Color Image Processing

Special Issue Editor
Edoardo Provenzi

MDPI • Basel • Beijing • Wuhan • Barcelona • Belgrade

Special Issue Editor
Edoardo Provenzi
Université de Bordeaux
France

Editorial Office
MDPI
St. Alban-Anlage 66
Basel, Switzerland

This edition is a reprint of the Special Issue published online in the open access journal *Journal of Imaging* (ISSN 2313-433X) from 2017–2018 (available at: http://www.mdpi.com/journal/jimaging/special_issues/color_image_processing).

For citation purposes, cite each article independently as indicated on the article page online and as indicated below:

Lastname, F.M.; Lastname, F.M. Article title. *Journal Name* **Year**, *Article number*, page range.

First Edition 2018

ISBN 978-3-03842-957-9 (Pbk)
ISBN 978-3-03842-958-6 (PDF)

Cover photo courtesy of Edoardo Provenzi

Table of Contents

About the Special Issue Editor

Edoardo Provenzi, full professor of mathematics and applications at University of Bordeaux, France. Edoardo Provenzi received his Master degree in Theoretical Physics from the University of Milano, Italy, in 2000 and a PhD in Mathematics and Applications from the University of Genova, Italy, in 2004. After 10 years of post-doc work in Italy and Spain in the field of applied mathematics, studying image processing and computer vision, he was hired as Associate Professor at the University Paris Descartes, France, in 2014. In 2017 he was hired as a full Professor of Mathematics at the University of Bordeaux, in the signal and image processing group. His main research interests are variational principles applied to color image processing, the metrology of visual attributes and differential geometry of visual information.

Preface to "Color Image Processing"

This book presents the collection of papers that formed the Special Issue "Color Image Processing" in the Journal of Imaging.

It reflects some of the most modern trends in color image processing. Authors from many different research areas have contributed to the multidisciplinary character of this book.

This large variety of subjects is possible thanks to the intrinsic multifaceted nature of color, which is a sensation created by our brain after a complicated and not yet fully understood combination of physical, biological and neurophysiological mechanisms.

We hereafter describe in more detail the contributions in this book; whenever possible, we tried to maintain semantic coherence by sequentially presenting papers dealing with similar subjects, as is the case for the first four, which are characterized by a rigorous analysis of color space geometry.

"RGB Color Cube-Based Histogram Specification for Hue-Preserving Color Image Enhancement", by Kohei Inoue, Kenji Hara and Kiichi Urahama, deals with the problem of saturation enhancement in a hue-preserving environment. By making use of the RGB space geometry, the authors were able to build a model free of tunable parameters.

The analysis of color space geometry is also the basis of the second paper, "Analytical Study of Colour Spaces for Plant Pixel Detection", by Pankaj Kumar and Stanley J. Miklavcic, where the properties of several color space features were analyzed for the segmentation of plants.

A fine mathematical analysis of the YUV space is presented in "Exemplar-Based Face Colorization Using Image Morphing", by Johannes Persch, Fabien Pierre and Gabriele Steidl, for the purpose of colorization via morphing techniques.

Finally, the quaternion framework for color images is the workspace for the development of watermarking in the paper "Image Fragile Watermarking through Quaternion Linear Transform in Secret Space", by Marco Botta, Davide Cavagnino and Victor Pomponiu.

The paper "Histogram-Based Color Transfer for Image Stitching", by Qi-Chong Tian and Laurent D. Cohen, presents a novel technique for the minimization of artifacts in the stitching process based on histogram-based color transfer.

In "Robust Parameter Design of Derivative Optimization Methods for Image Acquisition Using a Color Mixer", by HyungTae Kim, KyeongYong Cho, Jongseok Kim, KyungChan Jin and SeungTaek Kim, the so-called auto-lighting algorithm is analyzed. This is a process able to maximize the image quality of industrial machine vision by adjusting multiple LEDs, usually called color mixers. The authors propose a method to overcome the time-consuming gradient-based methods used so far.

A modern treatment of color processing must of course incorporate the very popular machine learning techniques. These were used in the form of convolutional neural networks (CNN) in "Improving CNN-Based Texture Classification by Color Balancing", by Simone Bianco, Claudio Cusano, Paolo Napoletano and Raimondo Schettini, where they show how suitable color balancing can help CNN performance in texture recognition.

Machine learning techniques are also used in "Automatic Recognition of Speed Limits on Speed-Limit Signs by Using Machine Learning", by Shigeharu Miyata, to detect speed limit signs (characterized by color) in a complex environment.

A variant of the classical Retinex algorithm of Land is introduced in "Color Consistency and Local Contrast Enhancement for a Mobile Image-Based Change Detection System", by Marco Tektonidis and David Monnin, to improve the performance of mobile change detection systems.

First-order statistics are used in "Improved Color Mapping Methods for Multiband Nighttime Image Fusion Maarten", by A. Hogervorst and Alexander Toet, to enhance the performance of color mapping methods in dark images by giving them a daylight look.

The paper "Illusion and Illusoriness of Color and Coloration", by Baingio Pinna, Daniele Porcheddu and Katia Deiana, provides a phenomenological analysis of chromatic illusion, underlying the difference between the perceptual and cognitive phenomena associated with this kind of illusion.

Finally, the contribution "The Academy Color Encoding System (ACES): A Professional Color-Management Framework for Production, Post-Production and Archival of Still and Motion Pictures", by Walter Arrighetti, provides a complete account of color encoding in professional cinema.

We believe that the large spectrum of subjects discussed in the papers can be both an inspiration for young researchers who wish to gain a broad vision of the modern research into color, as well as a state-of-the-art reference for established scholars in this domain.

Edoardo Provenzi

Special Issue Editor

Journal of
Imaging

MDPI

Article

RGB Color Cube-Based Histogram Specification for Hue-Preserving Color Image Enhancement

Kohei Inoue *, Kenji Hara and Kiichi Urahama

Department of Communication Design Science, Kyushu University, Fukuoka 815-8540, Japan; hara@design.kyushu-u.ac.jp (K.H.); urahama@design.kyushu-u.ac.jp (K.U.)
* Correspondence: k-inoue@design.kyushu-u.ac.jp

Received: 1 June 2017; Accepted: 27 June 2017; Published: 1 July 2017

Abstract: A large number of color image enhancement methods are based on the methods for grayscale image enhancement in which the main interest is contrast enhancement. However, since colors usually have three attributes, including hue, saturation and intensity of more than only one attribute of grayscale values, the naive application of the methods for grayscale images to color images often results in unsatisfactory consequences. Conventional hue-preserving color image enhancement methods utilize histogram equalization (HE) for enhancing the contrast. However, they cannot always enhance the saturation simultaneously. In this paper, we propose a histogram specification (HS) method for enhancing the saturation in hue-preserving color image enhancement. The proposed method computes the target histogram for HS on the basis of the geometry of RGB (rad, green and blue) color space, whose shape is a cube with a unit side length. Therefore, the proposed method includes no parameters to be set by users. Experimental results show that the proposed method achieves higher color saturation than recent parameter-free methods for hue-preserving color image enhancement. As a result, the proposed method can be used for an alternative method of HE in hue-preserving color image enhancement.

Keywords: color image enhancement; hue-preservation; histogram equalization; histogram specification; RGB color cube

1. Introduction

Color image enhancement is a challenging task in digital image processing with broad applications including human perception, machine vision applications, image restoration, image analysis, image compression, image understanding and pattern recognition [1], underwater image enhancement and image enhancement of low light scenes [2]. Sharo and Raimond surveyed the existing color image enhancement methods such as histogram equalization (HE), fuzzy-based methods and other optimization techniques [3]. Saleem and Razak also surveyed color image enhancement techniques using spatial filtering [4]. Suganya et al. analyzed the performance of various enhancement techniques based on noise ratio, time delay and quality [5].

In color image enhancement, preserving the hue of an input image is frequently required to preserve the appearance of the objects in the image. Bisla surveyed hue-preserving color image enhancement techniques [6]. Zhang et al. proposed a method for hue-preserving and saturation scaling color image enhancement using optimal linear transform [7]. Aashima and Verma proposed a hue-preserving and gamut problem-free color image enhancement technique, and compared it with a discrete cosine transform-based method [8]. Porwal et al. also proposed an algorithm for hue-preserving and gamut problem-free color image enhancement [9]. Chien and Tseng proposed a set of formulae for the color transformation between RGB (red, green and blue) and exact HSI (hue, saturation and intensity), and used it for color image enhancement [10]. Gorai and Ghosh considered image enhancement as an optimization problem and solved it using particle swarm optimization [11]. Taguchi reviewed color systems and color image enhancement methods, and introduced an improved HSI color space [12]. Menotti et al. proposed two fast hue-preserving HE methods based on 1D and 2D histograms of RGB color space for color image contrast enhancement [13]. Pierre et al. [14] introduced a variational model for the enhancement of color images, and compared their method with the state-of-the-art methods including Nikolova and Steidl's method [15], which is based on their strict ordering algorithm for exact HS [16].

Almost all of the above hue-preserving color image enhancement methods are based on the pioneering work of Naik and Murthy [17], where a scheme is proposed to avoid gamut problem arising during the process of enhancement of the intensity of color images using a general hue-preserving contrast enhancement function, in which HE is a typical example for intensity transformation. Han et al. also proposed the equivalent method from a viewpoint of 3D color HE [18]. However, Naik and Murthy's method cannot increase the saturation of colors to be enhanced. To overcome this problem, Yang and Lee [19] proposed a modified hue-preserving gamut mapping method that outputs higher saturation than Naik and Murthy's method. Yang and Lee's method divides the range of luminance into three parts corresponding to dark, middle and bright colors, and handles the input colors in different manners, that is, for dark and bright colors, their saturation is enhanced first, and then, Naik and Murthy's method is applied to the saturation-enhanced colors. On the other hand, for the remaining colors with middle luminance, Naik and Murthy's method is applied to the original colors directly. Therefore, the saturation of the middle luminance colors cannot be improved as well as Naik and Murthy's method.

In this paper, we propose a parameter-free HS method for hue-preserving color image enhancement based on the geometry of RGB color space. The proposed method can improve the color saturation in both Naik and Murthy's and Yang and Lee's methods. Experimental results show that the proposed HS method applied to Naik and Murthy's and Yang and Lee's methods improves the color saturation compared with the conventional Naik and Murthy's and Yang and Lee's methods using HE.

The rest of this paper is organized as follows: Section 2 first defines the saturation of a color, and then summarizes Naik and Murthy's and Yang and Lee's methods. Section 3 describes the detailed procedures of HS for color image enhancement, where HE is also summarized, and then RGB color cube-based HS method is proposed. Section 4 shows experimental results of hue-preserving color image enhancement. Finally, Section 5 discusses the results and the utility of the proposed method.

2. Hue-Preserving Color Image Enhancement

In this section, we briefly summarize previous hue-preserving color image enhancement methods proposed by Naik and Murthy [17] and Yang and Lee [19] after the description of color saturation.

Let $\boldsymbol{p} = [r, g, b]^T$ be a point in RGB color space or an RGB color vector, where r, g and b denote red, green and blue values, respectively, and satisfy $0 \leq r \leq 1$, $0 \leq g \leq 1$ and $0 \leq b \leq 1$, and the superscript T denotes the matrix transpose. Then, the intensity of $\boldsymbol{p}$ is given by $l = r + g + b$ [17] satisfying $l \in [0, 3]$, and the saturation of $\boldsymbol{p}$ is the perpendicular distance from the intensity axis to $\boldsymbol{p}$ [20] as follows:

$$S(\boldsymbol{p}) = \sqrt{\boldsymbol{p}^T \left[I - \frac{\mathbf{1}}{\|\mathbf{1}\|} \left(\frac{\mathbf{1}}{\|\mathbf{1}\|} \right)^T \right] \boldsymbol{p}} = \sqrt{\|\boldsymbol{p}\|^2 - \left(\frac{\mathbf{1}^T \boldsymbol{p}}{\|\mathbf{1}\|} \right)^2} = \sqrt{\frac{(r-g)^2 + (g-b)^2 + (b-r)^2}{3}}, \quad (1)$$

where $\mathbf{1} = [1, 1, 1]^T$, I is the 3×3 identity matrix, and $\| \cdot \|$ denotes the Euclidean norm.

Lemma 1. *The saturation $S(\boldsymbol{p})$ of $\boldsymbol{p}$ given by (1) has the following properties:*

$$S(\beta \boldsymbol{p}) = \beta S(\boldsymbol{p}), \quad (2)$$

where β *is a positive number, and*

$$S(\mathbf{1}-\gamma(\mathbf{1}-\boldsymbol{p}))=\gamma S(\boldsymbol{p}) \tag{3}$$

for $0<\gamma<1$.

The proof of Lemma 1 is given in Appendix A.

2.1. Naik and Murthy's Method

Let $\alpha(l)=f(l)/l$ for $l>0$, where $f(l)$ is a function of l for transforming the original intensity into the modified one in the same range as its domain, i.e., $f(l)\in[0,3]$. Then, Naik and Murthy considered a hue-preserving transformation of the form

$$\boldsymbol{p}'=\alpha(l)\boldsymbol{p}, \tag{4}$$

where $\boldsymbol{p}'=[r',g',b']^T$ denotes the transformed color vector of $\boldsymbol{p}$. The value of $\alpha(l)$ will be greater than 1 when $f(l)>l$. In such a case, the element value of $\boldsymbol{p}'$ may exceed 1 and thus result in a gamut problem. To overcome this problem, Naik and Murthy proposed a gamut problem-free procedure as follows:

Naik and Murthy's method

1. Case (i) If $\alpha(l)\le 1$, then compute $\boldsymbol{p}'=\alpha(l)\boldsymbol{p}$.
2. Case (ii) If $\alpha(l)>1$, then perform the following procedure:
   (1) Transform the RGB color vector $\boldsymbol{p}$ to CMY (cyan, magenta and yellow) color vector $\boldsymbol{q}=[c,m,y]^T$, where $c=1-r$, $m=1-g$ and $y=1-b$.
   (2) Find $\bar{l}=c+m+y=3-l$.
   (3) Find $\overline{f(l)}=3-f(l)$, $\bar{\alpha}(l)=\overline{f(l)}/\bar{l}$. Note that $\bar{\alpha}(l)=[3-f(l)]/(3-l)<1$ since $f(l)>l$.
   (4) Compute $\boldsymbol{q}'=\bar{\alpha}(l)\boldsymbol{q}$.
   (5) Transform the CMY color vector $\boldsymbol{q}'$ to RGB color vector $\boldsymbol{p}'=\mathbf{1}-\boldsymbol{q}'$.

Note that, in Step (3) in Case (ii), we cannot compute $\bar{\alpha}(\bar{l})$ when $l=3$ because it results in division by zero. To avoid such difficulties, we set that $\boldsymbol{p}'=\mathbf{1}$ if $l=3$, and $\boldsymbol{p}'=\mathbf{0}=[0,0,0]^T$ if $l=0$. We find that Case (ii) can be concisely written as follows:

Case (ii) If $\alpha(l)>1$, then compute $\boldsymbol{p}'=\mathbf{1}-\bar{\alpha}(l)(\mathbf{1}-\boldsymbol{p})$ for $\bar{\alpha}(l)=[3-f(l)]/(3-l)$.

Additionally, it has been proved that Naik and Murthy's method does not increase the saturation, that is, $S(\boldsymbol{p}')\le S(\boldsymbol{p})$ [21].

2.2. Yang and Lee's Method

Yang and Lee also pointed out that the color saturation of the resulting images by Naik and Murthy's method is low, and proposed a hue-preserving gamut mapping method, the resulting images of which show higher saturation than that of Naik and Murthy's.

Although Yang and Lee defined the luminance of $\boldsymbol{p}$ as $n=(r+g+b)/3=l/3$ instead of the intensity l to describe algorithms in their paper [19], we would like to use l rather than n in this paper consistently. Using l, we can describe Yang and Lee's method as follows:

Yang and Lee's method

1. Case (I) If $l\le 1$, then compute $\tilde{\boldsymbol{p}}=\boldsymbol{p}/l$, whose intensity is 1. Apply Naik and Murthy's method to $\tilde{\boldsymbol{p}}$ as follows:
   (1) Case (I-i) If $f(l)\le 1$, then compute $\boldsymbol{p}''=f(l)\tilde{\boldsymbol{p}}$.
   (2) Case (I-ii) If $f(l)>1$, then compute $\boldsymbol{p}''=\mathbf{1}-[3-f(l)](\mathbf{1}-\tilde{\boldsymbol{p}})/2$.

2. Case (II) If $1 < l \leq 2$, then apply Naik and Murthy's method to $\boldsymbol{p}$ as follows:

   (1) Case (II-i) If $\alpha(l) \leq 1$, then compute $\boldsymbol{p}'' = \boldsymbol{p}' = \alpha(l)\boldsymbol{p}$.
   (2) Case (II-ii) If $\alpha(l) > 1$, then compute $\boldsymbol{p}'' = \boldsymbol{p}' = \boldsymbol{1} - \bar{\alpha}(l)(\boldsymbol{1} - \boldsymbol{p})$.

3. Case (III) If $l > 2$, then transform $\boldsymbol{p}$ into the CMY color vector $\boldsymbol{q} = \boldsymbol{1} - \boldsymbol{p}$, and then lower the intensity of $\boldsymbol{q}$ to 2 as $\tilde{\boldsymbol{q}} = \boldsymbol{q}/(3-l)$ to have the RGB color vector $\tilde{\boldsymbol{p}} = \boldsymbol{1} - \tilde{\boldsymbol{q}}$. Apply Naik and Murthy's method to $\tilde{\boldsymbol{p}}$ as follows:

   (1) Case (III-i) If $f(l) \leq 2$, then compute $\boldsymbol{p}'' = f(l)\tilde{\boldsymbol{p}}/2$.
   (2) Case (III-ii) If $f(l) > 2$, then compute $\boldsymbol{p}'' = \boldsymbol{1} - [3 - f(l)](\boldsymbol{1} - \tilde{\boldsymbol{p}})$.

We have the following lemma:

Lemma 2. *The saturation given by Yang and Lee's method is greater than or equal to that given by Naik and Murthy's method, that is,* $S(\boldsymbol{p}'') \geq S(\boldsymbol{p}')$.

The proof of Lemma 2 is given in Appendix A.

3. Histogram Specification for Color Image Enhancement

Let $P = [\boldsymbol{p}_{ij}]$ be a digital color image, where $\boldsymbol{p}_{ij} = [r_{ij}, g_{ij}, b_{ij}]^T$ denotes the RGB color vector at the position (i, j) of a pixel in P for $i = 1, 2, \ldots, m$ and $j = 1, 2, \ldots, n$, where m and n denote the numbers of pixels in the vertical and horizontal directions in P, respectively. Suppose that P is a 24-bit true color image. Then, each element of $\boldsymbol{p}_{ij}$ is an integer between 0 and 255, i.e., $r_{ij}, g_{ij}, b_{ij} \in \{0, 1, \ldots, 255\}$, and the intensity of $\boldsymbol{p}_{ij}$ is given by $\bar{l}_{ij} = r_{ij} + g_{ij} + b_{ij} \in \{0, 1, \ldots, L\}$, where $L = 255 \times 3 = 765$.

3.1. Histogram Equalization

Let $\boldsymbol{h} = [h_0, h_1, \ldots, h_L]$ be the histogram of the intensity $\bar{l}_{ij}$ of $\boldsymbol{p}_{ij}$ in P. Then, the $\bar{l}$th element of $\boldsymbol{h}$ is given by $h_{\bar{l}} = \sum_{i=1}^{m} \sum_{j=1}^{n} \delta_{\bar{l}, \bar{l}_{ij}}$, where $\delta_{\bar{l}, \bar{l}_{ij}}$ denotes the Kronecker delta; $\delta_{\bar{l}, \bar{l}_{ij}} = 1$ if $\bar{l} = \bar{l}_{ij}$ and 0, otherwise. Let $\boldsymbol{H} = [H_0, H_1, \ldots, H_L]$ be the cumulative histogram of $\boldsymbol{h}$, where the $\bar{l}$th element of $\boldsymbol{H}$ is given by $H_{\bar{l}} = \sum_{k=0}^{\bar{l}} h_k$. Then, the intensity transformation function $f(l)$ for HE is given by

$$f^{E}(l) = \text{round}\left(\frac{L}{H_L} H_{\bar{l}}\right), \tag{5}$$

where 'round( )' operator rounds a given argument toward the nearest integer, $H_L = \sum_{k=0}^{L} h_k = mn$, and $\bar{l} = \text{round}(Ll/3)$ for $l \in [0, 3]$. The histogram-equalized intensity image of P is given by $P^E = [p_{ij}^E]$ where

$$p_{ij}^{E} = f^{E}\left(\frac{3}{L}\bar{l}_{ij}\right) \in \{0, 1, \ldots, L\}. \tag{6}$$

3.2. Histogram Specification

Let $\tilde{\boldsymbol{h}} = [\tilde{h}_0, \tilde{h}_1, \ldots, \tilde{h}_L]$ be a target histogram into which we want to transform the original histogram of intensity, and let $\tilde{\boldsymbol{H}} = [\tilde{H}_0, \tilde{H}_1, \ldots, \tilde{H}_L]$ be the cumulative histogram of $\tilde{\boldsymbol{h}}$, where the $\bar{l}$th element of $\tilde{\boldsymbol{H}}$ is given by $\tilde{H}_{\bar{l}} = \sum_{k=0}^{\bar{l}} \tilde{h}_k$. Then, the intensity transformation function $f(l)$ for HS is given by

$$f^{S}(l) = \arg\min_{k} \left\{ \left| H_L \tilde{H}_k - \tilde{H}_L H_{\bar{l}} \right| \right\}, \tag{7}$$

where $\tilde{H}_L = \sum_{k=0}^{L} \tilde{h}_k$, and $\bar{l} = \text{round}(Ll/3)$ for $l \in [0, 3]$. The histogram-specified intensity image of P is given by $P^S = [p_{ij}^S]$ where

$$p_{ij}^{S} = f^{S}\left(\frac{3}{L}\bar{l}_{ij}\right) \in \{0, 1, \ldots, L\}. \tag{8}$$

3.3. RGB Color Cube-Based Histogram Specification

In this subsection, we propose a parameter-free HS method named RGB color-cube based HS. As described in Section 2, the saturation of a color in RGB color space is defined as the perpendicular distance between the intensity axis and a point corresponding to the color. The locus of the perpendicular line around the intensity axis forms an equiintensity plane. Let us consider the cross section of the equiintensity plane and RGB color cube as shown in Figure 1, where the cross sections are painted in light blue, and Figure 1a–c show three cases of the value of the intensity l found in the equation of the plane, $\mathbf{1}^T \boldsymbol{p} = \|\mathbf{1}\| l$, that is, $0 \le l \le 1$, $1 \le l \le 2$ and $2 \le l \le 3$, respectively. In these figures, the triangles drawn by red broken and solid lines denote the cross sections of $l = 1$ and $l = 2$, respectively.

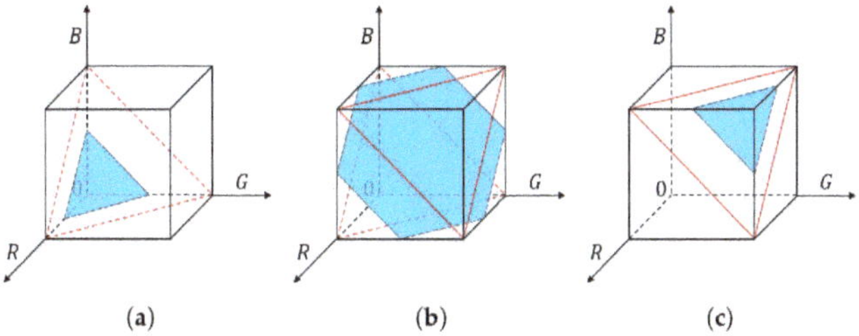

Figure 1. Cross sections of RGB color cube with equiintensity planes, $\mathbf{1}^T \boldsymbol{p} = \|\mathbf{1}\| l$, where $\boldsymbol{p}$ denotes a point on the plane: (**a**) $0 \le l \le 1$; (**b**) $1 \le l \le 2$; (**c**) $2 \le l \le 3$.

The area of the cross section indicates the variety of saturation for a given intensity l. Let $a(l)$ be the area of the cross section for an intensity l. Then, we have the following analytic form of $a(l)$:

$$a(l) = \begin{cases} a_1(l) = \frac{\sqrt{3}}{2} l^2 & \text{if } 0 \le l \le 1, \\ a_2(l) = \frac{3\sqrt{3}}{4} - \sqrt{3}\left(l - \frac{3}{2}\right)^2 & \text{if } 1 \le l \le 2, \\ a_3(l) = \frac{\sqrt{3}}{2}(3 - l)^2 & \text{if } 2 \le l \le 3, \end{cases} \tag{9}$$

which is continuous at $l = 1$ and $l = 2$, that is, $a_1(1) = a_2(1) = \sqrt{3}/2$ and $a_2(2) = a_3(2) = \sqrt{3}/2$. Additionally, the derivative function of $a(l)$ is given by

$$a'(l) = \begin{cases} a_1'(l) = \sqrt{3} l & \text{if } 0 \le l \le 1, \\ a_2'(l) = 2\sqrt{3}\left(\frac{3}{2} - l\right) & \text{if } 1 \le l \le 2, \\ a_3'(l) = \sqrt{3}(l - 3) & \text{if } 2 \le l \le 3, \end{cases} \tag{10}$$

which is also continuous at $l = 1$ and $l = 2$, that is, $a_1'(1) = a_2'(1) = \sqrt{3}$ and $a_2'(2) = a_3'(2) = -\sqrt{3}$. The integral of $a(l)$ is given by

$$A(l) = \int_0^l a(x)dx = \begin{cases} A_1(l) = \int_0^l a_1(x)dx = \frac{\sqrt{3}}{6} l^3 & \text{if } 0 \le l \le 1, \\ A_2(l) = A_1(1) + \int_1^l a_2(x)dx = \frac{\sqrt{3}}{6}\left(3 - 2l^3 + 9l^2 - 9l\right) & \text{if } 1 \le l \le 2, \\ A_3(l) = A_2(2) + \int_2^l a_3(x)dx = \frac{\sqrt{3}}{6}\left[6 - (3 - l)^3\right] & \text{if } 2 \le l \le 3, \end{cases} \tag{11}$$

which is also continuous at $l = 1$ and $l = 2$, that is, $A_1(1) = A_2(1) = \sqrt{3}/6$ and $A_2(2) = A_3(2) = 5\sqrt{3}/6$. Figure 2 shows the graphs of $a(l)$ and $A(l)$, where the vertical and horizontal axes denote the function value of $a(l)$ or $A(l)$ and the intensity l, respectively, and the blue and green lines denote the functions $a(l)$ and $A(l)$, respectively.

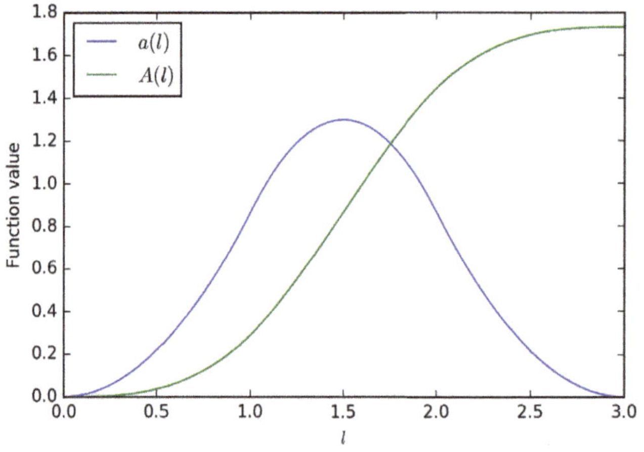

Figure 2. Area of cross section of RGB color cube and equiintensity plane, $a(l)$, and its integral, $A(l)$.

We propose to use $a(l)$ and $A(l)$ as the substitution of the target histogram and its cumulative one in HS, respectively. The detailed procedure is as follows.

Let $\tilde{\boldsymbol{h}}^C = [\tilde{h}_0^C, \tilde{h}_1^C, \ldots, \tilde{h}_L^C]$ be the target histogram for the proposed RGB color cube-based HS. Then, the $\tilde{l}$th element $\tilde{h}_{\tilde{l}}$ of $\tilde{\boldsymbol{h}}^C$ is given by

$$\tilde{h}_{\tilde{l}}^C = a\left(\frac{3}{L}\tilde{l}\right) \tag{12}$$

for $\tilde{l} = 0, 1, \ldots, L$. By means of the cumulation of $\tilde{\boldsymbol{h}}^C$, we have the cumulative histogram $\tilde{\boldsymbol{H}}^C$. In another way, since we have the analytic form of the integral of $a(l)$ as $A(l)$, we can also compute $\tilde{\boldsymbol{H}}^C = [\tilde{H}_0^C, \tilde{H}_1^C, \ldots, \tilde{H}_L^C]$ from $A(l)$ directly as follows:

$$\tilde{H}_{\tilde{l}}^C = A\left(\frac{3}{L}\tilde{l}\right) \tag{13}$$

for $\tilde{l} = 0, 1, \ldots, L$. Then, the intensity transformation function $f(l)$ for RGB color cube-based HS is given by

$$f^C(l) = \arg\min_k \left\{ \left| H_L \tilde{H}_k^C - \tilde{H}_L^C H_{\tilde{l}} \right| \right\}, \tag{14}$$

where $\tilde{H}_L^C = A(3) = \sqrt{3}$, and $\tilde{l} = \mathrm{round}(Ll/3)$ for $l \in [0, 3]$. The histogram-specified intensity image of P is given by $P^C = [p_{ij}^C]$ where

$$p_{ij}^C = f^C\left(\frac{3}{L}\tilde{l}_{ij}\right) \in \{0, 1, \ldots, L\}. \tag{15}$$

The above intensity transformation functions, $f^E(l)$, $f^S(l)$ and $f^C(l)$, can be used instead of $f(l)$ in Naik and Murthy's and Yang and Lee's methods.

3.4. Conditions for Saturation Improvement

The above RGB color cube-based HS can be used in both Naik and Murthy's and Yang and Lee's methods as well as the conventional HE. In this subsection, we summarize the conditions for improving color saturation by the proposed HS compared with HE in the two methods.

3.4.1. Naik and Murthy's Method

Let $\alpha^E(l) = f^E(l)/l$, $\alpha^C(l) = f^C(l)/l$, and $\boldsymbol{p}'^C$ be the enhanced color of $\boldsymbol{p}$ by Naik and Murthy's method with the proposed HS. Then, we have $S(\boldsymbol{p}'^C) \geq S(\boldsymbol{p}')$ under the following conditions:

$$\begin{cases} f^E(l) \leq f^C(l) \leq l, & \text{if} \quad \alpha^E(l) \leq 1 \quad \text{and} \quad \alpha^C(l) \leq 1, \\ f^E(l) \leq l \leq f^C(l) \leq f^E(l) + 3\left[1 - \alpha^E(l)\right], & \text{if} \quad \alpha^E(l) \leq 1 \quad \text{and} \quad \alpha^C(l) > 1, \\ \frac{3 - f^E(l)}{3 - l} l \leq f^C(l) \leq l < f^E(l), & \text{if} \quad \alpha^E(l) > 1 \quad \text{and} \quad \alpha^C(l) \leq 1, \\ l < f^C(l) \leq f^E(l), & \text{if} \quad \alpha^E(l) > 1 \quad \text{and} \quad \alpha^C(l) > 1. \end{cases} \tag{16}$$

3.4.2. Yang and Lee's Method

Let $\boldsymbol{p}''^C$ be the enhanced color of $\boldsymbol{p}$ by Yang and Lee's method with the proposed HS. Then, we have $S(\boldsymbol{p}''^C) \geq S(\boldsymbol{p}'')$ under the following conditions for three cases of l:

If $l \leq 1$ (Case I in Yang and Lee's method), then we have the following conditions:

$$\begin{cases} f^E(l) \leq f^C(l), & \text{if} \quad f^E(l) \leq 1 \quad \text{and} \quad f^C(l) \leq 1, \\ f^C(l) \leq 3 - 2f^E(l), & \text{if} \quad f^E(l) \leq 1 \quad \text{and} \quad f^C(l) > 1, \\ \frac{3 - f^E(l)}{2} \leq f^C(l), & \text{if} \quad f^E(l) > 1 \quad \text{and} \quad f^C(l) \leq 1, \\ f^C(l) \leq f^E(l), & \text{if} \quad f^E(l) > 1 \quad \text{and} \quad f^C(l) > 1. \end{cases} \tag{17}$$

If $1 < l \leq 2$ (Case II in Yang and Lee's method), then we have the same conditions as Equation (16) because Yang and Lee's method coincides with Naik and Murthy's method.

If $l > 2$ (Case III in Yang and Lee's method), then we have the following conditions:

$$\begin{cases} f^E(l) \leq f^C(l), & \text{if} \quad f^E(l) \leq 2 \quad \text{and} \quad f^C(l) \leq 2, \\ f^C(l) \leq 3 - \frac{f^E(l)}{2}, & \text{if} \quad f^E(l) \leq 2 \quad \text{and} \quad f^C(l) > 2, \\ 2\left[3 - f^E(l)\right] \leq f^C(l), & \text{if} \quad f^E(l) > 2 \quad \text{and} \quad f^C(l) \leq 2, \\ f^C(l) \leq f^E(l), & \text{if} \quad f^E(l) > 2 \quad \text{and} \quad f^C(l) > 2. \end{cases} \tag{18}$$

4. Experimental Results

In this section, we show the experimental results of hue-preserving color image enhancement, and demonstrate that the proposed method improve the color saturation in comparison with Naik and Murthy's and Yang and Lee's methods using HE.

Figure 3 shows input and output images for hue-preserving color image enhancement, where the first top row shows the original input images, and the second to fifth rows show the corresponding output images. The original images in the top row are collected from the Standard Image Data-BAse (SIDBA) [22]. The second row shows the results by Naik and Murthy's method, which uses HE for intensity transformation to enhance the contrast. However, the color saturation has faded in all images. As a result, the output images become close to their contrast-enhanced grayscale images. Moreover, we can see that the 2nd (Airplane) and the 6th (Girl) images become noisy by contrast overenhancement caused by HE. The third row shows the results of the proposed HS used in Naik and Murthy's method instead of HE, where the color saturation is recovered and the noise is suppressed compared with the second row. The fourth row shows the results of Yang and Lee's method with HE, where the 4th (Couple) and 6th (Girl) images have improved saturation and are more colorful than the second row of Naik and Murthy's method. However, the other images are similar to that of Naik and Murthy's method. The fifth row shows the results of the proposed HS used in Yang and Lee's method instead of HE, where the saturation is improved compared with the third and fourth rows.

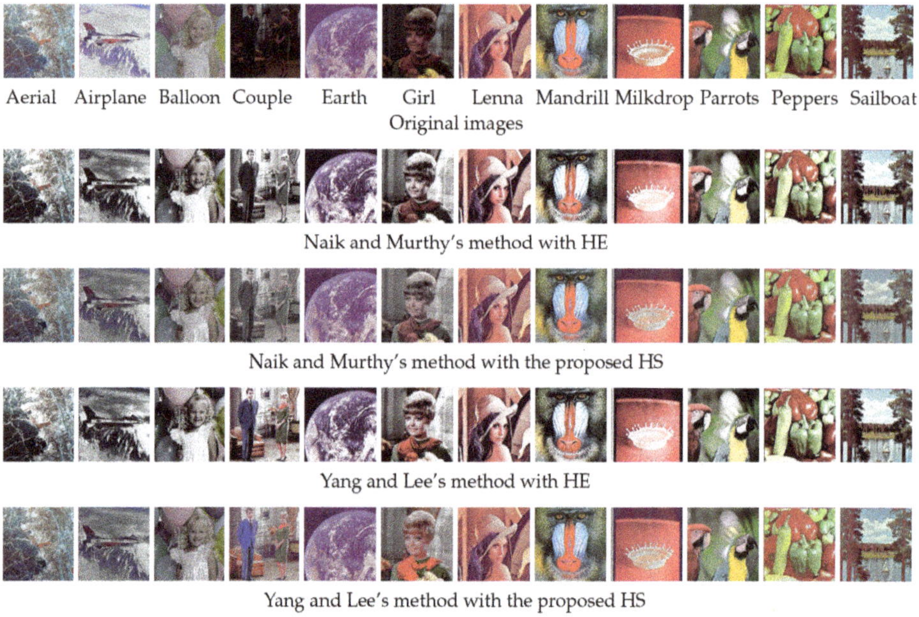

Figure 3. Results of hue-preserving color image enhancement.

Figure 4 shows the saturation images whose pixel values are given by the saturation values $S(\boldsymbol{p})$ in Equation (1). The order of the images are the same as that of Figure 3. The images in the second row are not brighter than that in the first row, which demonstrates visually that Naik and Murthy's method cannot increase the saturation from the original images. The third row shows the saturation images by Naik and Murthy's method with the proposed HS, which can improve the saturation—for example, we can see brighter regions in 7th (Lenna) to 10th (Parrots) images than the corresponding images in the second row. The fourth row shows the saturation images by Yang and Lee's method with HE, which achieves higher saturation than Naik and Murthy's method in the second row. The fifth row shows the saturation images by Yang and Lee's method with the proposed HS, which further increases the saturation compared with the fourth row.

Figure 5 shows the difference maps between the enhanced and original saturation images. These maps are generated by the following procedure: let $\boldsymbol{p}$ and $\boldsymbol{p}^{\text{enh}}$ be the corresponding pixels of an original color image and its enhanced one, respectively. Then, we compute the difference of their saturations as $d = S(\boldsymbol{p}^{\text{enh}}) - S(\boldsymbol{p})$, and set the pixel color in the difference map by $(\mathbf{1}/2) + [d, d, 0]^T$ if $d \geq 0$, and $(\mathbf{1}/2) - [0, d, d]^T$ otherwise. That is, cyan and green in the difference map mean a decrease and increase in saturation, respectively, and gray $(\mathbf{1}/2)$ means neutral.

The top row in Figure 5 shows the difference maps between Naik and Murthy's results with HE and the original images, where we can see a number of deep cyan regions, which mean a decrease in saturation from the original images. On the other hand, the second row shows the results by Naik Murthy's method with the proposed HS, where the cyan regions are diluted compared with the top row. The third row shows the results by Yang and Lee's method with HE, which gives similar results to Naik and Murthy's method in the top row except for the 4th (Couple) and 6th (Girl) images, in which we can see the yellow regions that mean the increase in saturation. The bottom row shows the results by Yang and Lee's method with the proposed HS, where the cyan regions are diluted as well as the second row, and yellow regions are made deeper and broader than the third row.

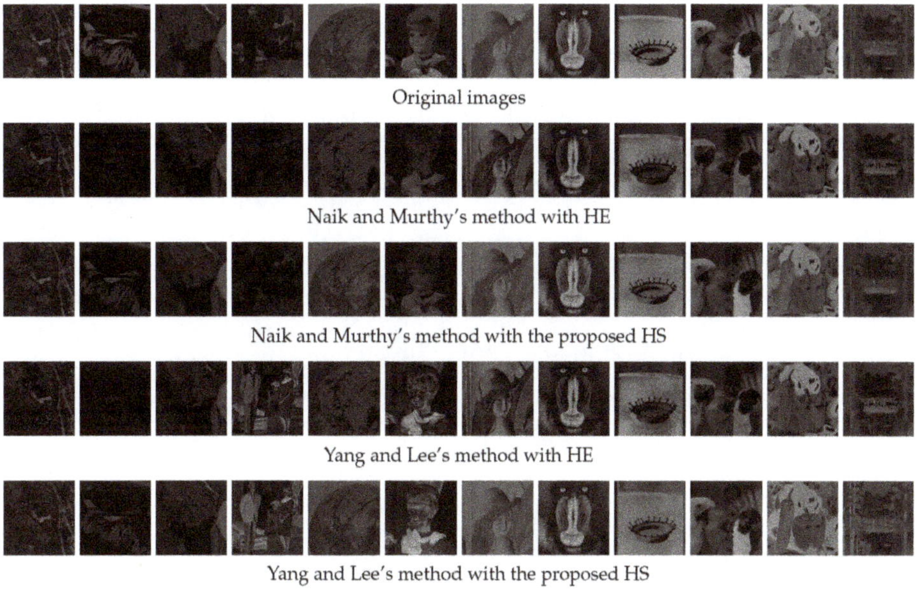

Figure 4. Saturation images.

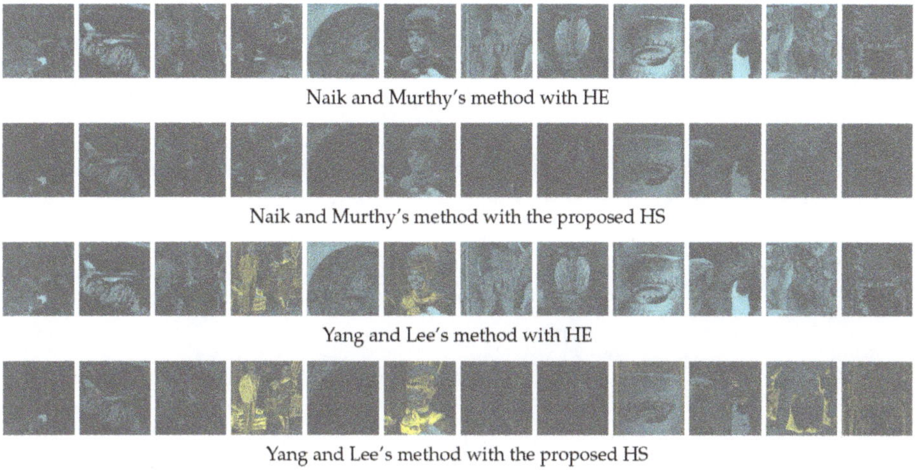

Figure 5. Difference maps of saturation images.

Figure 6 shows the mean saturation value per pixel for each image in Figure 4, where the vertical and horizontal axes denote the mean saturation and the names of images, respectively. Compared with the original images denoted by cyan bars, the Naik and Murthy's results denoted by light green have decreased mean saturation. The proposed HS improves the mean saturation as shown by the yellow bars; however, the improvement is limited to the values of the original images by the fact that Naik and Murthy's method does not increase the saturation [21]. Yang and Lee's results denoted by the orange bars indicate the values greater than or equal to Naik and Murthy's results (light green bars), as stated in Lemma 2. The proposed method also improves the mean saturation for Yang and Lee's method as shown in the red bars.

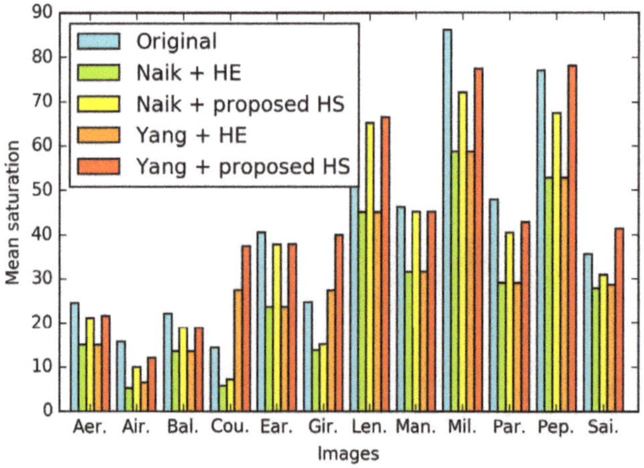

Figure 6. Mean saturation.

The total mean saturation is summarized in Table 1, where twelve mean values of each color bar are averaged to get the values in the table. Naik and Murthy's method denoted by Naik + HE in the table decreases the total mean saturation from the value of the original images. The proposed method (Naik + Proposed HS) increases the total mean saturation from Naik + HE. Yang and Lee's method (Yang + HE) achieves greater value than Naik + HE, which demonstrates the claim in Lemma 2, and the proposed method (Yang + Proposed HS) also improves it.

Table 1. Total mean saturation. In this table, 'Original' means the original images, 'Naik+HE' means Naik and Murthy's method with HE, 'Naik+Proposed HS' means Naik and Murthy's method with the proposed HS, 'Yang+HE' means Yang and Lee's method with HE, and 'Yang+Proposed HS' means Yang and Lee's method with the proposed HS.

Original	Naik + HE	Naik + Proposed HS	Yang + HE	Yang + Proposed HS
41.89	26.84	35.98	29.97	43.31

Figure 7 shows the results on the INRIA Holidays dataset [23,24], where the top row shows the original images, and the second to fifth rows show the results by Naik and Murthy's method with HE and the proposed HS, and Yang and Lee's method with HE and the proposed HS, respectively. The number under each image denotes the mean saturation value. The proposed HS improves the mean saturation of almost all examples except for an image with the mean saturation value 7.59, which is lower than 8.32 given by Naik and Murthy's method with HE.

Figure 7. Results on the INRIA (Institut National de Recherche en Informatique et en Automatique) dataset with the mean saturation values.

5. Discussion

In the above experimental results, we have compared four hue-preserving color image enhancement methods: Naik + HE, Naik + Proposed HS, Yang + HE and Yang + Proposed HS. First, we confirmed the fact that Naik and Murthy's method does not increase the saturation of original colors experimentally. Next, we also experimentally confirmed the claim in Lemma 2, that is, Yang and Lee's method can improve the saturation compared with Naik and Murthy's method. Moreover, we demonstrated that the proposed HS method can improve the saturation compared with the conventional HE method used in both Naik and Murthy's and Yang and Lee's methods.

The target histogram for the proposed HS method is derived from the geometric shape of RGB color space, that is, a cube with a side length of 1, and has an analytic expression that can be integrated to obtain the cumulative target histogram used in the proposed HS. As a result, the proposed HS method has no additional assumptions or parameters. Therefore, there is no need for users to be bothered with any parameter settings. Additionally, the proposed HS method can suppress the contrast overenhancement that frequently occurs when HE is used.

Consequently, the proposed HS method can be used for an alternative method of HE because it is a parameter-free method as well as HE, and can enhance the color saturation compared with the conventional hue-preserving color image enhancement methods, while it can suppress the contrast overenhancement that occurs in HE frequently.

Acknowledgments: This work was supported by JSPS KAKENHI Grant Number JP16H03019.

Author Contributions: All of the authors contributed extensively to this work presented in the paper. Kohei Inoue has coordinated the work and participated in all section. Kiichi Urahama is an advisor guiding the designing of image processing methods. Kenji Hara provides suggestions to improve the algorithm and revise the whole paper.

Conflicts of Interest: The authors declare no conflict of interest.

Abbreviations

The following abbreviations are used in this manuscript:

RGB	red, green and blue
CMY	cyan, magenta and yellow
HSI	hue, saturation and intensity
HE	histogram equalization
HS	histogram specification

Appendix A. Proofs of Lemmas

In this section, we prove Lemmas 1 and 2 described above.

Proof of Lemma 1. By the definition of $S(\boldsymbol{p})$ in (1), it follows that

$$S(\beta\boldsymbol{p}) = \sqrt{(\beta\boldsymbol{p})^T\left[I - \frac{\mathbf{1}}{\|\mathbf{1}\|}\left(\frac{\mathbf{1}}{\|\mathbf{1}\|}\right)^T\right](\beta\boldsymbol{p})} = \beta\sqrt{\boldsymbol{p}^T\left[I - \frac{\mathbf{1}}{\|\mathbf{1}\|}\left(\frac{\mathbf{1}}{\|\mathbf{1}\|}\right)^T\right]\boldsymbol{p}} = \beta S(\boldsymbol{p}), \tag{A1}$$

and thus Equation (2) holds. Similarly, we find that

$$S(\mathbf{1} - \gamma(\mathbf{1} - \boldsymbol{p})) = S((1-\gamma)\mathbf{1} + \gamma\boldsymbol{p}) = \sqrt{\|(1-\gamma)\mathbf{1} + \gamma\boldsymbol{p}\|^2 - \left(\frac{\mathbf{1}^T[(1-\gamma)\mathbf{1} + \gamma\boldsymbol{p}]}{\|\mathbf{1}\|}\right)^2} \tag{A2}$$

$$= \sqrt{(1-\gamma)^2\|\mathbf{1}\|^2 + 2(1-\gamma)\gamma\mathbf{1}^T\boldsymbol{p} + \gamma^2\|\boldsymbol{p}\|^2 - \left[(1-\gamma)^2\|\mathbf{1}\|^2 + 2(1-\gamma)\gamma\mathbf{1}^T\boldsymbol{p} + \gamma^2\left(\frac{\mathbf{1}^T\boldsymbol{p}}{\|\mathbf{1}\|}\right)^2\right]} \tag{A3}$$

$$= \gamma\sqrt{\|\boldsymbol{p}\|^2 - \left(\frac{\mathbf{1}^T\boldsymbol{p}}{\|\mathbf{1}\|}\right)^2} = \gamma S(\boldsymbol{p}), \tag{A4}$$

hence Equation (3) holds. □

Proof of Lemma 2. In Case (I), for $l \le 1$, there are three cases of the position of $f(l)$ corresponding to the left, middle and right: $f(l) \le l \le 1$, $l < f(l) \le 1$ and $l \le 1 < f(l)$. For $f(l) \le l \le 1$, we have $\mathbf{p}'' = f(l)\tilde{\mathbf{p}} = f(l)\mathbf{p}/l = \alpha(l)\mathbf{p} = \mathbf{p}'$, from which it follows that $S(\mathbf{p}'') = S(\mathbf{p}')$. For $l < f(l) \le 1$, we have $\mathbf{p}' = \mathbf{1} - \bar{\alpha}(l)(\mathbf{1} - \mathbf{p})$ for $\bar{\alpha}(l) < 1$, whose saturation is given by $S(\mathbf{p}') = \bar{\alpha}(l)S(\mathbf{p})$ from Equation (3) in Lemma 1. We also have $\mathbf{p}'' = \alpha(l)\mathbf{p}$ for $\alpha(l) > 1$, whose saturation is given by $S(\mathbf{p}'') = \alpha(l)S(\mathbf{p})$ from Equation (2) in Lemma 1. Therefore, we have that $S(\mathbf{p}'') - S(\mathbf{p}') = [\alpha(l) - \bar{\alpha}(l)]S(\mathbf{p}) = 3[f(l) - l]S(\mathbf{p})/[l(3-l)] > 0$ or $S(\mathbf{p}'') > S(\mathbf{p}')$. For $l \le 1 < f(l)$, we have $S(\mathbf{p}') = \bar{\alpha}(l)S(\mathbf{p})$ and $\mathbf{p}'' = \mathbf{1} - [3 - f(l)](\mathbf{1} - \tilde{\mathbf{p}})/2$ for $\tilde{\mathbf{p}} = \mathbf{p}/l$, whose saturation is given by $S(\mathbf{p}'') = [3 - f(l)]S(\tilde{\mathbf{p}})/2 = S(\mathbf{p}/l)/2 = [3 - f(l)]S(\mathbf{p})/(2l)$ by Lemma 1. Therefore, we have that

$$S(\mathbf{p}'') - S(\mathbf{p}') = \left(\frac{3 - f(l)}{2l} - \frac{3 - f(l)}{3 - l}\right) S(\mathbf{p}) = \frac{3[3 - f(l)](1 - l)}{2l(3 - l)} S(\mathbf{p}) \ge 0 \tag{A5}$$

or $S(\mathbf{p}'') \ge S(\mathbf{p}')$.

In Case (II), since Yang and Lee's method outputs the same result as Naik and Lee's method, we have $S(\mathbf{p}'') = S(\mathbf{p}')$ immediately.

In Case (III), for $2 < l$, there are three cases of the position of $f(l)$ corresponding to the left, middle and right: $f(l) \le 2 < l$, $2 < f(l) \le l$ and $2 < l < f(l)$. For $f(l) \le 2 < l$, we have $\mathbf{p}' = \alpha(l)\mathbf{p}$, whose saturation is given by $S(\mathbf{p}') = \alpha(l)S(\mathbf{p})$. We also have $\mathbf{p}'' = f(l)\tilde{\mathbf{p}}/2$ for $\tilde{\mathbf{p}} = \mathbf{1} - (\mathbf{1} - \mathbf{p})/(3 - l)$, whose saturation is given by $S(\mathbf{p}'') = f(l)S(\tilde{\mathbf{p}})/2 = f(l)S(\mathbf{p})/[2(3 - l)]$. Therefore, we have that

$$S(\mathbf{p}'') - S(\mathbf{p}') = \left(\frac{f(l)}{2(3 - l)} - \frac{f(l)}{l}\right) S(\mathbf{p}) = \frac{3f(l)(l - 2)}{2(3 - l)l} S(\mathbf{p}) > 0 \tag{A6}$$

or $S(\mathbf{p}'') > S(\mathbf{p}')$. For $2 < f(l) \le l$, we have $S(\mathbf{p}') = \alpha(l)S(\mathbf{p})$ and $\mathbf{p}'' = \mathbf{1} - [3 - f(l)](\mathbf{1} - \tilde{\mathbf{p}})$, whose saturation is given by $S(\mathbf{p}'') = [3 - f(l)]S(\tilde{\mathbf{p}}) = [3 - f(l)]S(\mathbf{p})/(3 - l) = \bar{\alpha}(l)S(\mathbf{p})$. Therefore, we have

$$S(\mathbf{p}'') - S(\mathbf{p}') = \left(\frac{3 - f(l)}{3 - l} - \frac{f(l)}{l}\right) S(\mathbf{p}) = \frac{3[l - f(l)]}{(3 - l)l} S(\mathbf{p}) \ge 0 \tag{A7}$$

or $S(\mathbf{p}'') \ge S(\mathbf{p}')$. For $2 < l < f(l)$, we have $\mathbf{p}' = \mathbf{1} - \bar{\alpha}(l)(\mathbf{1} - \mathbf{p})$, whose saturation is given by $S(\mathbf{p}') = \bar{\alpha}(l)S(\mathbf{p})$, and $\mathbf{p}'' = \mathbf{1} - [3 - f(l)](\mathbf{1} - \tilde{\mathbf{p}})$, whose saturation is given by $S(\mathbf{p}'') = [3 - f(l)]S(\tilde{\mathbf{p}}) = [3 - f(l)]S(\mathbf{p})/(3 - l) = \bar{\alpha}(l)S(\mathbf{p}) = S(\mathbf{p}')$.

Consequently, $S(\mathbf{p}'') \ge S(\mathbf{p}')$ holds in any case. □

References

1. Anjana, N.; Jezebel Priestley, J.; Nandhini, V.; Elamaran, V. Color Image Enhancement using Edge Based Histogram Equalization. *Indian J. Sci. Technol.* **2015**, *8*, 1–6.
2. Mudigonda, S.P.; Mudigonda, K.P. Applications of Image Enhancement Techniques—An Overview. *MIT Int. J. Comput. Sci. Inf. Technol.* **2015**, *5*, 17–21.
3. Sharo, T.A.; Raimond, K. A Survey on Color Image Enhancement Techniques. *IOSR J. Eng.* **2013**, *3*, 20–24.
4. Saleem, S.A.; Razak, T.A. Survey on Color Image Enhancement Techniques using Spatial Filtering. *Int. J. Comput. Appl.* **2014**, *94*, 39–45.
5. Suganya, P.; Gayathri, S.; Mohanapriya, N. Survey on Image Enhancement Techniques. *Int. J. Comput. Appl. Technol. Res.* **2013**, *2*, 623–627.
6. Bisla, S. Survey on Hue-Preserving Color Image Enhancement without Gamut Problem. *Int. J. Sci. Res.* **2017**, *6*, 1911–1914.
7. Zhang, X.; Xin, Y.; Huang, H.; Xie, N. Color Image Enhancement Using Optimal Linear Transform with Hue Preserving and Saturation Scaling. *Int. J. Intell. Eng. Syst.* **2015**, *8*, 1–9.

8. Aashima; Verma, N. Hue Preserving Improvement in Quality of Colour Image. *Int. J. Adv. Res. Eng. Technol.* **2016**, *4*, 1–5.
9. Porwal, D.; Alam, Md. S.; Porwal, A. HUE Preserving Color Image Enhancement without GAMUT Problem using newly proposed Algorithm. *Int. J. Eng. Comput. Sci.* **2013**, *2*, 3450–3454.
10. Chien, C.-L.; Tseng, D.-C. Color Image Enhancement with Exact HSI Color Model. *Int. J. Innov. Comput. Inf. Control* **2011**, *7*, 6691–6710.
11. Gorai, A.; Ghosh, A. Hue-preserving color image enhancement using particle swarm optimization. In Proceedings of the IEEE Recent Advances in Intelligent Computational Systems (RAICS), Trivandrum, India, 22–24 September 2011; pp. 563–568.
12. Taguchi, A. Color Systems and Color Image Enhancement Methods. *ECTI Trans. Comput. Inf. Technol.* **2016**, *10*, 97–110.
13. Menotti-Gomes, D.; Najman, L.; Facon, J.; De Araújo, A.A. Fast Hue-Preserving Histogram Equalization Methods for Color Image Contrast Enhancement. *Int. J. Comput. Sci. Inf. Technol.* **2012**, *4*, 243–259.
14. Pierre, F.; Aujol, J.-F.; Bugeau, A.; Steidl, G.; Ta, V.-T. Hue-preserving perceptual contrast enhancement. In Proceedings of the IEEE International Conference Image Process, Phoenix, AZ, USA, 25–28 September 2016.
15. Nikolova, M.; Steidl, G. Fast Hue and Range Preserving Histogram Specification: Theory and New Algorithms for Color Image Enhancement. *IEEE Trans. Image Process.* **2014**, *23*, 4087–4100.
16. Nikolova, M.; Steidl, G. Fast Ordering Algorithm for Exact Histogram Specification. *IEEE Trans. Image Process.* **2014**, *23*, 5274–5283.
17. Naik, S.K.; Murthy, C.A. Hue-preserving color image enhancement without gamut problem. *IEEE Trans. Image Process.* **2003**, *12*, 1591–1598.
18. Han, J.-H.; Yang, S.; Lee, B.-U. A novel 3-d color histogram equalization method with uniform 1-d gray scale histogram. *Trans. Image Process.* **2011**, *20*, 506–512.
19. Yang, S.; Lee, B. Hue-preserving gamut mapping with high saturation. *Electronics Letters* **2013**, *49*, 1221–1222.
20. Gonzalez, R. C.; Woods, R. E. *Digital Image Processing*, 3rd ed.; Prentice-Hall, Inc.: Upper Saddle River, NJ, USA, 2006.
21. Inoue, K.; Hara, K.; Urahama, K. On hue-preserving saturation enhancement in color image enhancement. *IEICE Trans. Fundam.* **2015**, *98*, 927–931.
22. The Standard Image Data-BAse (SIDBA). Available online: http://www.ess.ic.kanagawa-it.ac.jp/app_images_j.html (accessed on 30 June 2017).
23. Jegou, H.; Douze, M.; Schmid, C. Hamming Embedding and Weak Geometry Consistency for Large Scale Image Search. In Proceedings of the 10th European Conference on Computer Vision, Marseille, France, 12–18 October 2008.
24. The INRIA Holidays Dataset. Available online: http://lear.inrialpes.fr/~jegou/data.php (accessed on 30 June 2017).

Journal of
Imaging

MDPI

Article

Analytical Study of Colour Spaces for Plant Pixel Detection

Pankaj Kumar *,†, and Stanley J. Miklavcic †

School of Information Technology and Mathematical Sciences; Phenomics and Bioinformatics Research Centre, University of South Australia, Adelaide SA 5001, Australia; stan.miklavcic@unisa.edu.au

* Correspondence: pankaj.kumar@unisa.edu.au; Tel.: +61-8-830-23196

† These authors contributed equally to this work.

Received: 26 September 2017; Accepted: 12 February 2018; Published: 16 February 2018

Abstract: Segmentation of regions of interest is an important pre-processing step in many colour image analysis procedures. Similarly, segmentation of plant objects in digital images is an important preprocessing step for effective phenotyping by image analysis. In this paper, we present results of a statistical analysis to establish the respective abilities of different colour space representations to detect plant pixels and separate them from background pixels. Our hypothesis is that the colour space representation for which the separation of the distributions representing object and background pixels is maximized is the best for the detection of plant pixels. The two pixel classes are modelled by Gaussian Mixture Models (GMMs). In our statistical modelling we make no prior assumptions on the number of Gaussians employed. Instead, a constant bandwidth mean-shift filter is used to cluster the data with the number of clusters, and hence the number of Gaussians, being automatically determined. We have analysed the following representative colour spaces: RGB, rgb, HSV, $Ycbcr$ and $CIE\text{-}Lab$. We have analysed the colour space features from a two-class variance ratio perspective and compared the results of our model with this metric. The dataset for our empirical study consisted of 378 digital images (and their manual segmentations) of a variety of plant species: Arabidopsis, tobacco, wheat, and rye grass, imaged under different lighting conditions, in either indoor or outdoor environments, and with either controlled or uncontrolled backgrounds. We have found that the best segmentation of plants is found using HSV colour space. This is supported by measures of Earth Mover Distance (EMD) of the GMM distributions of plant and background pixels.

Keywords: plant phenotyping; plant pixel classification; colour space; Gaussian Mixture Model; Earth Mover Distance; variance ratio; plant segmentation

1. Introduction

Compared with the growing interest in plant phenotyping using computer vision and image analysis, plant phenotyping by visual inspection is slow and subjective, relying as it does on human evaluation. Two of the aims of digital imaging and image analysis are (a) the removal of any degree of subjectivity associated with an individual human's perception, and (b) the expedition of the analysis procedure. This is especially important for the high throughput assessment of the phenotypic manifestations of genetic expressions in plants.

Another and particular application of digital imaging in an agricultural setting is the detection and identification of weeds for the purpose of spot spraying of herbicide. Spot spraying, as opposed to blanket spraying, is more economical and environmentally less detrimental. It is worth considering that successful application of spot spraying may also depend on weed size (volume of herbicide) as well as weed identification (type of herbicide). Consequently, being able to estimate weed volume or biomass by 3D reconstruction from digital images is potentially beneficial. In [1], An et al. presented a novel method that used plant segmentation from images for 3D plant morphology quantification and phenotyping. Plant segmentation was used by An et al. in [2] to measure phenotypic traits such as leaf length and rosette area in 2D images. Plant pixel detection by a Gaussian Mixture Model (GMM) was used by Kovalchuk et al. in [3] for the automatic detection of plot canopy coverage and analysis of different genotypes. Thus, it can be noted that an important basic precursor to both detection, identification and 3D reconstruction [4,5] is the process of plant segmentation. That is, the binary classification of pixels into plant and non-plant groups.

There are many approaches to segmentation. These fall into one of two camps: supervised or unsupervised segmentation [4,6]. However, active contours or level sets and fuzzy logic can also be used for object segmentation [7,8]. All these different methods of segmentation will benefit from the study presented in this paper. A colour space which enhances the ability to separate plant pixels from non-plant pixels will improve the performance of any segmentation method based on colour.

Many segmentation methods are based on colour distinction. To achieve optimal plant segmentation, however, the natural question to first pose is which colour space is the more effective for the detection of plant pixels? Is there a suitable transformation of {Red Green Blue} (RGB) colour space to a representation that will make plant pixel detection more accurate and more reliable? Can a suitable representation be found that will improve the degree to which plant pixel detection is independent of illumination condition? Does the contrast between plant and background naturally get enhanced in certain colour space irrespective of illumination condition? These are some of the questions we address in this paper. Similar questions have been raised and answered for skin pixel segmentation [9–11], shadow and traffic object detection [12–14] and image segmentation by graph cut [15,16]. It has been shown in [17] that the choice of colour space does influence object recognition.

A related study seeking to improve plant segmentation by colour analysis was carried out by Golzarian et al. [18] using colour indices. Colour indices, individually however, do not provide a complete representation of a colour space. Individual indices are scalar-valued variables obtained by a linear manipulation of the components of the three dimensional colour space vector of a pixel. The individual colour indices considered by Golzarian et al. [18] were g, DGR, EGI, $MEGI$, NDI, Hue, and the hue channel of HSV colour space. Their results showed that hue achieved the least amount of type II error with a small loss of plant pixel. Our results and conclusions differ somewhat. We attribute the difference in conclusions to the fact that our study is more encompassing as our larger dataset includes a greater number of lighting conditions and a larger number of plant species. Thus, in contrast to Golzarian et al.'s findings, our results show that HSV is overall best suited for segmentation of plants under the majority of lighting conditions. An important aspect which has not been addressed in this study is how color balancing would affect the plant pixel detection. Color balancing was shown in [19] to affect texture classification studies.

This paper is organized as follows. In Section 2.1 we introduce briefly the different colour spaces we experimented with and their mathematical relationship to each other. Then, in Section 2.2 we outline our method for discriminating pixels into one of two classes using GMMs and evaluating the separability of the classes by computing their class distance using Earth Mover Distance (EMD) and variance ratios. EMD [20] has a long history for use in image processing and analysis. Rubner et al. used EMD on cluster signature of images for image retrieval and for object tracking by Zhao et al. in [21] and by Kumar et al. in [22]. The details of our dataset are provided in Section 2.3. We present our results and discuss those in Section 3, and finally conclude the paper with summary comments in Section 4.

2. Methods

2.1. Colour Representation

Colour spaces allow for different representations of intensity and colour information in colour images. Past research activities on colour representation, psycho-visual perception of colour, video signal transmission and computer graphics have given rise to many colour spaces having different desirable properties. Here, we briefly review the five well-known colour spaces *RGB*, *rgb*, *HSV*, *Ycbcr* and *CIE-Lab* that we shall utilize and we summarize how they are related to the common *RGB* colour space. A consideration of less well-known spaces such as the colour derivative spaces, as described in Gevers et al. [23], and opponent colour space as mentioned in [24] by Gevers and Stokman, could be made the subject of subsequent study.

2.1.1. *RGB*

Red, green and blue are the familiar primary colours and it is now accepted that their different practical combinations are capable of generating almost all possible colour shades. This colour space has been the basis for the design of *CRT*s, television and computer screens. Most still cameras and scanners save their images in this colour space. However, the high correlation between channels as well as the mixing of chrominance and luminance information makes *RGB* space a sub-optimal choice for colour-based detection schemes.

2.1.2. Normalized *rgb*

$$r = \frac{R}{R+G+B}, \quad g = \frac{G}{R+G+B}, \quad b = \frac{B}{R+G+B} \tag{1}$$

This is a colour space in which intensity information is normalized, which in turn leads to a reduced dependence on the luminance information. The normalization property, however, introduces a redundancy amongst the three components. For instance, no additional information is available in b since $b = 1 - r - g$. In such a case, the components, r and g, are referred to as pure colours due to the absence of a dependence on the brightness of the source *RGB*. A mention of *rg* space can also be found in [23].

2.1.3. *HSV*

This colour space specifies any colour in terms of three quantities: *Hue*, *Saturation*, and *Value*. It was introduced to satisfy user need to specify colour properties numerically. Hue defines the dominant colour of a pixel, Saturation measures the colourfulness of a pixel in proportion to its brightness, and Value is related to colour luminance. *HSV* is non-linearly related to *RGB* via the following set of equations

$$\begin{aligned} H &= \begin{cases} \theta & \text{if } B \leq G, \\ 2\pi - \theta & \text{if } B > G. \end{cases} \\ &\quad \text{where} \quad \theta = \cos^{-1}\frac{1/2[(R-G)+(R-B)]}{[(R-G)^2+(R-B)(G-B)]^{1/2}} \\ S &= 1 - \frac{3}{R+G+B}[\min(R,G,B)] \\ V &= 1/3(R+G+B) \end{aligned} \tag{2}$$

A polar co-ordinate representation of *HSV* results in a cyclic colour space. *HSV* colour space was recently used for adaptive skin classification in [25]. This colour space is similar to the colour space representations, *HIS*, *HLS*, and *HCI*.

2.1.4. *Ycbcr*

This colour space is utilized in most image compression standards such as *JPEG*, *H*.261, *MPEG*, and television studios (video cameras also usually save in this format). Pixel intensity is represented by Y luminance, computed as a weighted sum of *RGB* values; the matrix of weights which transforms the *RGB* pixel value to *Ycbcr* is given in Equation (3). The chrominance component of the pixel information is contained in the *cb* and *cr* channels. The colour space is characterized by a simple but explicit separation of luminance and chrominance components. It is similar to YIQ and YUV color spaces and linearly related to RGB as follows

$$\begin{bmatrix} Y \\ cr \\ cb \end{bmatrix} = \begin{bmatrix} 0.299 & 0.587 & 0.114 \\ 0.711 & -0.587 & -0.114 \\ -0.299 & -0.587 & 0.886 \end{bmatrix} \times \begin{bmatrix} R \\ G \\ B \end{bmatrix} \tag{3}$$

where the matrix elements are fixed.

2.1.5. CIE-Lab

This colour space, originally proposed by G. Wyszecki [26], to approximate perceptually-uniform colour space information has been standardized by the Commission Internationale de L' Eclairage (CIE). By "perceptually-uniform" one means that it was designed to approximate human vision. The L-channel contains information about pixel intensity/brightness, while a and b store the colour information. The CIE-Lab colour space is non-linearly related to CIE-XYZ.

The RGB to Lab conversion is achieved by a transformation , $\mathbf{M}$: $\begin{bmatrix} L \\ a \\ b \end{bmatrix} = \mathbf{M} \begin{bmatrix} R \\ G \\ B \end{bmatrix}$. Standard methods exist for specifying the transformation $\mathbf{M}$ when the co-ordinates of the RGB system and reference white has been specified. One such $\mathbf{M}$ has been used in [14] for $sRGB$, $D65$ device-dependent colour space. Related colour spaces are CIE-LUV and CIE-LCH. More details of the different colour spaces could be found in [27].

2.2. *Evaluation of Colour Space Representations*

To evaluate the suitability of a colour space representation for the detection of plant pixels, we differentiate background from foreground pixels based on their relative position within a GMM which has been constructed using the respective colour space information possessed by the pixels. The Gaussian Mixture Model is a function of a random variable, $\mathbf{z}$, which in our case is the feature vector comprising the information contained in the three pixel colour channels:

$$gmm(\mathbf{z}, \boldsymbol{\phi}) = \sum_{k=1}^{K} w_k g(\mathbf{z} : \mu_k, \boldsymbol{\Sigma}_k) \tag{4}$$

The model parameter set $\boldsymbol{\phi}$ is the set $\{w_k, \mu_k, \boldsymbol{\Sigma}_k\}_{k=1}^{K}$ where K is the number of Gaussian distributions in gmm and each g is of the form

$$g(\mathbf{z} : \mu_k, \boldsymbol{\Sigma}_k) = \frac{1}{\sqrt{2\pi\boldsymbol{\Sigma}_k}} \exp\left(-\frac{1}{2}(\frac{\| \mathbf{z} - \mu_k \|}{\boldsymbol{\Sigma}_k})^2\right). \tag{5}$$

In applying the GMM using the expectation maximization (EM) algorithm, there arises the fundamental problem of how to predetermine the number of Gaussian functions to include in the GMM. We ameliorate this problem by using a mean-shift algorithm to cluster both the background pixel and foreground plant pixel data. The same fixed bandwidth is used to cluster both sets of pixel data. Each cluster is then modelled as a Gaussian distribution with a diagonal co-variance matrix,

$$\boldsymbol{\Sigma}_k = \begin{bmatrix} \sigma_{11} & 0 & 0 \\ 0 & \sigma_{22} & 0 \\ 0 & 0 & \sigma_{33} \end{bmatrix}. \tag{6}$$

This is utilized to reduce computational load at the cost of an insignificant loss of accuracy. We denote the background GMM be gmm_{bg} and foreground plant pixel GMM be gmm_{fg}. A distance function based on Earth Mover Distance (EMD) is used as a measure of the distance between gmm_{bg} and gmm_{fg}. The EMD can be considered as a measure of dissimilarity between two multi-dimensional distributions. The greater the EMD between two distributions the more dissimilar they are. It was used by Rubner et al. in [28] for colour- and texture-based image retrieval. It was also employed by Kumar and Dick in [22] to track targets in an image sequence. In the present case, we apply the EMD measure to the multi-dimensional Gaussian mixture distributions that correspond, respectively, to background and foreground/plant pixel colour in a given colour space. We then consider the relative success of the EMD-based clustering in the different colour spaces to compare the effectiveness of the spaces. The greater the EMD value between the background and foreground GMMs in a given colour space, the better is that colour space for separating plant pixel from non-plant pixel. Using an EMD as a quantifier we aim to discover in which colour space the distance between the two distributions is maximal. We then explore how this distance varies with plant type and imaging condition.

Computing the EMD is based on a solution of the transportation problem [29]. In our case the two distributions are the two GMMs corresponding to the two classes of plant pixel and background pixel:

$$\begin{aligned} gmm_{bg} &= \textstyle\sum_{l=1}^{L} w_l g(\mathbf{z} : \boldsymbol{\mu}_l, \boldsymbol{\Sigma}_l) \quad and \\ gmm_{fg} &= \textstyle\sum_{k=1}^{K} w_k g(\mathbf{z} : \boldsymbol{\mu}_k, \boldsymbol{\Sigma}_k). \end{aligned} \tag{7}$$

Here, gmm_{bg} has L Gaussian in its model and gmm_{fg} has K. To compute the EMD between them, L need not be equal to K.

2.2.1. EMD on GMMs

In this section we provide an overview of the use of EMD as a measure of separation/distance between two GMM distributions. Let the two distributions, gmm_{bg} and gmm_{fg}, be characterized by their weights, means, and variances, $(w_l, \boldsymbol{\mu}_l, \boldsymbol{\Sigma}_l)_{l=1}^{L}$ and $(w_k, \boldsymbol{\mu}_k, \boldsymbol{\Sigma}_k)_{k=1}^{K}$. The EMD is used to compute the distance between these cluster signatures. Cluster signatures are characterised by weights; a signature differs from a distribution in the sense that the weights are not normalized. Also, cluster signatures do not have a cluster spread associated with them, since the GMMs have variances associated with each Gaussian. In our case the weights $w_k\{k = 1...K\}$ are normalized, *i.e.*, $\sum_{k=1}^{K} w_k = 1$. EMD can also be used to compute the dissimilarity between unnormalized cluster signatures. However, in our application the weights are normalized. The EMD is defined in terms of an optimal flow f_{kl} which minimizes the following

$$EMD(gmm_{fg}, gmm_{bg}) = \sum_{k=1}^{K} \sum_{l=1}^{L} f_{kl} d_{kl}, \tag{8}$$

where $d_{lk} = D(g(\mathbf{z} : \boldsymbol{\mu}_k, \boldsymbol{\Sigma}_k), g(\mathbf{z} : \boldsymbol{\mu}_l, \boldsymbol{\Sigma}_l))$ is a measure of dissimilarity/distance between Gaussians $g(\mathbf{z} : \boldsymbol{\mu}_k, \boldsymbol{\Sigma}_k)$ and $g(\mathbf{z} : \boldsymbol{\mu}_l, \boldsymbol{\Sigma}_l)$, and is also referred to as ground distance (GD). The computed flow after an optimization process satisfies the following constraints

$$\begin{aligned} f_{kl} &\geq 0, \quad for\ 1 \leq l \leq L,\ 1 \leq k \leq K \\ \textstyle\sum_{l=1}^{L} f_{kl} &\leq w_k, \quad for\ 1 \leq k \leq K \\ \textstyle\sum_{k=1}^{K} f_{kl} &\leq w_l, \quad for\ 1 \leq l \leq L \\ \textstyle\sum_{k=1}^{K}\sum_{l=1}^{L} f_{kl} &= min(\textstyle\sum_{k=1}^{K} w_k, \textstyle\sum_{l=1}^{L} w_l) = 1 \end{aligned} \tag{9}$$

The formulation of EMD is slightly different when the weights are not normalized. For computing the ground distance d_{lk} we need a distance measure between two Gaussians $g(\mathbf{z} : \boldsymbol{\mu}_l, \boldsymbol{\Sigma}_l)$ and $g(\mathbf{z} : \boldsymbol{\mu}_k, \boldsymbol{\Sigma}_k)$. We propose here to use a modified Mahalanobis distance to compute the ground distance d_{lk} between Gaussians l and k:

$$d_{lk} = (\boldsymbol{\mu}_l - \boldsymbol{\mu}_k)^T[(\boldsymbol{\Sigma}_l + \boldsymbol{\Sigma}_k)/2]^{-1}(\boldsymbol{\mu}_l - \boldsymbol{\mu}_k). \tag{10}$$

The Mahalanobis distance, formally introduced by P.C. Mahalanobis in [30], is a measure of similarity/difference of a multivariate data point $\mathbf{z} = (z_1, z_2, ..., z_N)$ with a known Gaussian distribution in the same dimension

$$D_M(\mathbf{z}) = \sqrt{(\mathbf{z}-\boldsymbol{\mu})^T \boldsymbol{\Sigma}^{-1} (\mathbf{z}-\boldsymbol{\mu})} \tag{11}$$

This distance is different from Euclidean distance in that it scales down the distance by the standard deviation of the distribution. The intuition behind this distance is that the distance of a data point to a normal distribution is inversely proportional to the latter's spread. This is an important concept in cluster analysis. The distance of a data point to a cluster is not just the Euclidean distance of the data point to the cluster centre. It also depends inversely on the spread of the cluster. The same intuitive notion extends to the distance between two Gaussians, as suggested by the distance function in Equation (10). The distance between two Gaussians with similar differences in mean values increases as their standard deviation decreases. A problem with the distance function in Equation (10) is that it becomes unbounded as the variance of the Gaussian goes to zero. However, this phenomena is of theoretical interest only, since for real life data sets it is seldom necessary to model something with a zero variance normal distribution. Furthermore, a zero variance normal distribution has no physical meaning. Other distance functions like Bhattacharyya distance, Hellinger distance, Kulback Leibler etc., could also have been modified to formulate a distance function between two multivariate Gaussians and in future studies we will consider this aspect of the problem.

2.2.2. Two-Class Variance Ratio

To study the discriminative power of different colour spaces with respect to segmenting plant pixels from background pixels, we compare the results of the present approach to results of augmented variance ratio (AVR). AVR has been used for feature ranking and as a preprocessing step in feature subset selection [31,32], and for online selection of discriminative feature tracking [33]. AVR is defined as the ratio of the inter-class variance to the intra-class variance of features. We use this variance ratio to measure the power of different colour spaces to discriminate plant pixel from background pixel. It is well known that linear discriminant analysis (LDA) and the variance ratio are inappropriate for separating multi-modal class distributions. The plant pixel colours and background pixel colours are generally multi-model. Therefore, we use the log-likelihood ratio, a non-linear transformation, to transform the features of a pixel i

$$L(i) = log\frac{max(pp(i), \delta)}{max(bg(i), \delta)} \tag{12}$$

The parameter δ is set to a small value, *e.g.*, 0.001, to avoid creating a divergence (divide by zero or logarithm of zero). The vectors $pp(i)$ and $bg(i)$ are the class-conditional probability distributions (normalized histograms) of plant and background pixels, respectively, learnt from a training data set. This log-likelihood ratio transforms the class distributions into a uni-modal form, making the use of the variance ratio appropriate for measuring the discriminative power of the colour space feature. The variance of $L(i)$ for class $pp(i)$ is

$$\begin{aligned} var(L;pp) &= E[L^2(i)] - (E[L(i)])^2 \\ &= \textstyle\sum_i pp(i)L^2(i) - [\sum_i pp(i)L(i)]^2 \end{aligned} \tag{13}$$

Similarly, the variance ratio for the background class is

$$var(L;bg) = \sum_i bg(i)L^2(i) - [\sum_i bg(i)L(i)]^2 \tag{14}$$

The variance ratio now is

$$VR(L;pp,bg) = \frac{var(L;(pp+bg)/2)}{var(L;pp)+var(L;bg)} \tag{15}$$

The denominator ensures that the colour space for which the within-class variance is smaller will be more discriminative, while the numerator favours the feature space in which the between-class variance is larger.

2.3. Dataset and Experiments

The data set contains images of Arabidopsis and tobacco plants grown in growth chambers which have been taken under controlled lighting conditions. A distinct subset of the data set consists of wheat and rye grass images which have been taken in the field and thus subject to different lighting conditions. Figure 1 shows the two imaging platforms used for imaging some of the plants used in this study. The platform on the left is for imaging indoor plants while the platform on the right is for imaging outdoor plants. After imaging, the plant regions were manually selected and segmented for this study. Arabidopsis images have two types of backgrounds: one black and one red. Plant images were taken both indoors and outdoors in order to capture as great a variety of illumination and background conditions as possible. The complete data set comprises 378 images.

(a) (b)

Figure 1. This figure shows the two imaging platforms we have build in house for imaging plants. (**a**) is the platform for imaging plants indoor and (**b**) is the platform for imaging outdoor plants growing in field conditions.

For segmentation of images in different colour spaces we used mean-shift clustering and region fusion. We selected the cluster related to leaves in a semi-supervised way and undertook a two-pass mean-shift clustering. The first pass was to determine the leaf area and separate it from the background. The second pass was to cluster individual leaves and separate them into different leaf areas. Different sets of parameters were used for the two passes of the clustering algorithm as described in [6].

3. Results

The results of applying the proposed EMD measure, Equation (8), on the GMMs of foreground and background pixels from training data are shown in Tables 1 and 2. These tables also show the variance ratios given by Equation (15) for the two classes of pixels in different colour channels. The *HSV* colour space had the highest scoring of all systems in terms of EMD distance. There appears to be no clear uniformally high performer amongst the different colour spaces, in terms of variance ratio. The segmentation results shown in Figure 2 and in Table 3 show that segmentation based on the HSV colour space was nevertheless superior in two out of three different scenarios. Table 3 shows that segmentation in terms of percentages using the FGBGDice code provided by the LSC challenge dataset. Finally, we show the segmentation results for the test data set in *CIE-Lab* colour space. The reason for using the *CIE-Lab* colour space for this comparison is that this colour space was recommended by the authors of [6].

(a) (b) (c) (d) (e) (f) (g)

Figure 2. This figure shows the results of segmentation using the method of mean-shift clustering, used in the paper, for different colour spaces. Image in column (**a**) are the original images of Arabidopsis and Tobacco plants. One set of Arabidopsis plant has a contrasting red background and the other has black background. The images in column (**b**) are the ground truth segmentation results. The ground truth segmentation were generated by manual labelling of the image data. In columns (**c**–**g**) are the segmentation in different colour spaces *RGB*, *HSV*, *CIE-Lab*, normalized *rgb*, and *YCbCr*, respectively

Table 1. Table for different plant types and the computed Earth Mover Distance (EMD) on the Gaussian Mixture Models (GMM) models in different colour spaces and their comparison with variance ratio in different colour spaces. EMD based distance are higher for *HSV* on both plant types, where as variance ratios are higher for normalized *rgb* colour space.

Plant Type	Colour Space	EMD Distance	Variance Ratio
Arabidopsis	*RGB*	282.46	1.17
	HSV	**847.24**	1.09
	CIE-Lab	246.14	2.19
	nrgb	264.67	**2.23**
	YCbCr	155.01	1.67
Tobacco	*RGB*	228.65	0.76
	HSV	**1389.56**	0.99
	CIE-Lab	415.17	0.43
	nrgb	347.63	**1.11**
	YCbCr	235.15	0.47

Table 2. Table for different imaging senarios for same plant type. Their computed EMDs on the GMM models in different colour spaces and their comparsion with variance ratio in different colour spaces. EMD based distance are higher for *HSV* on both background types, where as variance ratios are higher for normalized *rgb* colour space for contrasting red background and *CIE-Lab* for the black background.

Background Type	Colour Space	EMD Distance	Variance Ratio
Contrasting Green-Red	*RGB*	230.36	1.07
	HSV	**401.81**	0.78
	CIE-Lab	182.77	1.77
	nrgb	39.43	**1.88**
	YCbCr	178.47	1.57
Green-Black	*RGB*	282.16	1.27
	HSV	**404.70**	2.13
	CIE-Lab	272.88	**9.17**
	nrgb	257.87	2.24
	YCbCr	181.57	3.62

Table 3. Percentage foreground background segmentation results in the different colour spaces for there different datasets. A1's are Arabidopsis plants with a red background, A2's are Arabidopsis plants with black background and A3's are tobacoo plants imaged in controlled growth chambers.

Plant type	Percentage Foreground Background Segmentation				
	RGB	*HSV*	*CIE-Lab*	*rgb*	*YCbCr*
A1	96.32 %	96.67%	93.40%	10.87%	94.25%
A2	90.53%	98.51%	95.54%	49.69%	97.83%
A3	64.8%	89.6%	57.23%	19.56%	51.79%

Table 4. Overall plant and leaf segmention results using the method of mean-shift clustering as described in Section 2.3.

Plant Type	Plant Segmentation		Leaf Segmentation	
	Mean	Std	Mean	Std
A1	92.14%	2.82 %	47.14%	11.14%
A2	93.31%	2.41 %	55.16%	13.15%
A3	76.52%	35.32%	34.03%	22.35%

The segmentation results of separating plant leaves from the background obtained on the test dataset for Arabidopsis images were quite reasonable, achieving mean values of 0.9215 with a standard deviation of 0.0282 on A1 test images (see Table 4) and a mean of 0.93313 with a standard deviation of 0.0241 for test images of A2 (see Table 4). Set A1 are Arabidopsis plants with red background imaged indoors under controlled lighting conditions. Set A2 comprised images of Arabidopsis plants with black background also imaged indoors. Set A3 was composed of images of tobacco plants at different stages of development. Some errors in plant and background segmentation were mainly due to the presence of green moss on the background soil. The leaf segmentation results were not as good as the plant segmentation results as the algorithm was designed mainly for background-foreground segmentation. The results can be improved with the use of shape priors for leaf segmentation. The accuracy of plant segmentation for the tobacco test data set was quite poor, which could have resulted for one of two reasons. Firstly, the colours of some of the tobacco plant leaves in the test data set were quite different from what was typically found in the training data set. Secondly, the illumination present in the tobacco images was not as intense as that applied to the Arabidopsis plants. Consequently, some of the darker regions of the tobacco plants have been classified as background.

Plant pixel detection and segmentation is certainly affected by the choice of the colour space being used for image analysis. Hence, the choice of colour space should be given careful consideration for plant phenotyping purposes. In this study where we considered and analysed five different colour spaces, better plant pixel detection was achieved using the HSV colour space for almost all plant types and under different illumination conditions. This can be attributed to the fact that HSV is a perceptual colour space. Usually, best results of detection and segmentation are obtained in perceptual colour spaces. This outcome is supported by Golzarian et al.'s study in [18], where the authors obtained the least amount of type II error with a small loss of plant pixels. However, here we have shown that the segmentation in HSV colour space gives better results under a greater variety of illumination conditions and for a greater range of plant species.

4. Conclusions

In this paper we have presented a method for dynamically selecting the suitability of a feature space (colour space in this case) for segmenting plant pixels in digital images which have both classes of plant pixels and background pixels modelled by Gaussian Mixture Models. For the data set of plants imaged under controlled lighting conditions, the proposed method of colour space selection seems to be more effective than the variance ratio method. The HSV colour space clearly performs better for tobacco plants and is one of the higher quality segmentation performers for Arabidopsis plant images under two different scenarios. This conclusion extends to plants imaged either in field-like conditions where no lighting control is possible, or close to field-like conditions where there is a mix of ambient and controlled lighting. It is well known that the choice colour space influences the performance of image analysis, and the use of perceptual spaces generally provide more satisfying results, whatever database is being considered. Our experimental results on plant pixel detection under different illumination condition supports this prevalent hypothesis. In addition, the separability analysis based on EMD of GMMs for different colour spaces reveals the same phenomena. Our use of different distance functions to measure the separation of Gaussian distributions, and hence $GMMs$, has the added benefit of providing better analytical understanding of the results on which our conclusion is based. In future work we would like to experiment with how color balancing [19] affects the segmentation and detection of plant pixels.

Acknowledgments: The authors are grateful for the financial support from the Australian Research Council through its Linkage program project grants LP140100347. Thanks to Dr. Joshu Chopin for discussions and help with latex.

Author Contributions: Pankaj Kumar and Stanley J. Miklavcic conceived the experiments; Pankaj Kumar designed and performed the experiments, and analysed the data; Pankaj Kumar and Stanley J. Miklavcic wrote the paper; Stanley J. Miklavcic contributed with funding, computational resources and project management.

Conflicts of Interest: The authors declare no conflicts of interest.

References

1. An, N.; Welch, S.M.; Markelz, R.C.; Baker, R.L.; Palmer, C.M.; Ta, J.; Maloof, J.N.; Weinig, C. Quantifying time-series of leaf morphology using 2D and 3D photogrammetry methods for high-throughput plant phenotyping. *Comput. Electr. Agric.* **2017**, *135*, 222–232.
2. An, N.; Palmer, C.M.; Baker, R.L.; Markelz, R.C.; Ta, J.; Covington, M.F.; Maloof, J.N.; Welch, S.M.; Weinig, C. Plant high-throughput phenotyping using photogrammetry and imaging techniques to measure leaf length and rosette area. *Comput. Electr. Agric.* **2016**, *127*, 376–394.
3. Kovalchuk, N.; Laga, H.; Cai, J.; Kumar, P.; Parent, B.; Lu, Z.; Miklavcic, S.J.; Haefele, S.M. Phenotyping of plants in competitive but controlled environments: A study of drought response in transgenic wheat. *Funct. Plant Biol.* **2016**, *44*, 290–301.

4. Kumar, P.; Cai, J.; Miklavcic, S.J. High-throughput 3D modelling of plants for phenotypic analysis. In *Proceedings of the 27th Conference on Image and Vision Computing New Zealand*; ACM: New York, NY, USA, 2012; pp. 301–306.
5. Kumar, P.; Connor, J.N.; Miklavcic, S.J. High-throughput 3D reconstruction of plant shoots for phenotyping. In Proceedings of the 2014 13th International Conference on Automation Robotics and Computer Vision (ICARCV), Singapore, 10–12 December 2014.
6. Comaniciu, D.; Meer, P. Mean Shift: A robust approach toward feature space analysis. *IEEE Trans. Pattern Anal. Mach. Intell.* **2002**, *24*, 603–619.
7. Golzarian, M.R.; Cai, J.; Frick, R.A.; Miklavcic, S.J. Segmentation of cereal plant images using level set methods, a comparative study. *Int. J. Inf. Electr. Eng.* **2011**, *1*, 72–78.
8. Valliammal, N.; Geethalakshmi, S.N. A novel approach for plant leaf image segmentation using fuzzy clustering. *Int. J. Comput. Appl.* **2012**, *44*, 10–20.
9. Phung, S.L.; Bouzerdoum, A.; Chai, D. Skin segmentation using color pixel classification: Analysis and comparison. *IEEE Trans. Pattern Anal. Mach. Intell.* **2005**, *27*, 148–154.
10. Jones, M.J.; Rehg, J.M. Statistical color models with application to skin detection. *J. Comput. Vis.* **2002**, *46*, 81–96.
11. Vezhnevets, V.; Sazonov, V.; Andreeva, A. A survey on pixel-based skin color detection techniques. *Proc. Graph.* **2003**, *3*, 85–92.
12. Prati, A.; Mikić, I.; Trivedi, M.M.; Cucchiara, R. Detecting moving shadows: Formulation, algorithms and evaluation. *IEEE Trans. Pattern Anal. Mach. Intell.* **2003**, *25*, 918–923.
13. Fleyeh, H. Color detection and segmentation for road and traffic signs. *Cybern. Intell. Syst.* **2004**, *2*, 809–814.
14. Kumar, P.; Sengupta, K.; Lee, A. A comparative study of different color spaces for foreground and shadow detection for traffic monitoring system. In Proceedings of the The IEEE 5th International Conference onIntelligent Transportation Systems, Singapore, 6 September 2002; pp. 100–105.
15. Khattab, D.; Ebied, H.M.; Hussein, A.S.; Tolba, M.F. Color image segmentation based on different color space models using automatic GrabCut. *Sci. World J.* **2014**, *127*, 1–10.
16. Wang, X.; Hansch, R.; Ma, L.; Hellwich, O. Comparison of different color spaces for image segmentation using Graph-Cut. In Proceedings of the International Conference on Computer Vision Theory and Applications (VISAPP), Lisbon, Portugal, 5–8 January 2014; pp. 301–308.
17. Muselet, D.; Macaire, L. Combining color and spatial information for object recognition across illumination changes. *Pattern Recognit. Lett.* **2007**, *28*, 1176–1185.
18. Golzarian, M.R.; Lee, M.K.; Desbiolles, J.M.A. Evaluation of color indices for improved segmentation of plant images. *Trans. ASABE* **2012**, *55*, 261–273.
19. Bianco, S.; Cusano, C.; Napoletano, P.; Schettini, R. Improving CNN-Based Texture Classification by Color Balancing. *J. Imaging* **2017**, *3*, 33.
20. Levina, E.; Bickel, P. The earth mover distance is the Mallows distance: Some insights from statistics. In Proceedings of the IEEE International Conference on Computer Vision, Vancouver, BC, Canada, 7–14 July 2001; Volume 2, pp. 251–256.
21. Zhao, Q.; Brennan, S.; Tao, H. Differential EMD tracking. In Proceedings of the IEEE International Conference on Computer Vision, Rio de Janeiro, Brazil, 14–21 October 2007; pp. 1–8.
22. Kumar, P.; Dick, A. Adaptive earth mover distance-based Bayesian multi-target tracking. *Comput. Vis. IET* **2013**, *7*, 246–257.
23. Gevers, T.; Weijer, J.V.D.; Stokman, H. *Color Feature Detection: An Overview; Color Image Processing: Methods and Applications*; CRC Press: Boca Raton, FL, USA, 2006.
24. Gevers, T.; Stokman, H. Robust histogram construction from color invariants for object recognition. *IEEE Trans. Pattern Anal. Mach. Intell.* **2004**, *26*, 113–118.
25. Bianco, S.; Gasparini, F.; Schettini, R. Adaptive Skin Classification Using Face and Body Detection. *IEEE Trans. Image Process.* **2015**, *24*, 4756–4765.
26. Wyszecki, G.; Stiles, W.S. *Color Science: Concepts and Methods, Quantitative Data and Formulae*; Wiley: Hoboken, NJ, USA, 2000; Chapter 6.
27. Busin, L.; Vandenbroucke, N.; Macaire, L. Color spaces and image segmentation. *Adv. Imaging Electr. Phys.* **2009**, *151*, 65–168.

28. Rubner, Y.; Tomasi, C.; Guibas, L.J. The earth mover distance as a metric for image retrieval. *Int. J. Comput. Vis.* **2000**, *40*, 99–121.
29. Hitchcock, F.L. The distribution of a product from several sources to numerous localities. *J. Math. Phys.* **1941**, *20*, 224–230.
30. Mahalanobis, P.C. On the generalised distance in statistics. *Proc. Natl. Inst. Sci. India* **1936**, *2*, 49–55.
31. Liu, Y.; Schmidt, K.L.; Cohn, J.F.; Mitra, S. Facial asymmetry quantification for expression invariant human identification. *Comput. Vis. Image Underst.* **2003**, *91*, 138–159.
32. Liu, Y.; Teverovskiy, L.; Carmichael, O.; Kikinis, R.; Shenton, M.; Carter, C.; Stenger, V.; Davis, S.; Aizenstein, H.; Becker, J.; et al. Discriminative MR image feature analysis for automatic schizophrenia and Alzheimer's disease classification. In *Medical Image Computing and Computer-Assisted Intervention—MICCAI 2004*; Barillot, C.; Haynor, D.; Hellier, P., Eds.; Springer: Berlin/Heidelberg, Germany, 2004; Volume 3216, pp. 393–401.
33. Collins, R.T.; Liu, Y.; Leordeanu, M. Online selection of discriminative tracking features. *IEEE Trans. Pattern Anal. Mach. Intell.* **2005**, *27*, 1631–1643.

Journal of
Imaging

MDPI

Article

Exemplar-Based Face Colorization Using Image Morphing

Johannes Persch [1], Fabien Pierre [2,*] and Gabriele Steidl [1,3]

1 Department of Mathematics, Technische Universität Kaiserslautern, Paul-Ehrlich-Str. 31, 67663 Kaiserslautern, Germany; persch@mathematik.uni-kl.de (J.P.); steidl@mathematik.uni-kl.de (G.S.)
2 Laboratoire Lorrain de Recherche en Informatique et ses Applications, UMR CNRS 7503, Université de Lorraine, INRIA projet Magrit, 54500 Vandoeuvre-les-Nancy, France
3 Fraunhofer ITWM, Fraunhofer-Platz 1, 67663 Kaiserslautern, Germany
* Correspondence: fabien.pierre@univ-lorraine.fr

Received: 30 May 2017; Accepted: 19 October 2017; Published: 31 October 2017

Abstract: Colorization of gray-scale images relies on prior color information. Exemplar-based methods use a color image as source of such information. Then the colors of the source image are transferred to the gray-scale target image. In the literature, this transfer is mainly guided by texture descriptors. Face images usually contain few texture so that the common approaches frequently fail. In this paper, we propose a new method taking the geometric structure of the images rather their texture into account such that it is more reliable for faces. Our approach is based on image morphing and relies on the YUV color space. First, a correspondence mapping between the luminance Y channel of the color source image and the gray-scale target image is computed. This mapping is based on the time discrete metamorphosis model suggested by Berkels, Effland and Rumpf. We provide a new finite difference approach for the numerical computation of the mapping. Then, the chrominance U,V channels of the source image are transferred via this correspondence map to the target image. A possible postprocessing step by a variational model is developed to further improve the results. To keep the contrast special attention is paid to make the postprocessing unbiased. Our numerical experiments show that our morphing based approach clearly outperforms state-of-the-art methods.

Keywords: image colorization; morphing; face images

1. Introduction

Colorization consists in adding color information to a gray-scale images. This technique is used for instance by the cinema industry to make old productions more attractive. As usual we can consider a gray-scale image as luminance channel Y of an RGB image [1–4]. The Y channel is defined as a weighted average of the RGB channels, see, e.g., [5]:

$$Y = 0.299R + 0.587G + 0.114B.$$

In addition to the luminance channel, two chrominance channels, called U and V, enable to recover the RGB image. Recovering an RGB image from the luminance channel alone is an ill-posed problem and requires additional information [3,6]. This information is provided in the literature in two different ways, namely by *manual* [1,4,7] or *exemplar-based* [3,6,8] methods. In the first one the user augments the image by some color strokes as basis for the algorithm to compute the color of each pixel. The colorization of a complex scene by a manual prior can be a tedious work for the user [2]. In the second approach a color image is used as a source of information. Here the results strongly depend on the choice of the image. Therefore it is often called *semi-automatic* [1].

In this paper, we focus on the exemplar-based methods. A common back-bones of these techniques is the matching of the images. First, the color *source* image is transformed to a gray-scale image, referred to as *template* which is compared with the input gray-scale image called *target*. The main issue of the exemplar-based colorization consists in matching the pixels of the template and the target gray-scale images. The basic hypothesis in the literature is that the color content is similar in similar texture patches [8]. Then the main challenge is the choice of appropriate texture descriptors.

In the literature, the exemplar-based methods come along with various matching procedures. For instance, Gupta et al. [6] use SURF and Gabor features, Irony et al. [2] use DCT, Welsh et al. [8] use standard-deviation, etc. Some of them are done after an automatic segmentation, whereas others use local information. The seminal paper on exemplar-based colorization by Welsh et al. [8] was inspired by the texture synthesis of Efros and Leung [9] and uses basic descriptors for image patches (intensity and standard-deviation) to describe the local texture. Pierre et al. [3] proposed an exemplar-based framework based on various metrics between patches to produce a couple of colorization results. Then a variational model with total variation like regularization is applied to choose between the different results in one pixel with a spatial regularity assumption. The approach of Irony et al. [2] is built on the segmentation of the images by a mean-shift algorithm. The matching between segments of the images is computed from DCT descriptors which analyse the textures. The method of Gupta et al. [6] is rather similar. Here, an over-segmentation (SLIC, see, e.g., [10]) is used instead of the mean-shift algorithm. The comparison between textures in done by SURF and Gabor features. Chen et al. [11] proposed a segmentation approach based on Bayesian image matching which can also deal with smooth images including faces. The authors pointed out that the colorization of faces is a particular hard problem. However, their approach uses a manual matching between objects to skirt the problem of smooth parts. Charpiat et al. [12] ensured spatial coherency without segmenting, but their method involves many complex steps. The texture discrimination is mainly based on SURF descriptors. In the method of Chia et al. [13], the user has manually to segment and label the objects and the algorithm finds similar segments in a set of images available in the internet. Recently a convolutional neural network (CNN) has been used for colorization by Zhang et al. [14] with promising results. Here the colorization is computed from a local description of the image. However, no regularization is applied to ensure a spatial coherence. This produces "halo effects" near strong contours. All the described methods efficiently distinguish textures and possibly correct them with variational approaches, but fail when similar textures have to be colorized with different colors. This case arises naturally for face images. Here the smooth skin is considered nearly as a constant part. Thus, as we show in this paper, when the target image contains constant parts outside the face, the texture-based methods fail.

In this paper,we propose a new technique for the colorization of face images guided by image morphing. Our framework relies on the hypothesis that the global shape of faces is similar. The matching of the two images is performed by computing a morphing map between the target and the template image. Image morphing is a generic term for smooth image transition which is an old problem in image processing and computer vision. For example in feature based morphing only specific features are mapped to each other and the whole deformation is then calculated by interpolation. Such method [15] was used by the film industry for example in the movie *Willow*. For an overview of this and similar techniques see also [16,17]. A special kind of image morphing, the so-called metamorphosis was proposed by Miller, Trouvé and Younes [18–20]. The metamorphosis model can be considered as an extension of the flow of diffeomorphism model and its large deformation diffeomorphic metric mapping framework [21–25] in which each image pixel is transported along a flow determined by a diffeomorphism. As an extension the metamorphosis model allows the variation of image intensities along trajectories of the flow. Shooting methods for the metamorphosis model were developed e.g., in [26]. For a metamorphosis regression model and corresponding shooting methods we refer to [27]. A comprehensive overview over the topic is given in the book [28] as well as the review article [29], for a historic account see also [30]. In this paper, we build up on the time discrete metamorphosis model by Berkels, Effland and Rumpf [31]. In contrast to these authors we apply a finite difference approach for the computation of the morphing maps. This involves the solution of a chain of registration problems in each iteration step. There exists a rich literature on registration problems, e.g., [32–35], and we refer to the books of Modersitzki [36,37] for an overview.

Having the morphing map available, the chrominance channels can be transported by this map to the target image, while preserving its luminance channel. This gives very good results and outperforms state-of-the-art methods. For some images we can further improve the quality by applying a variational post-processing step. Our variational model incorporates a total variation like regularization term which takes the edges of the target image into account. This was also proposed by one of the authors of this paper in [3]. The method is accomplished by adapting a procedure of Deledalle et al. [38] to keep the contrast of the otherwise biased total variation method.

The outline of the paper is as follows: in Section 2 we sketch the ideas of the morphing approach. In particular, we show how the morphing map is computed with an alternating minimization algorithm and describe our finite difference approach. Details of the computation are shifted to the Appendix. Section 3 deals with the color transfer. Having the morphing map at hand we explain the transfer of the chrominance values. In particular we address the necessary scalings. Sometimes it is useful to apply a variational model with a modified total variation regularization as post-processing step to remove possible spatial inconsistencies. Such a procedure is developed in Section 4. Numerical experiments demonstrate the very good performance of our algorithm in Section 5. The paper ends with conclusions in Section 6.

2. Image Morphing

Our colorization method is based on the time discrete image metamorphosis model [31]. We briefly explain the model used also in the numerical part of [31] in Section 2.1 and describe our numerical realization for digital images by a finite difference approach in Section 2.2. Note that in [31] a finite element approach for the spatial discretization was proposed without a detailed description. Finite element methods are highly flexible and can be also applied, e.g., for shape metamorphosis. However, having the rectangular structure of the image grid in mind, we propose to use finite differences for the spatial discretization. Then, in Step 1 of our alternating algorithm, we can build up on registration methods proposed e.g., by Haber and Modersitzki [34].

2.1. Morphing Model Based on [31]

Let $\Omega \subset \mathbb{R}^2$ be an open, bounded domain with Lipschitz continuous boundary. We are given a gray-value template image $I_{\text{temp}} : \Omega \to \mathbb{R}$ and a target image $I_{\text{tar}} : \Omega \to \mathbb{R}$ which are supposed to be continuously differentiable and compactly supported. For $K \geq 2$ set

$$I_0 := I_{\text{temp}}, \quad I_K := I_{\text{tar}}.$$

In our application, the template image will be the luminance channel of the color source image and the target image the gray-scale image we want to colorize. We want to find a sequence of $K-1$ images $\mathbf{I}$ together with a sequence of mappings $\boldsymbol{\varphi}$ on Ω, i.e.,

$$\mathbf{I} := (I_1, \ldots, I_{K-1}), \quad \boldsymbol{\varphi} := (\varphi_1, \ldots, \varphi_K),$$

such that

$$I_k \approx I_{k-1} \circ \varphi_k \quad \text{for all } x \in \Omega,$$

see Figure 1, and the deformations φ_k have a small linearized elastic potential defined below. To this end, we suppose for $k = 1, \ldots, K$ that φ_k is related to the displacement v_k by

$$\varphi_k(\mathbf{x}) = \mathbf{x} - v_k(\mathbf{x}) = \begin{pmatrix} x - v_{k,1}(\mathbf{x}) \\ y - v_{k,2}(\mathbf{x}) \end{pmatrix}, \quad \mathbf{x} = (x, y)^{\mathrm{T}} \in \Omega, \tag{1}$$

and set $\mathbf{v} := (v_1, \ldots, v_K)$. The (Cauchy) strain tensor of the displacement $v = (v_1, v_2)^{\mathrm{T}} : \Omega \to \mathbb{R}^2$ is defined by

$$\varepsilon(v) := \frac{1}{2}(\nabla v + \nabla v^{\mathrm{T}}) = \begin{pmatrix} \partial_x v_1 & \frac{1}{2}(\partial_y v_1 + \partial_x v_2) \\ \frac{1}{2}(\partial_y v_1 + \partial_x v_2) & \partial_y v_2 \end{pmatrix},$$

where ∇v denotes the Jacobian of v. The linearized elastic potential is given by

$$\mathcal{S}(v) := \int_\Omega \mu \,\mathrm{trace}\left(\varepsilon^{\mathrm{T}}(v)\varepsilon(v)\right) + \frac{\lambda}{2}\,\mathrm{trace}\left(\varepsilon(v)\right)^2 \,\mathrm{d}\mathbf{x}, \tag{2}$$

where $\mu, \lambda > 0$. Then we want to minimize

$$\mathcal{J}(\mathbf{I}, \mathbf{v}) := \sum_{k=1}^{K} \int_\Omega |I_k - I_{k-1} \circ \varphi_k|^2 + \mathcal{S}(v_k)\,\mathrm{d}\mathbf{x}, \quad \text{subject to} \quad I_0 = I_{\mathrm{temp}},\ I_K = I_{\mathrm{tar}}. \tag{3}$$

This functional was used in the numerical part of [31]. Note that the term $\mathcal{S}(v)$ may be accomplished by a higher order derivative

$$\int_\Omega |D^m v_k(x)|^2 \mathrm{d}x, \quad m > 2, \tag{4}$$

which ensures in the time continuous setting that φ is indeed a diffeomorphism, see [22]. In the time discrete setting the additional term does not guaranty that $\varphi_k, k = 1, \ldots, K$, is a diffeomorphism. Hence we do not include (4) into our functional (3), as the linearized elastic potential $\mathcal{S}$ is sufficient for our purposes.

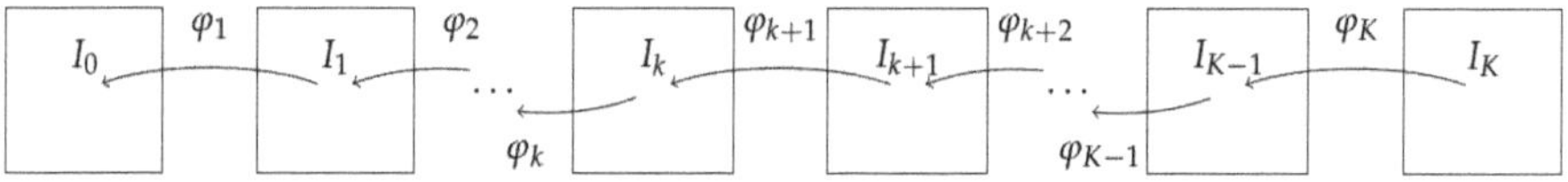

Figure 1. Illustration of the image path and the diffeomorphism path, where $I_0 := I_{\mathrm{temp}}$ and $I_K := I_{\mathrm{tar}}$ are the given template and target images.

The minimizer $(\mathbf{I}, \mathbf{v})$ of the functional provides us with both a sequence of images $\mathbf{I}$ along the approximate geodesic path and a sequence of displacements $\mathbf{v}$ managing the transport of the gray values through this image sequence. For finding a minimizer of (3) we alternate the minimization over $\mathbf{I}$ and $\mathbf{v}$:

1. Fixing $\mathbf{I}$ and minimizing over $\mathbf{v}$ leads to the following K single registration problems:

$$\operatorname*{arg\,min}_{v_k} \mathcal{J}_{\mathbf{I}}(v_k) := \int_\Omega |I_k - I_{k-1} \circ \varphi_k|^2 + \mathcal{S}(v_k)\,\mathrm{d}\mathbf{x}, \quad k = 1, \ldots, K, \tag{5}$$

   where φ_k is related to v_k by (1).
2. Fixing $\mathbf{v}$, resp., $\boldsymbol{\varphi}$ leads to solving the following image sequence problem

$$\operatorname*{arg\,min}_{\mathbf{I}} \mathcal{J}_{\boldsymbol{\varphi}}(\mathbf{I}) := \sum_{k=1}^{K} \int_\Omega |I_k - I_{k-1} \circ \varphi_k|^2 \,\mathrm{d}\mathbf{x}. \tag{6}$$

   This can be done via the linear system of equations arising from Euler-Lagrange equation of the functional which we describe in the Appendix A.

Note that Miller et al. [39] considered for two *given* image sequences $\mathbf{I} := (I_j)_{j=1}^N$ and $\mathbf{J} := (J_j)_{j=1}^N$ (e.g., related to corresponding image patches) the registration problem

$$\underset{v}{\arg\min} \int_\Omega \sum_{j=1}^N |I_j - J_j \circ \varphi|^2 + \mathcal{S}(v)\, d\mathbf{x},$$

where φ is related to v by (1). In contrast to our problem these authors search for the same mapping v and the N template-target pairs are known.

2.2. Space Discrete Morphing Model

When dealing with digital images we have to work in a spatially discrete setting. Let

$$\mathcal{G}_p := \{1, \ldots, n_1\} \times \{1, \ldots, n_2\}$$

be the (primal) image grid, i.e., $I_{\text{ref}} : \mathcal{G}_p \to \mathbb{R}$ and $I_{\text{temp}} : \mathcal{G}_p \to \mathbb{R}$. We discretize the integrals on the integration domain $\bar{\Omega} := [\frac{1}{2}, n_1 + \frac{1}{2}] \times [\frac{1}{2}, n_2 + \frac{1}{2}]$ by the midpoint quadrature rule, i.e., with pixel values defined on $\mathcal{G}_p$. For discretizing the operators in (2) we work as usual on staggered grids. For the application of mimetic grid techniques in optical flow computation see also [40]. Let

$$\mathcal{G}_d := \{\tfrac{3}{2}, \ldots, n_1 - \tfrac{1}{2}\} \times \{\tfrac{3}{2}, \ldots, n_2 - \tfrac{1}{2}\}.$$

be the (inner) dual grid, i.e., $\mathcal{G}_p$ shifted by $\frac{1}{2}$ in each direction, and

$$\mathcal{G}_1 := \{\tfrac{3}{2}, \ldots, n_1 - \tfrac{1}{2}\} \times \{1, \ldots, n_2\}, \quad \mathcal{G}_2 := \{1, \ldots, n_1\} \times \{\tfrac{1}{2}, \ldots, n_2 - \tfrac{3}{2}\}.$$

We start by considering the registration problems in the first alternating step (5) and turn to the image sequence problem in the second step (6) with (A2) afterwards.

Solution of the Registration Problems

Let us fix $k \in \{1, \ldots, K\}$ in (5) and abbreviate $T := I_{k-1}$, $R := I_k$ as template and reference image, resp., and $v := v_k$. Then we have to find a minimizer of the continuous registration problem

$$\mathcal{J}(v) := \int_\Omega |R(\mathbf{x}) - T(\mathbf{x} - v(\mathbf{x}))|^2 + \mathcal{S}(v)\, d\mathbf{x} \tag{7}$$

where the elastic potential in (2) can be rewritten as

$$\mathcal{S}(v) = \mu \int_\Omega (\partial_x v_1)^2 + \frac{1}{2}(\partial_y v_1 + \partial_x v_2)^2 + (\partial_y v_2)^2\, d\mathbf{x} + \frac{\lambda}{2} \int_\Omega (\partial_x v_1 + \partial_y v_2)^2\, d\mathbf{x}.$$

For the spatial discrete setting we assume that $T, R : \mathcal{G}_p \to \mathbb{R}^s$ are given. Further, we consider $v = (v_1, v_2)^{\mathrm{T}}$ with $v_1 : \mathcal{G}_1 \to \mathbb{R}$ and $v_2 : \mathcal{G}_2 \to \mathbb{R}$. In contrast to [34] we assume no flow over the image boundary and set

$$\begin{aligned} v_1(\tfrac{3}{2}, y) &= v_1(n_1 - \tfrac{1}{2}, y) = 0, \quad y \in \{1, \ldots, n_2\}, \\ v_2(x, \tfrac{3}{2}) &= v_2(x, n_2 - \tfrac{1}{2}) = 0, \quad x \in \{1, \ldots, n_1\}. \end{aligned}$$

See Figure 2 for an illustration. We approximate $\partial_x v_1$ for $\mathbf{x} = (x, y)^{\mathrm{T}} \in \mathcal{G}_p$ by

$$(D_{1,x} v_1)(x, y) := \begin{cases} 0, & x = 1, \\ v_1(x + \frac{1}{2}, y) - v_1(x - \frac{1}{2}, y), & x = 2, \ldots, n_1 - 1, \\ 0, & x = n_1, \end{cases}$$

and $\partial_y v_1$ for $x \in \{1, \dots n_1 - 1\}$ and $y \in \{1, \dots, n_2 - 1\}$ by

$$(D_{1,y}v_1)(x + \tfrac{1}{2}, y + \tfrac{1}{2}) = \Big\{ \; v_1(x + \tfrac{1}{2}, y + 1) - v_1(x + \tfrac{1}{2}, y), \quad x = 1, \dots, n_1 - 1,$$

and similarly for the derivatives with respect to v_2. Then we obtain

$$\mathcal{S}(v) = \sum_{\mathbf{x} \in \mathcal{G}_p} \mu \left((D_{1,x}v_1)^2(\mathbf{x}) + (D_{2,y}v_2)^2(\mathbf{x}) \right) + \frac{\lambda}{2} \left((D_{1,x}v_1)(\mathbf{x}) + (D_{2,y}v_2)(\mathbf{x}) \right)^2 \tag{8}$$
$$+ \sum_{\mathbf{x} \in \mathcal{G}_d} \frac{\mu}{2} \left((D_{1,y}v_1)(\mathbf{x}) + (D_{2,x}v_2)(\mathbf{x}) \right)^2.$$

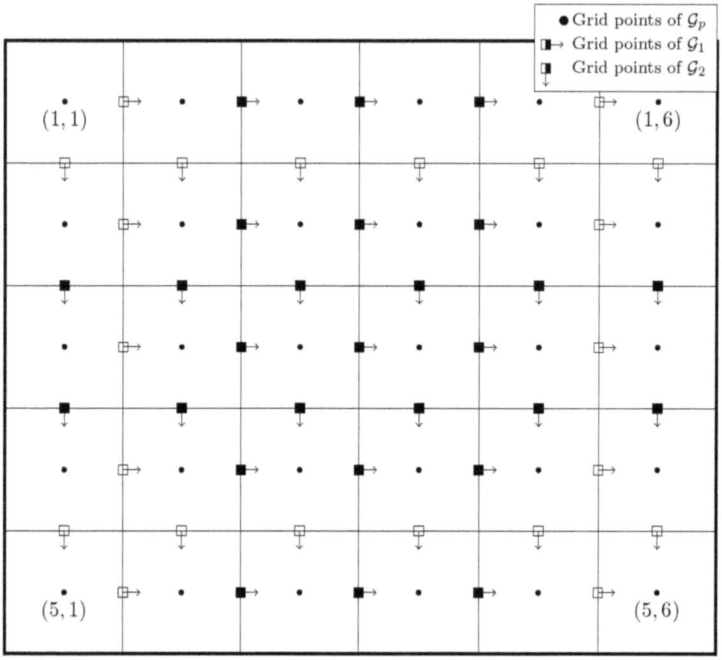

Figure 2. Illustration of the grids, where empty boxes mean zero movement.

To discretize the data term in (7) we need to approximate $T(\mathbf{x} - v(\mathbf{x}))$. Since v is not defined on $\mathcal{G}_p$ we use instead of v its averaged version $Pv := \mathcal{G}_p \to \mathbb{R}^2$ given by

$$(P_1v_1)(x, y) := \begin{cases} 0, & x = 1 \\ \frac{1}{2} \left(v_1(x - \frac{1}{2}, y) + v_1(x + \frac{1}{2}, y) \right), & x = 2, \dots, n_1 - 1, \\ 0, & x = n_1 \end{cases}$$

and similarly for P_2v_2 in y-direction. Let the continuous function $\tilde{T}$ on $\Omega := [1, n_1] \times [1, n_2]$ be constructed from given values of T on $\mathcal{G}_p$ by interpolation. Here we restrict our attention to bilinear interpolations, see (A3), but more sophisticated approaches are possible. In summary, the discrete registration problem reads

$$\arg\min_v \mathcal{J}(v) := \sum_{\mathbf{x} \in \mathcal{G}_p} |R(\mathbf{x}) - \tilde{T}(\mathbf{x} - (Pv)(\mathbf{x}))|^2 + \mathcal{S}(v),$$

where $\mathcal{S}(v)$ is given by (8). We solve this nonlinear least squares problem by a Gauss-Newton like method described in the Appendix B.

Multilevel Strategy

As usual in optical flow and image registration we apply a coarse-to-fine strategy. First we iteratively smooth and downsample our given images. On the coarsest level we perform a registration to obtain a deformation on the coarse grid. We apply a bilinear interpolation v of this deformation to construct intermediate images on the finer level by

$$I_k(x) = \left(T + \frac{k}{\tilde{K}}(R \circ \varphi^{-1} - T)\right)\left(x - \tfrac{k}{\tilde{K}}v(x)\right), \quad \varphi(x) = x - v(x), \tag{9}$$

where T, R are the start and end images at the new level. Then we use the alternating algorithm on this level with $\tilde{K} \leq K$ images to obtain deformations and intermediate images on this level. Going to the finer level by bilinear interpolation, we construct more intermediate images by interpolating between neighboring images with (9). We repeat this process until we reach the maximal number K of images and finest level. The multilevel strategy of the algorithm is sketched in Algorithm 1.

Algorithm 1 Morphing Algorithm (informal).

```
T_0 := T, R_0 := R, Ω_0 := Ω
create image stack (T_0, ..., T_l), (R_0, ..., R_l) on (Ω_0, ..., Ω_l) by smoothing and downsampling
solve (3) for T_l, R_l with K = 1, for ṽ
l → l − 1
use bilinear interpolation to get v on Ω_l from ṽ
obtain K̃_l images I_l^(0) from T_l, R_l, v by (9)
while l ≥ 0 do
    find image path Ĩ_l and deformation path ṽ_l minimizing (3) with initialization I_l^(0)
    l → l − 1
    if l > 0 then
        use bilinear interpolation to get I_l and v_l on Ω_l
        for k = 1, ..., K̃_l do
            calculate K̃_l intermediate images between I_{l,k−1}, I_{l,k} with v_{l,k} using (9)
I := I_0
```

3. Face Colorization

In this section, we describe a method to colorize a gray-scale image based on the morphing map between the luminance channel of a source image (template) and the present image (target). The idea consists in transferring of the chrominance channels from the source image to the target one. To this end, we work in the YUV color space. For the importance of the chrominance channels in color imaging, see, e.g., [41,42]. While in RGB images the color channels are highly correlated, the YUV space shows a suitable decorrelation between the luminance channel Y and the two chrominance channels U, V. The transfer from the RGB space to YUV space is given by (see, e.g., [5])

$$\begin{pmatrix} Y \\ U \\ V \end{pmatrix} = \begin{pmatrix} 0{,}299 & 0{,}587 & 0{,}114 \\ -0{,}14713 & -0{,}28886 & 0{,}436 \\ 0{,}615 & -0{,}51498 & -0{,}10001 \end{pmatrix} \begin{pmatrix} R \\ G \\ B \end{pmatrix}. \tag{10}$$

Most of the colorization approaches are based on the hypothesis that the target image is the luminance channel of the desired image. Thus, the image colorization process is based on the computation of the unknown chrominance channels.

3.1. Luminance Normalization

The first step of the algorithm consists in transforming the RGB source image to the YUV image by (10). The range of the target gray-value image and the Y channel of the source image may differ making the meaningful comparison between these images not possible.

To tackle this issue, most of state-of-the art methods use a technique called *luminance remapping* which was introduced in [43]. This affine mapping between images which aims to fit the average and the standard deviation of the target and the template images is defined as

$$I_{\text{temp}} := \sqrt{\frac{\text{var}(I_{\text{tar}})}{\text{var}(I_Y)}} \, (I_Y - \text{mean}(I_Y)) + \text{mean}(I_{\text{tar}}),$$

where mean is the average of the pixel values, and var is the empirical variance.

3.2. Chrominance Transfer by the Morphing Maps

Next we compute the morphing map between the two gray-scale images I_{temp} and I_{tar} with model (3). This results in the deformation sequence φ which produces the resulting map from the template image to the target one by concatenation

$$\Phi = \varphi_1 \circ \varphi_2 \circ \ldots \circ \varphi_K.$$

Due to the discretization of the images, the map Φ is defined, for images of size $n_1 \times n_2$, on the discrete grid $\mathcal{G} := \{1, \ldots, n_1\} \times \{1, \ldots, n_2\}$:

$$\Phi : \mathcal{G} \to [1, n_1] \times [1, n_2], \quad x \mapsto \Phi(x),$$

where $\Phi(x)$ is the position in the source image which corresponds to the pixel $x \in \mathcal{G}$ in the target image. Now we colorize the target image by computing its chrominance channels, denoted by $(U_{\text{tar}}(x), V_{\text{tar}}(x))$ at position x as

$$(U_{\text{tar}}(x), V_{\text{tar}}(x)) := (U(\Phi(x)), V(\Phi(x))). \tag{11}$$

The chrominance channels of the target image are defined on the image grid $\mathcal{G}$, but usually $\Phi(x) \notin \mathcal{G}$. Therefore the values of the chrominance channels at $\Phi(x)$ have to be computed by interpolation. In our algorithm we use just bilinear interpolation which is defined for $\Phi(x) = (p, q)$ with $(p, q) \in [i, i+1] \times [j, j+1]$, $(i, j) \in \{1, \ldots, n_1 - 1\} \times \{1, \ldots, n_2 - 1\}$ by

$$U(\Phi(x)) = U(p, q) := (i + 1 - p, p - i) \begin{pmatrix} U(i,j) & U(i,j+1) \\ U(i+1,j) & U(i+1,j+1) \end{pmatrix} \begin{pmatrix} j + 1 - q \\ q - j \end{pmatrix}.$$

Finally, we compute a colorized RGB image from its luminance $I_{\text{tar}} = Y_{\text{tar}}$ and the chrominance channels (11) by the inverse of (10):

$$\begin{pmatrix} R(x) \\ G(x) \\ B(x) \end{pmatrix} = \begin{pmatrix} 1 & 0 & 1,13983 \\ 1 & -0,39465 & -0,58060 \\ 1 & 2,03211 & 0 \end{pmatrix} \begin{pmatrix} Y_{\text{tar}}(x) \\ U_{\text{tar}}(x) \\ V_{\text{tar}}(x) \end{pmatrix} \tag{12}$$

for all $x \in \mathcal{G}$. Figure 3 (Image segements with unified background of George W. Bush https://commons.wikimedia.org/wiki/File:George-W-Bush.jpeg and Barack Obama https://commons.wikimedia.org/wiki/File:Barack_Obama.jpg both in public domain.) summarizes our color transfer method.

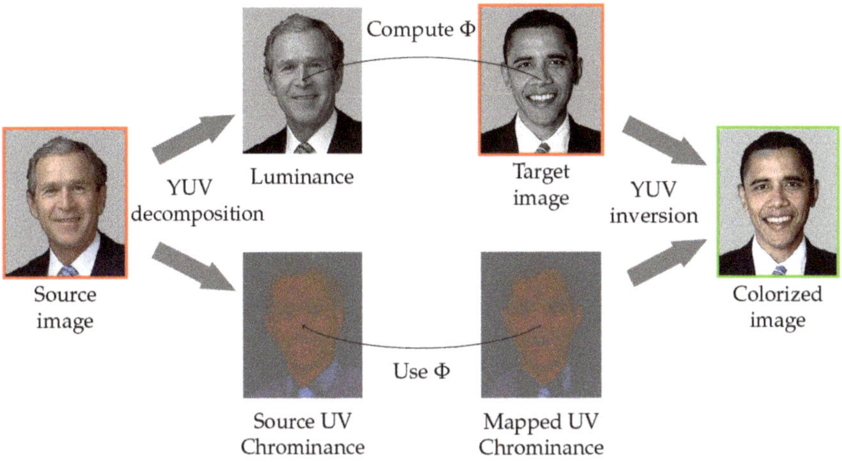

Figure 3. Overview of the color transfer. The mapping Φ is computed from Model (3) between the luminance channel of the source image and the target one. From this map, the chrominances of the source image are mapped. Finally, from these chrominances and the target image the colorization result is computed.

4. Variational Methods for Chrominance Postprocessing

Sometimes the color transfer computed from the morphing map can be disturbed by artifacts. To improve the results, post-processing steps are usually applied in the literature.

Variational approaches are frequently applied in image colorization either directly or as a post-processing step, see, e.g., [1,3,44]. For instance, the technique of Gupta et al. [6] uses the chrominance diffusion approach of Levin et al. [1].

In this paper, we propose a variational method with a total variation based regularization as a post-processing step to remove possible artifacts. We build up on the model [3] suggested by one of the authors. This variational model uses a functional with a specific regularization term to avoid "halo effects". More precisely, we consider the minimizer of

$$\hat{u} = (\hat{U}, \hat{V}) = \underset{(U,V)}{\arg\min} \, \mathrm{TV}_{Y_{\mathrm{tar}}}(U, V) + \alpha \int_\Omega |U(x) - U_{\mathrm{tar}}(x)|^2 + |V(x) - V_{\mathrm{tar}}(x)|^2 dx, \tag{13}$$

with

$$\mathrm{TV}_{Y_{\mathrm{tar}}}(U, V) := \int_\Omega \sqrt{\gamma |\nabla Y_{\mathrm{tar}}|^2 + |\nabla U|^2 + |\nabla V|^2} \, dx.$$

The first term in (13) is a coupled total variation term which enforces the chrominance channels to have a contour at the same location as the target gray-value image. The data fidelity term is the classical squared L_2-norm of the differences of the given and the desired chrominance channels. Note that the model in [3] contains an additional box constraint.

The parameter γ manages the coupling of the chrominance channels with the luminance one. It has been shown in [3] that a parameter value around 25 can be used for most images. The parameter α is related to the object size in the images. If some colors on large objects leak on small ones, this parameter has to be increased. On the other hand, if some artifacts are still present in the result, the value of α has to be decreased. We apply the primal-dual Algorithm 2 to find the minimizer of the strictly convex model (13). It uses an update on the step time parameters τ and σ, as proposed by Chambolle and Pock [45], as well as a relaxation parameter θ to speed-up the convergence. Here we use the abbreviation $b := (U_{\mathrm{tar}}, V_{\mathrm{tar}})$ and $u := (U, V)$. Further, p is the dual variable which is pixel-wise in $\mathbb{R}^6$. The parameters τ and σ are intern time step sizes. The operator div stands for the discrete divergence and ∇ for the discrete gradient. Further, the proximal mapping $P_{\mathcal{B}}$ is given pixel-wise, for $p \in \mathbb{R}^6$ by

$$P_{\mathcal{B}}(p) = \frac{\hat{p}}{\max\left(1, \|\hat{p}\|_2^2\right)}, \quad \text{where} \quad \hat{p} := p - \sigma \begin{pmatrix} 0 \\ 0 \\ 0 \\ 0 \\ \partial_x Y \\ \partial_y Y \end{pmatrix}. \tag{14}$$

Algorithm 2 Minimization of (13).

1: $u^0 \leftarrow b, \overline{u}^0 \leftarrow u^0$
2: $p^0 \leftarrow \nabla u^0$
3: $\sigma \leftarrow 0.001, \quad \tau \leftarrow 20$
4: **for** $n > 0$ **do**
5: $\quad p^{n+1} \leftarrow P_{\mathcal{B}}(p^n + \sigma \nabla \overline{u}^n)$
6: $\quad u^{n+1} \leftarrow \dfrac{u^n + \tau\left(\operatorname{div}(p^{n+1}) + \alpha b\right)}{1 + \tau\alpha}$
7: $\quad \theta = 1/\sqrt{1+\tau\alpha}$
8: $\quad \tau = \theta\tau \quad \sigma = \sigma/\theta$
9: $\quad \overline{u}^{n+1} \leftarrow u^{n+1} + \theta(u^{n+1} - u^n)$
10: $\hat{u} \leftarrow u^{+\infty}$.

As mentioned in the work of Deledalle et al. [38], the minimization of the TV-L_2 model produces a biased result. This bias causes a lost of contrast in the case of gray-scale images, whereas it is visible as a lost of colorfulness in the case of model (13). The authors of [38] describe an algorithm to remove such bias. In this paper, we propose to modify this method for our model (13) in order to enhance the result of Algorithm 2.

The CLEAR method of [38] relies on the *refitting estimator* $\mathcal{R}_{\hat{x}}(y)$ of the data y from the biased estimation $\hat{x}(y)$:

$$\mathcal{R}_{\hat{x}}(y) \in \underset{h \in \mathcal{H}_{\hat{x}}}{\arg\min} \|h(y) - y\|_2^2 \tag{15}$$

where $\mathcal{H}_{\hat{x}}$ is defined as the set of maps $h : \mathbb{R}^n \to \mathbb{R}^p$ satisfying, $\forall y \in \mathbb{R}^n$:

$$h(y) = \hat{x}(y) + \rho J_{\hat{x}(y)}(y - \hat{x}(y)), \text{ with } \rho \in \mathbb{R},$$

where $J_{\hat{x}(y)}$ is the Jacobian of the biased estimator with respect to the data y:

$$J_{\hat{x}(y)} d = \lim_{\varepsilon \to 0} \frac{\hat{x}(y + \varepsilon d) - \hat{x}(y)}{\varepsilon}.$$

$J_{\hat{x}(y)}$ contains some structural information of the biased estimator $\hat{x}$ such as the jumps. A closed formula for ρ can be given:

$$\rho = \begin{cases} \dfrac{\left\langle J_{\hat{x}(y)}(\delta) | \delta \right\rangle}{\|J_{\hat{x}(y)}(\delta)\|_2^2} & \text{if } J_{\hat{x}(y)}(\delta) \neq 0, \\ 1 & \text{otherwise.} \end{cases}$$

where $\delta = y - \hat{x}(y)$. With Equation (15), the refitting process is as closer as possible to the original data.

The final algorithm is summarized in Algorithm 3. Note that it uses the result $\hat{u}$ of Algorithm 1 as an input. The proximal mapping $\Pi_p(\tilde{p})$ within the algorithm is defined pixel-wise, for variables $p \in \mathbb{R}^6$ and $\tilde{p} \in \mathbb{R}^6$, as

$$\Pi_p(\tilde{p}) = \begin{cases} \hat{\hat{p}} & \text{if } \|\hat{p}\| < 1 \\ \frac{1}{\|\hat{p}\|}\left(\hat{\hat{p}} - \frac{\langle \hat{p}, \hat{\hat{p}} \rangle}{\|\hat{p}\|^2}\hat{p}\right) & \text{otherwise,} \end{cases}$$

where $\hat{p}$ and $\hat{\hat{p}}$ are defined as in (14).

Algorithm 3 Debiasing of Algorithm 2.

$u^0 = b, \overline{u}^0 = b$

$\delta \leftarrow b - \hat{u}$

$\tilde{u}^0 = \delta, \overline{\tilde{u}}^0 = \delta$

$p^0 \leftarrow \nabla u, \tilde{p}^0 \leftarrow \nabla \tilde{u}$

$\sigma \leftarrow 0.001, \quad \tau \leftarrow 20$

for $n \geq 0$ **do**

$\quad p^{n+1} \leftarrow P_B(p^n + \sigma \nabla \overline{u}^n)$

$\quad \tilde{p}^{n+1} \leftarrow \Pi_{p^n + \sigma \nabla \overline{u}^n}\left(\tilde{p}^n + \sigma \nabla \overline{\tilde{u}}^n\right)$

$\quad u^{n+1} \leftarrow \dfrac{u^n + \tau\left(\mathrm{div}(p^{n+1}) + \alpha b\right)}{1 + \tau\alpha}$

$\quad \tilde{u}^{n+1} \leftarrow \dfrac{\tilde{u}^n + \tau\left(\mathrm{div}(\tilde{p}^{n+1}) + \alpha \delta\right)}{1 + \tau\alpha}$

$\quad \theta = 1/\sqrt{1 + \tau\alpha}$

$\quad \tau = \theta\tau \quad \sigma = \sigma/\theta$

$\quad \overline{u}^{n+1} \leftarrow u^{n+1} + \theta(u^{n+1} - u^n)$

$\quad \overline{\tilde{u}}^{n+1} \leftarrow \tilde{u}^{n+1} + \theta(\tilde{u}^{n+1} - \tilde{u}^n)$

$\rho \leftarrow \begin{cases} \dfrac{\langle \tilde{u}^{+\infty} | \delta \rangle}{\|\tilde{u}^{+\infty}\|_2^2} & \text{if } \tilde{u}^{+\infty} \neq 0 \\ 1 & \text{otherwise.} \end{cases},$

$u_{\text{debiased}} \leftarrow \hat{u} + \rho \tilde{u}^{+\infty}.$

The results obtained at the different steps of the work-flow are presented for a particular image in Figure 4. First we demonstrate in Figure 4c that simply transforming the R, G, B channels via the morphing map gives no meaningful results. Letting our morphing map act to the chrominance channels of our source image and applying (12) with the luminance of our target image we get , e.g., Figure 4d which is already a good result. However, the forehead of Obama contains an artifact; a gray unsuitable color is visible here. After a post-processing of the chrominance channels by our variational method the artifacts disappear as can be seen in Figure 4e.

(a) Source (b) Target (c) RGB morphing (d) (UV) mapping (e) Final result

Figure 4. Illustration of the colorization steps of our algorithm. (**a**) The color source image; (**b**) The gray value target image; (**c**) The transport of the R, G, B channel via the morphing map is not suited for colorization; (**d**) The result with our morphing method is already very good; (**e**) It can be further improved by our variational post-processing.

5. Numerical Examples

In this section, we compare our method with

- the patch-based algorithm of Welsh et al. [8],
- the patch-based method of Pierre et al. [3], and
- the segmentation approach of Gupta et al. [6].

We implemented our morphing algorithm in Matlab 2016b and used the Matlab intern function for the bilinear interpolation. The parameters are summarized in Table 1. Here $\lambda = \mu$ and K are the parameters for the morphing step. The parameters γ and α appear in the variational model for post-processing and were only applied in three images. The number of deformations K depends on the size and details of the image, i.e., large detailed images need a larger K as small image having not so much details. Usually $K = 24$ was a good choice for our experiments. The parameter λ controls the smoothness of the deformation, i.e., a small value leads to large movements which might not be reasonable, while for large values the deformations become zero and the image metamorphosis is just a blending. In our experiments, a value of magnitude 10^{-2} lead to good results.

Table 1. Parameters for our numerical experiments.

Image	μ	K	γ	α
Figure 5-1. row	0.025	24	50	0.005
Figure 5-2. row	0.05	24	25	0.005
Figure 5-3. row	0.05	12	-	-
Figure 5-4. row	0.05	24	-	-
Figure 6-1. row	0.005	32	-	-
Figure 6-2. row	0.0075	18	50	0.05
Figure 6-3. row	0.04	24	-	-
Figure 7	0.0075	18	-	-
Figure 8	0.01	25	-	-
Figure 9-1. row	0.005	25	-	-
Figure 9-2. row	0.01	25	-	-
Figure 9-3. row	0.01	25	-	-

Figure 5. Comparison of our approach with state-of-the-art methods on photographies. In contrast to these methods our model is not based on texture comparisons, but on the morphing of the shapes. Therefore it is able to handle faces images, where the background has frequently a similar texture as the skin.

First we compare portraits in Figure 5 (Image segments of Steve Jobs https://commons.wikimedia.org/wiki/File:Steve_Jobs_Headshot_2010-CROP.jpg image courtesy of Matthew Yohe, Albert Einstein https://commons.wikimedia.org/wiki/File:Einstein_Portr_05936.jpg in public domain, Catherine Deneuve https://commons.wikimedia.org/wiki/File:Catherine_Deneuve_2010.jpg image courtesy of Georges Biard and Renée Deneuve https://commons.wikimedia.org/wiki/File:Ren%C3%A9e_Deneuve.jpg in public domain.) starting with the modern photographies in the first row. The approach of Welsh et al. [8] is based on a patch matching between images. The patch comparison is done with basic texture descriptors (intensity of the central pixel and standard-deviation of the patches). Since the background, as well as the skin are smooth, the distinction between them is unreliable if their intensities are similar. Moreover, since no regularization is used after the color transfer, some artifacts occur. For instance, some blue points appear on Obama's face, see Figure 5, first row. The approach of Pierre et al. [3] is based on more sophisticated texture features for patches and applies a variational model with total variation like regularization. With this approach the artifacts mentioned above are less visible. Nevertheless, the forehead of Obama is purple which is unreliable. The method of Gupta et al. [6] uses texture descriptors after an over-segmentation, see, e.g., SLIC [10]. The texture descriptors are based on SURF and Gabor features. In the case of the Obama image, the descriptors are not able to distinguish the skin and other smooth parts, leading to a background color different from the source image. Our method is able to colorize the second image in a more reasonable way, i.e., face and background color are different and the tie gets a blue color. However, our methods is not perfect so far. For example part of the forehead of Obama becomes gray is due to the gray hair of Bush, which has a gray value closer to the forehead as to the hair of Obama.

The second and the third rows of Figure 5 focus on the colorization of old photographies. This challenging problem is a real application of image colorization which is sought, e.g., by the cinema industry. Note that the texture of old images are disturbed by the natural grain which is not the case in modern photography. Thus, the texture comparison is unreliable for this application. This issue is visible in all the comparison methods. For the portrait of Einstein the background is not colorized with the color of the source. Moreover, the color of the skin is different from those of the person in the source image. For the picture of Deneuve, the color of her lips is not transferred to the target image (Deneuve's mother) with the state-of-the-art texture-based algorithms. With our morphing approach, the shapes of the faces are mapped. Thus, the lips, as well as the eyes and the skin are well colorized with a color similar to the source image. In the last row we have two images of the same person. Here the state-of-the-art-texture-based methods give unlikely results, especially the methods of Welsh et al. and Gupta et al. lead to a non smooth colorization of the background and the face, respectively. Our method provides a reasonable result with only small artifacts around the glasses.

Source. Target. Welsh et al. Gupta et al. Pierre et al. Our.

Figure 6. Results including painted faces. Only our morphing method is able to colorize the target images in an authentic way.

In Figure 6 (Image segments of self-portraits of Vincent van Gogh https://commons.wikimedia.org/wiki/File:Vincent_Willem_van_Gogh_102.jpg and https://commons.wikimedia.org/wiki/File:Vincent_van_Gogh_-_Self-Portrait_-_Google_Art_Project.jpg both in public domain, a photography of a woman https://pixabay.com/en/woman-portrait-face-model-canon-659352/ licensed CC0, a drawing of a woman https://pixabay.com/en/black-and-white-drawing-woman-actor-1724363/ licensed CC0, a color image of Marilyn Monroe https://www.flickr.com/photos/7477245@N05/5272564106 created by Luiz Fernando Reis, and a drawing of Marilyn Monroe https://pixabay.com/en/marilyn-monroe-art-draw-marilyn-885229/ licensed CC0), we provide results including painted images. Note that we use the same Van Gogh self-portraits as in [31]. Due to the low contrast of the ear and suit to the background we add here the same segmentation information as in [31], which means our images are two dimensional during the calculation for the results shown in the first row of Figure 6. In these examples the similarity of the shapes between the source and target images is again more reliable than the matching of the textures so that only our morphing approach produces suitable results. Consider in particular the lips of the woman in the second and third row. A non post-processed result for the woman in the second row is shown in Figure 7. Comparing the two images we see that only small details change but most of the colorization is done by the morphing.

Figure 7. Color transport along the image path.

Figure 7 visualizes the path of the color transfer. As the morphing approach calculates at the same time an image and mapping path, we can not only show the final colorization result, but also the way the color is transported along this path. We illustrate this by the image in second row of Figure 6. The UV channels are transported via the mapping path of the Y channels of the source and the target images, where every second image along the path is shown. We see that even though the right part of the images undergoes large transformations, the eyes and the mouth are moved to the correct places. Note that the final image does not contain a post-processing, in contrast to those in the second row of Figure 6. However, the result is already quite good.

Figure 8 is included to evaluate the quality of our colorization method. We took two photographs of the same person in different poses. Assuming that the color of the person does not change due to the movement, we have a ground truth target image and can do a quantitative comparison using the PSNR. We see that the visual superiority of our result is confirmed by the much higher PSNR.

In Figure 9 we considered three RGB image pairs. Using the luminance of one of the images as target and the other one as source and conversely we colorized the target image and computed the PSNR. In Table 2 the PSNR values of the colorization with our method and the state-of-the-art methods are shown. The performance of our method is very convincing for all these example images.

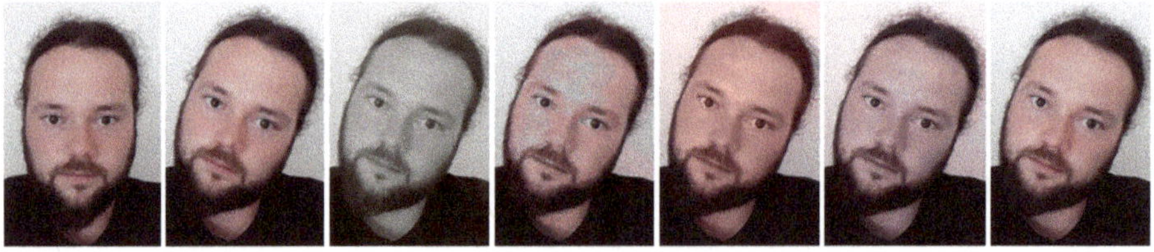

Figure 8. Results on a color image turned into a gray-scale one for a quantitative comparison. The qualitative comparisons with the state-of-the-art methods are confirmed by the PSNR measures.

Table 2. Comparison of the different PSNR values for the image pairs in Figure 9.

	Gray	Welsh et al. [8]	Gupta et al. [6]	Pierre et al. [3]	Our
Figure 9-1. row 1. pair	24.8023	20.0467	26.3527	33.7694	44.7808
Figure 9-1. row 2. pair	24.5218	23.9513	25.9457	34.4231	45.4682
Figure 9-2. row 1. pair	24.3784	22.6729	27.5586	32.0119	41.1413
Figure 9-2. row 2. pair	23.7721	23.2375	25.9375	30.1398	39.4254
Figure 9-3. row 1. pair	24.5950	30.3985	24.3112	31.5263	42.3861
Figure 9-3. row 2. pair	24.3907	27.7816	25.6207	31.8982	42.4092

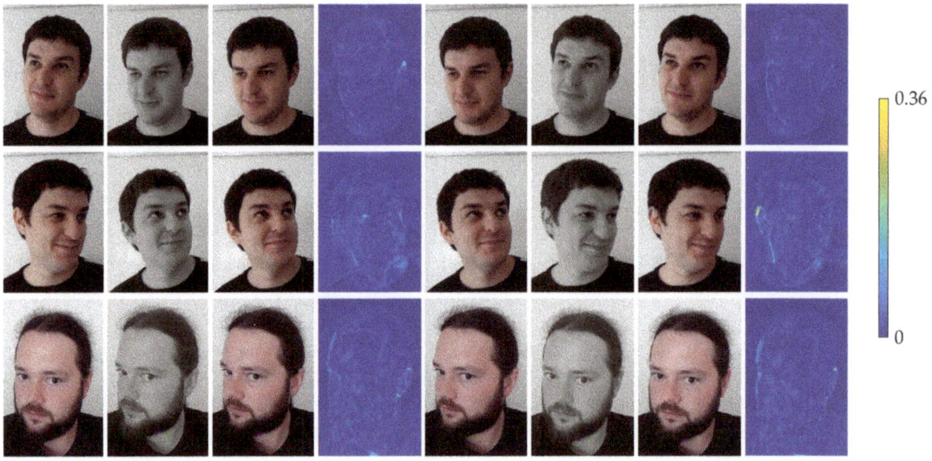

Figure 9. Multiple colorizations of known RGB-images with difference plots measured in Euclidean distance in $\mathbb{R}^3$.

Remark (Limitations of our approach). *In all our experiments our method leads to good results and outperforms the state-of-the-art colorization methods. However, the method is not perfect and the face images can not be arbitrarily deformed. Since the morphing is only done in the Y channel, the illumination of the faces should be similar. For example there is a shadow on the left side of the images in the second row of Figure 5. Mirroring one of the images would lead to bad matching results. The same holds true for the features themselves, so we can not expect to match bright hair perfectly to dark hair, as we see for example in Figure 4d.*

Since the deformations are close to diffeomorphisms, features should not be excluded or appear. For example, the left eye in the last row of Figure 9 starts in the middle of the face and moves to the edge, so a part of the skin-color stays in the eye. Similarly in the second matching here the eye moves from the edge to the inside, so the skin obtains the color from the background.

By a proper initialization the faces should be aligned in such a way that there are no similar features atop of each other, e.g., a left eye in the source should not be atop of the right eye of the target. However with more prior knowledge we could use the given features as landmarks and perform a landmark matching, e.g., [46], for the initialization on the coarsest level.

6. Conclusions

In this paper, we propose an finite difference method to compute a morphing map between a source and a target gray-value image. This map enables us to transfer the color from the source image to the target one, based on the shapes of the faces. An additional variational post-processing step with a luminance guided total variation regularization and an update to make the result unbiased may be added to remove some possible artifacts. The results are very convincing and outperform state-of-the-art approaches on historical photographies.

Let us notice some special issues of our approach in view of an integration into a more global framework for an exemplar-based image and video colorization. First of all, our method works well on faces and on object with similar shapes, but when this hypothesis is not fulfilled, some artifacts can appear. Therefore, instead of using our algorithm on full images, a face detection algorithm can be used to focus on the face colorization. Let us remark that faces in image can be detected by efficient, recent methods, see, e.g., [47]. In future work, the method will be integrated into a complete framework for exemplar-based image and video colorization.

Second, the morphing map has to be computed between images with the same size. This issue can be easily solved with a simple interpolation of the images. Keeping the ratio between the width and the height of faces similar, the distortion produced by such interpolation is small enough to support our method.

Acknowledgments: Funding by the German Research Foundation (DFG) within the project STE 571/13-1 is gratefully acknowledged.

Author Contributions: All the three authors have substantially contributed to the paper. The paper is the outcome of many joint and fruitful discussions J. Persch and G. Steidl had with F. Pierre during the months the later was a Postdoc at the University at Kaiserslautern (Germany). Therefore it is not possible to split the contributions. However, J. Persch was the leading expert concerning morphing and F. Pierre concerning color image processing.

Conflicts of Interest: The authors declare no conflict of interest.

Appendix A. Solution of the Image Sequence Problem

Fixing $\mathbf{v}$, resp., $\boldsymbol{\varphi}$ leads to the image sequence problem (6). In the following we show how this problem can be solved via the linear system of equations arising from Euler-Lagrange equation of the functional. We mainly follow [31].

The terms in $\mathcal{J}_{\boldsymbol{\varphi}}(\mathbf{I})$ containing I_k are

$$\int_\Omega |I_k(\varphi_{k+1}(\mathbf{x})) - I_{k+1}(\mathbf{x})|^2 + |I_k(\mathbf{x}) - I_{k-1}(\varphi_k(\mathbf{x}))|^2 \, d\mathbf{x}.$$

Assuming that φ_{k+1} is a diffeomorphism, the variable transform $\mathbf{y} := \varphi_{k+1}(\mathbf{x})$ gives

$$\begin{aligned}&\int_\Omega |I_k(\mathbf{x}) - I_{k+1}\left(\varphi_{k+1}^{-1}(\mathbf{x})\right)|^2 |\det \nabla \varphi_{k+1}^{-1}\left(\varphi_{k+1}^{-1}(\mathbf{x})\right)| + |I_k(\mathbf{x}) - I_{k-1}(\varphi_k(\mathbf{x}))|^2 \, d\mathbf{x}\\ &= \int_\Omega |I_k(\mathbf{x}) - I_{k+1}\left(\varphi_{k+1}^{-1}(\mathbf{x})\right)|^2 |\det \nabla \varphi_{k+1}\left(\varphi_{k+1}^{-1}(\mathbf{x})\right)|^{-1} + |I_k(\mathbf{x}) - I_{k-1}(\varphi_k(\mathbf{x}))|^2 \, d\mathbf{x}\end{aligned}$$

Setting

$$a_k\left(\varphi_{k+1}^{-1}(\mathbf{x})\right) := |\det \nabla \varphi_{k+1}\left(\varphi_{k+1}^{-1}(\mathbf{x})\right)|^{-1},$$

the Euler-Lagrange equations read for $k = 1, \ldots, K-1$ as

$$
\begin{aligned}
(I_k(\mathbf{x}) - I_{k+1}(\varphi_{k+1}^{-1}(\mathbf{x})))a_k\left(\varphi_{k+1}^{-1}(\mathbf{x})\right) + I_k(\mathbf{x}) - I_{k-1}(\varphi_k(\mathbf{x})) &= 0,\\
-I_{k-1}(\varphi_k(\mathbf{x})) + \left(1 + a_k\left(\varphi_{k+1}^{-1}(\mathbf{x})\right)\right) I_k(\mathbf{x}) - a_k\left(\varphi_{k+1}^{-1}(\mathbf{x})\right) I_{k+1}(\varphi_{k+1}^{-1}(\mathbf{x})) &= 0.
\end{aligned} \tag{A1}
$$

We introduce

$$X_K(\mathbf{x}) := \mathbf{x}, \quad X_{k-1}(\mathbf{x}) := \varphi_k(X_k(\mathbf{x})), \quad k = 1, \ldots, K,$$

which can be computed for each $\mathbf{x} \in \Omega$ since the φ_k, $k = 1, \ldots, K$ are given. Then (A1) can be rewritten for $k = 1, \ldots, K-1$ as

$$-I_{k-1}(X_{k-1}(\mathbf{x})) + (1 + a_k(X_{k+1}(\mathbf{x}))\, I_k(X_k(\mathbf{x})) - a_k(X_{k+1}(\mathbf{x}))\, I_{k+1}(X_{k+1}(\mathbf{x})) = 0. \tag{A2}$$

In matrix-vector form, this reads with

$$F_0 = I_{\text{temp}}(X_0(\mathbf{x})), \quad F_K = I_{\text{ref}}(\mathbf{x}), \quad F_k := I_k(X_k(\mathbf{x})), \quad k = 1, \ldots, K-1$$

for fixed $\mathbf{x} \in \Omega$ and $F := (F_1, \ldots, F_{K-1})^{\mathrm{T}}$ as

$$AF = (F_0, 0, \ldots, 0, a_{K-1}F_K)^{\mathrm{T}}, \quad A := \operatorname{tridiag}(-1, 1 + a_k, -a_k)_{k=1}^{K-1}.$$

Assuming that $a_k > 0$ which is the case in practical applications, the matrix A is irreducible diagonal dominant and thus invertible.

Appendix B. Gauss-Newton Method for the Registation Problem

We consider the nonlinear least squares problem in (7)

$$\arg\min_{v} \mathcal{J}(v) := \sum_{\mathbf{x} \in \mathcal{G}_p} |R(\mathbf{x}) - \tilde{T}(\mathbf{x} - (Pv)(\mathbf{x}))|^2 + \mathcal{S}(v).$$

Let us first write $\mathcal{S}(v)$ in a convenient matrix-vector notation. Using $v_1 = \left(v_1(x + \frac{1}{2}, y)\right)_{x,y=1}^{n_1-1,n_2} \in \mathbb{R}^{n_1-1,n_2}$, $v_2 = \left(v_1(x, y + \frac{1}{2})\right)_{x,y=1}^{n_1,n_2-1} \in \mathbb{R}^{n_1,n_2-1}$ and

$$
D_{1,x} := \begin{pmatrix} 0 & & & & \\ -1 & 1 & & & \\ & & \ddots & & \\ 0 & & 0 & -1 & 1 \\ 0 & & & 0 & 0 \end{pmatrix} \in \mathbb{R}^{n_1 \times n_1 - 1}, \quad
D_{1,y} := \begin{pmatrix} -1 & 1 & & & & \\ 0 & -1 & 1 & & & \\ & & \ddots & & & \\ & & & -1 & 1 & 0 \\ 0 & & & 0 & -1 & 1 \end{pmatrix} \in \mathbb{R}^{n_2 - 2 \times n_2 - 1},
$$

and similarly $D_{2,x} \in \mathbb{R}^{n_1 - 2 \times n_1 - 1}$ and $D_{2,y} \in \mathbb{R}^{n_2 \times n_2 - 1}$ we obtain

$$\mathcal{S}(v) = \mu\left(\|D_{1,x}v_1\|_F^2 + \|v_2 D_{2,y}^{\mathrm{T}}\|_F^2\right) + \frac{\lambda}{2}\|D_{1,x}v_1 + v_2 D_{2,y}^{\mathrm{T}}\|_F^2 + \frac{\mu}{2}\|v_1 D_{1,y}^{\mathrm{T}} + D_{2,x}v_2\|_F^2$$

where $\|\cdot\|_F$ denotes the Frobenius norm of matrices. Reshaping the v_i, $i = 1,2$ columnwise into a vectors $\mathbf{v}_1 \in \mathbb{R}^{(n_1-1)n_2}$ and $\mathbf{v}_2 \in \mathbb{R}^{n_1(n_2-1)}$ (where we keep the notation) and using tensor products $\otimes$ of matrices

$$\mathbf{D}_{1,x} := I_{n_2} \otimes D_{1,x}, \quad \mathbf{D}_{2,x} := I_{n_2-1} \otimes D_{2,x}, \quad \mathbf{D}_{1,y} := D_{1,x} \otimes I_{n_1-1}, \quad \mathbf{D}_{1,y} := D_{2,y} \otimes I_{n_1}$$

the regularizing term can be rewritten as

$$\mathcal{S}(\mathbf{v}) = \|\mathbf{S}\mathbf{v}\|_2^2, \quad \mathbf{S} := \begin{pmatrix} \sqrt{\mu}\,\mathbf{D}_{1,x} & 0 \\ 0 & \sqrt{\mu}\,\mathbf{D}_{2,y} \\ \sqrt{\frac{\mu}{2}}\,\mathbf{D}_{1,y} & \sqrt{\frac{\mu}{2}}\,\mathbf{D}_{2,x} \\ \sqrt{\frac{\lambda}{2}}\,\mathbf{D}_{1,x} & \sqrt{\frac{\lambda}{2}}\,\mathbf{D}_{2,y} \end{pmatrix}, \quad \mathbf{v} = \begin{pmatrix} \mathbf{v}_1 \\ \mathbf{v}_2 \end{pmatrix}.$$

The nonlinear data term in the registration problem with the columnwise reshaped image $\mathbf{T}(\mathbf{v})$ of $\left(\tilde{T}\left(\mathbf{x} - (Pv)(\mathbf{x})\right)\right)_{\mathbf{x}\in\mathcal{G}_p}$ and $\mathbf{R}$ of $(R(\mathbf{x}))_{\mathbf{x}\in\mathcal{G}_p}$ reads

$$\mathcal{D}(\mathbf{v}) := \|\mathbf{R} - \mathbf{T}(\mathbf{v})\|^2.$$

We linearize the term at a previous iterate $\mathbf{v}^{(r)}$, i.e.,

$$\mathcal{D}(\mathbf{v}) \approx \|\mathbf{R} - \mathbf{T}(\mathbf{v}^{(r)}) + \underbrace{\nabla_v \mathbf{T}(\mathbf{v}^{(r)})}_{\mathbf{G}^{(r)}} \left(\mathbf{v} - \mathbf{v}^{(r)}\right)\|_2^2.$$

We comment on the computation of $\mathbf{G}^{(r)}$ below. Then we consider

$$\|\mathbf{R} - \mathbf{T}(\mathbf{v}^{(r)}) + \mathbf{G}^{(r)}\left(\mathbf{v} - \mathbf{v}^{(r)}\right)\|_2^2 + \|\mathbf{S}\mathbf{v}\|_2^2.$$

The gradient is given by

$$\left((\mathbf{G}^{(r)})^{\mathrm{T}}\mathbf{G}^{(r)} + \mathbf{S}^{\mathrm{T}}\mathbf{S}\right)\mathbf{v} - (\mathbf{G}^{(r)})^{\mathrm{T}}\mathbf{G}^{(r)}\mathbf{v}^{(r)} - (\mathbf{G}^{(r)})^{\mathrm{T}}(\mathbf{R} - \mathbf{T}(\mathbf{v}^{(r)})).$$

Setting the gradient to zero we obtain for $\mathbf{v}$

$$\begin{aligned} \mathbf{v} &:= \mathbf{v}^{(r)} - \left((\mathbf{G}^{(r)})^{\mathrm{T}}\mathbf{G}^{(r)} + \mathbf{S}^{\mathrm{T}}\mathbf{S}\right)^{-1}\left((\mathbf{G}^{(r)})^{\mathrm{T}}(\mathbf{T}(\mathbf{v}^{(r)}) - \mathbf{R}) + \mathbf{S}^{\mathrm{T}}\mathbf{S}\mathbf{v}^{(r)}\right) \\ &= \mathbf{v}^{(r)} - \left((\mathbf{G}^{(r)})^{\mathrm{T}}\mathbf{G}^{(r)} + \mathbf{S}^{\mathrm{T}}\mathbf{S}\right)^{-1}\nabla_v \mathcal{J}(\mathbf{v}^{(r)}) \end{aligned}$$

As we want to minimize $\mathcal{J}$, the next iterate is given by

$$\mathbf{v}^{(r+1)} = \mathbf{v}^{(r)} - \left(\tfrac{1}{2}\right)^l \left((\mathbf{G}^{(r)})^{\mathrm{T}}\mathbf{G}^{(r)} + \mathbf{S}^{\mathrm{T}}\mathbf{S}\right)^{-1}\nabla_v \mathcal{J}(\mathbf{v}^{(r)}),$$

where $l \in \mathbb{N}$ is the smallest number such that $\mathcal{J}(\mathbf{v}^{(k+1)}) < \mathcal{J}(\mathbf{v}^{(k)})$, if $\nabla_v \mathcal{J}(\mathbf{v}^{(r)})$ is not zero.

Finally we comment on the computation of

$$\mathbf{G}^{(r)} := \nabla_v \mathbf{T}(\mathbf{v}^{(r)}) = \begin{pmatrix} -P_1^{\mathrm{T}}\partial_x \tilde{T}(x - (P_1 v_1)(\mathbf{x}), y - P_2 v_2(\mathbf{x})) \\ -P_2^{\mathrm{T}}\partial_y \tilde{T}(x - (P_1 v_1)(\mathbf{x}), y - P_2 v_2(\mathbf{x})) \end{pmatrix}.$$

Since $\tilde{T}$ is computed by linear interpolation from $(T(\mathbf{x}))_{\mathbf{x}\in\mathcal{G}_p}$, Then we have for an arbitrary point $(p,q) \in [i, i+1] \times [j, j+1]$, $(i,j) \in \{1, \ldots, n_1 - 1\} \times \{1, \ldots, n_2 - 1\}$ that

$$\tilde{T}(p,q) := (i + 1 - p, p - i) \begin{pmatrix} T(i,j) & T(i,j+1) \\ T(i+1,j) & T(i+1,j+1) \end{pmatrix} \begin{pmatrix} j + 1 - q \\ q - j \end{pmatrix}. \tag{A3}$$

Then

$$\partial_x \bar{T}(p,q) = (j+1-q)\,(T(i+1,j) - T(i,j)) + (q-j)\,(T(i+1,j+1) - T(i,j)),$$
$$\partial_y \bar{T}(p,q) = (i+1-p)\,(T(i,j+1) - T(i,j)) + (p-i)\,(T(i+1,j+1) - T(i+1,j)).$$

Hence the derivatives can be calculated from the forward differences of T in x- resp. y-direction and appropriate weighting.

References

1. Levin, A.; Lischinski, D.; Weiss, Y. Colorization using optimization. *ACM Trans. Graph.* **2004**, *23*, 689–694.
2. Irony, R.; Cohen-Or, D.; Lischinski, D. Colorization by example. In Proceedings of the 16th Eurographics Conference on Rendering Techniques, Konstanz, Germany, 29 June 29–1 July 2005; pp. 201–210.
3. Pierre, F.; Aujol, J.F.; Bugeau, A.; Papadakis, N.; Ta, V.T. Luminance-chrominance model for image colorization. *SIAM J. Imaging Sci.* **2015**, *8*, 536–563.
4. Yatziv, L.; Sapiro, G. Fast image and video colorization using chrominance blending. *IEEE Trans. Image Process.* **2006**, *15*, 1120–1129.
5. Jack, K. *Video Demystified: A Handbook for the Digital Engineer*; Elsevier: Amsterdam, The Netherlands, 2011.
6. Gupta, R.K.; Chia, A.Y.S.; Rajan, D.; Ng, E.S.; Zhiyong, H. Image colorization using similar images. In Proceedings of the 20th ACM International Conference on Multimedia, Nara, Japan, 29 October–2 November 2012; pp. 369–378.
7. Horiuchi, T. Colorization algorithm using probabilistic relaxation. *Image Vis. Comput.* **2004**, *22*, 197–202.
8. Welsh, T.; Ashikhmin, M.; Mueller, K. Transferring color to greyscale images. *ACM Trans. Graph.* **2002**, *21*, 277–280.
9. Efros, A.A.; Leung, T.K. Texture synthesis by non-parametric sampling. In Proceedings of the 7th IEEE International Conference on Computer Vision, Kerkyra, Greece, 20–27 September 1999; Volume 2, pp. 1033–1038.
10. Achanta, R.; Shaji, A.; Smith, K.; Lucchi, A.; Fua, P.; Süsstrunk, S. SLIC superpixels compared to state-of-the-art superpixel methods. *IEEE Trans. Pattern Anal. Mach. Intell.* **2012**, *34*, 2274–2282.
11. Chen, T.; Wang, Y.; Schillings, V.; Meinel, C. Grayscale image matting and colorization. In Proceedings of the Asian Conference on Computer Vision, Jeju Island, Korea, 27–30 January 2004; pp. 1164–1169.
12. Charpiat, G.; Hofmann, M.; Schölkopf, B. Automatic image colorization via multimodal predictions. In *European Conference on Computer Vision*; Springer: Berlin/Heidelberg, Germany, 2008; pp. 126–139.
13. Chia, A.Y.S.; Zhuo, S.; Kumar, R.G.; Tai, Y.W.; Cho, S.Y.; Tan, P.; Lin, S. Semantic colorization with internet images. *ACM Trans. Graph.* **2011**, *30*, 156.
14. Zhang, R.; Isola, P.; Efros, A.A. Colorful image colorization. In *European Conference on Computer Vision*; Springer: Berlin/Heidelberg, Germany, 2016; pp. 1–16.
15. Smythe, D.B. *A Two-Pass Mesh Warping Algorithm for Object Transformation and Image Interpolation*; Technical report; ILM Technical Memo Department, Lucasfilm Ltd.: San Francisco, CA, USA, 1990.
16. Wolberg, G. *Digital Image Warping*; IEEE Computer Society Press: Los Alamitos, CA, USA, 1990; Volume 10662.
17. Wolberg, G. Image morphing: A survey. *Vis. Comput.* **1998**, *14*, 360–372.
18. Miller, M.I.; Younes, L. Group actions, homeomorphisms, and matching: A general framework. *Int. J. Comput. Vis.* **2001**, *41*, 61–84.
19. Trouvé, A.; Younes, L. Local geometry of deformable templates. *SIAM J. Math. Anal.* **2005**, *37*, 17–59.
20. Trouvé, A.; Younes, L. Metamorphoses through Lie group action. *Found. Comput. Math.* **2005**, *5*, 173–198.
21. Christensen, G.E.; Rabbitt, R.D.; Miller, M.I. Deformable templates using large deformation kinematics. *IEEE Trans. Image Process.* **1996**, *5*, 1435–1447.
22. Dupuis, P.; Grenander, U.; Miller, M.I. Variational problems on flows of diffeomorphisms for image matching. *Q. Appl. Math.* **1998**, *56*, 587–600.
23. Trouvé, A. An infinite dimensional group approach for physics based models in pattern recognition. *Int. J. Comput. Vis.* **1995**, *28*, 213-221.

24. Trouvé, A. Diffeomorphisms groups and pattern matching in image analysis. *Int. J. Comput. Vis.* **1998**, *28*, 213–221.
25. Beg, M.F.; Miller, M.I.; Trouvé, A.; Younes, L. Computing large deformation metric mappings via geodesic flows of diffeomorphisms. *Int. J. Comput. Vis.* **2005**, *61*, 139–157.
26. Richardson, C.L.; Younes, L. Metamorphosis of images in reproducing kernel Hilbert spaces. *Adv. Comput. Math.* **2016**, *42*, 573–603.
27. Hong, Y.; Joshi, S.; Sanchez, M.; Styner, M.; Niethammer, M. Metamorphic geodesic regression. *Med. Image Comput. Comput. Assist. Interv.* **2012**, *15*, 197–205.
28. Younes, L. *Shapes and Diffeomorphisms*; Springer: Berlin, Germany, 2010.
29. Miller, M.I.; Trouvé, A.; Younes, L. Hamiltonian systems and optimal control in computational anatomy: 100 years since D'Arcy Thompson. *Annu. Rev. Biomed. Eng.* **2015**, *17*, 447–509.
30. Miller, M.I.; Trouvé, A.; Younes, L. On the metrics and Euler-Lagrange equations of computational anatomy. *Annu. Rev. Biomed. Eng.* **2002**, *4*, 375–405.
31. Berkels, B.; Effland, A.; Rumpf, M. Time discrete geodesic paths in the space of images. *SIAM J. Imaging Sci.* **2015**, *8*, 1457–1488.
32. Christensen, G.E.; Johnson, H.J. Consistent image registration. *IEEE Trans. Med. Imaging* **2001**, *20*, 568–582.
33. Fischer, B.; Modersitzki, J. Curvature based image registration. *J. Math. Imaging Vis.* **2003**, *18*, 81–85.
34. Haber, E.; Modersitzki, J. A multilevel method for image registration. *SIAM J. Sci. Comput.* **2006**, *27*, 1594–1607.
35. Han, J.; Berkels, B.; Droske, M.; Hornegger, J.; Rumpf, M.; Schaller, C.; Scorzin, J.; Urbach, H. Mumford–Shah Model for one-to-one edge matching. *IEEE Trans. Image Process.* **2007**, *16*, 2720–2732.
36. Modersitzki, J. *Numerical Methods for Image Registration*; Oxford University Press: Oxford, UK, 2004.
37. Modersitzki, J. *FAIR: Flexible Algorithms for Image Registration*; SIAM: Philadelphia, PA, USA, 2009.
38. Deledalle, C.A.; Papadakis, N.; Salmon, J.; Vaiter, S. CLEAR: Covariant LEAst-square Refitting with applications to image restoration. *SIAM J. Imaging Sci.* **2017**, *10*, 243–284.
39. Miller, M.I.; Christensen, G.E.; Amit, Y.; Grenander, U. Mathematical textbook of deformable neuroanatomies. *Proc. Natl. Acad. Sci. USA* **1993**, *90*, 11944–11948,
40. Yuan, J.; Schnörr, C.; Steidl, G. Simultaneous higher order optical flow estimation and decomposition. *SIAM J. Sci. Comput.* **2007**, *29*, 2283–2304.
41. Bertalmio, M. *Image Processing for Cinema*; CRC Press: Boca Raton, FL, USA, 2014.
42. Nikolova, M.; Steidl, G. Fast Hue and Range Preserving Histogram Specification: Theory and New Algorithms for Color Image Enhancement. *IEEE Trans. Image Process.* **2014**, *23*, 4087–4100.
43. Hertzmann, A.; Jacobs, C.E.; Oliver, N.; Curless, B.; Salesin, D.H. Image analogies. In Proceedings of the 28th Annual Conference on Computer Graphics and Interactive Techniques, Los Angeles, CA, USA, 12–17 August 2001; pp. 327–340.
44. Peter, P.; Kaufhold, L.; Weickert, J. Turning diffusion-based image colorization into efficient color compression. *IEEE Trans. Image Process.* **2016**, *26*, 860-869.
45. Chambolle, A.; Pock, T. A first-order primal-dual algorithm for convex problems with applications to imaging. *J. Math. Imaging Vis.* **2011**, *40*, 120–145.
46. Joshi, S.C.; Miller, M.I. Landmark matching via large deformation diffeomorphisms. *IEEE Trans. Image Process.* **2000**, *9*, 1357–1370.
47. Chen, D.; Hua, G.; Wen, F.; Sun, J. Supervised transformer network for efficient face detection. In *European Conference on Computer Vision*; Springer: Cham, Switzerland, 2016; pp. 122–138.

Journal of
Imaging

MDPI

Article

Image Fragile Watermarking through Quaternion Linear Transform in Secret Space

Marco Botta [1,*], Davide Cavagnino [1] and Victor Pomponiu [2]

1 Dipartimento di Informatica, Università degli Studi di Torino, 10149 Turin, Italy; davide@di.unito.it
2 Agency for Science, Technology and Research, 1 Fusionopolis Way, 487372 Singapore, Singapore; victor.pomponiu@gmail.com
* Correspondence: marco.botta@unito.it; Tel.: +39-011-670-6789

Received: 14 June 2017; Accepted: 5 August 2017; Published: 11 August 2017

Abstract: In this paper, we apply the quaternion framework for color images to a fragile watermarking algorithm with the objective of multimedia integrity protection (Quaternion Karhunen-Loève Transform Fragile Watermarking (QKLT-FW)). The use of quaternions to represent pixels allows to consider the color information in a holistic and integrated fashion. We stress out that, by taking advantage of the host image quaternion representation, we extract complex features that are able to improve the embedding and verification of fragile watermarks. The algorithm, based on the Quaternion Karhunen-Loève Transform (QKLT), embeds a binary watermark into some QKLT coefficients representing a host image in a secret frequency space: the QKLT basis images are computed from a secret color image used as a symmetric key. A computational intelligence technique (i.e., genetic algorithm) is employed to modify the host image pixels in such a way that the watermark is contained in the protected image. The sensitivity to image modifications is then tested, showing very good performance.

Keywords: data hiding; fragile watermarking; image authentication; color image processing; quaternions; genetic algorithm (GA); Karhunen-Loève Transform (KLT)

1. Introduction

The protection of digital media is one fundamental topic in the present age, in which practically every kind of content is represented in digital form. Without an integrity guard system, the transmission via open and unsecured networks of digital assets could not be verified. Researchers have developed and are still devising various techniques to solve the problem. For example, digital signature is a method to ensure authenticity and proof of origin for a digital media; Message Authentication Codes are another method to authenticate the integrity of a digital media for a restricted set of entities. Both methods require appending a certain amount of information to the protected digital object.

Another effective solution to defend digital objects from various attacks is digital watermarking [1]. Watermarking techniques insert a signal in the digital object itself with various purposes: content authentication, content integrity, copyright protection, traitor tracing, etc.

Depending on the application requirements, various watermarking methods (not necessarily excluding one other) have been devised, every one having specific properties. We briefly recall these characteristics in the following.

A watermarking algorithm may be *robust* or *fragile*: the first kind is intended to survive (intentional) modifications of the digital object aimed at its removal (while maintaining an acceptable quality of the resulting object); fragile watermarks have the opposite purpose: being destroyed at the minimal modification of the digital object, and possibly localizing the modified area. Therefore, robust watermarks are useful for copyright protection and track of origin, whilst fragile ones may be used for authentication and integrity check purposes. Some fragile watermarks have been devised to accept minimal modifications (like high quality JPEG compression of an image), and are thus called semi-fragile; others have the ability, after detecting an altered area, to (partially) restore the original object.

Another property of watermarking methods is the capacity to recover the original (host) object from the watermarked one by authorized entities: an algorithm that possesses this ability is called *reversible*, and *non-reversible* (or *lossy*) otherwise.

The present work is devoted to fragile watermarking of color images, thus in the following we will restrict the subject to this type of data only.

The watermark signal may be inserted mainly into two different domains, namely the *spatial* and the *frequency domains*. The spatial domain refers to the pixels of the image, so the watermark signal is embedded by modifying the pixel values. On the contrary, the frequency domain generally refers to a transformed space, like the Fourier (DFT), Discrete Cosine (DCT), Karhunen-Loève (KLT) or Singular Value Decomposition (SVD), into which the pixels are projected producing a set of features (typically coefficients) used for watermark embedding. Embedding into frequency domains requires the inverse step to obtain pixels again: we note that some works ignore the possibility that the inverse operation returns floating point pixel values and the necessary rounding to integer values may, in general, remove part or all of the watermark. We already developed [2] and optimized a methodology based on Genetic Algorithms (GAs) to overcome this problem, using a modification of the approach suggested in [3]. The use of the KLT is peculiar with regard to other linear transforms like DFT and DCT because the kernel used by the KLT is not fixed but is instead computed from a set of samples (in our case, a set of subimages obtained from a secret key image): this allows the creation of a secret space of features that provide the necessary security for hiding a watermark signal.

In this paper, we consider bitmap images coded in RGB format having a bit-depth of 8 bits per channel. The objective is to adapt the methodology of a previously developed *fragile watermarking algorithm* for gray scale images to color images considering the holistic representation of the color information: we believe that the quaternion framework is a powerful representation, which could offer the optimal interplay between the robustness and imperceptibility components of the watermarking scheme and also improves on known ones [4]. In particular, [5] defines the computation of eigenvectors and eigenvalues for a set of color pixels, and proposes the Quaternion Karhunen-Loève Transform (QKLT). We think that the QKLT is a valuable tool that can be seemly incorporated into a fragile watermarking framework since the color information is processed as a whole and not as three independent channels, easing the protection of the integrity of an image even using a reduced number of watermark bits with regard to other approaches. The QKLT gives a formalization for the KLT of a color image combining the RGB values of a pixel. Moreover, the use of the QKLT defines a single approach in computing the secret space where the watermark is inserted, *differently* from the methods used in [4] which consider the three color channels as separate entities allowing many different KLT basis computation approaches that, even if correct and with very good performance, lack a motivation for the choice of one with regard to the other and lead to empirical combinations of the vectors of the three channels.

Thus, the main contribution of this work is the development of a fragile watermarking algorithm for color images which employs the channel color information as integrated producing very high quality images with a high level of security in detecting unauthorized modifications.

The quaternion framework we propose has been integrated into the modular watermarking architecture [4] by developing a new module that computes the suitable features for the watermarking process and brings improvements on various directions, as it will be detailed in the experimental results section. In particular, the use of quaternions allows to keep the transform space dimension limited to n^2 instead of $3n^2$ (where n^2 is the number of pixels per subimage) because quaternions incorporate the whole color information instead of keeping it separate and requiring to consider vectors three times bigger to take into account the correlation among color channels. The resulting algorithm has smaller running times and a higher sensitivity in detecting tampering, while maintaining the same high quality of the watermarked images as the other approaches presented in [4].

The main characteristics, novelties and improvements of the presented algorithm with regard to previous fragile watermarking algorithms are:

(a) representation of color pixels using quaternions, which translates in integrating the color information employing a linear transform that considers this *holistic* interpretation, and embedding with a *smaller impact* on the host image;
(b) *improved* running times for watermarking embedding with regard to a previous algorithm developed in the same framework;
(c) *increased* sensitivity to attacks;
(d) *very high quality* of the watermarked images, along with a good and *flexible localization potential* (satisfying almost all application contexts);
(e) *flexibility* in payload embedding for a chosen localization capability;
(f) application of the QKLT to the fragile watermarking domain.

The paper is organized as follows: firstly, we survey some works on the representation of color images in the quaternion framework and on image watermarking. Then, with the aim of making the paper as self-contained as possible, we devolve a section to briefly recall some basic concepts on quaternions, and afterwards we present the QKLT. The section on the watermarking algorithm QKLT-FW (Fragile Watermarking) presents the steps for embedding and extracting the watermark with the aim of detecting and localizing alterations to the image. Then, experimental results are shown along with a comparison with existing algorithms. In the final section we draw some conclusions on the developed algorithm and discuss the obtained performance.

Regarding notation, in this paper, we represent matrices, vectors and scalars by capital letters, bold lowercase letters and plain lowercase letters, respectively. All vectors are column-wise by default.

2. Related Works

2.1. Quaternion Signal Processing

In the last decade, quaternion signal processing (QSP) has started to be widely employed and several common signal processing transforms have been extended to the quaternion domain. For instance, the quaternion Fourier transform (QFT) was firstly introduced by Sangwine [6] and later extended with new results [7].

Other works proposed descriptors for the quaternion Singular Value Decomposition (QSVD), the quaternion Eigenvalue Decomposition (QEVD) and the QKLT.

The calculation and properties of the SVD for quaternion matrices (generated by vector-array signals) were extensively studied in [8].

According to the previous works, the main purpose of using the quaternion counterpart of these tools is that it can naturally account for the correlation among color channels, providing a holistic representation [9] of color images. Thus, the quaternion theory treats a color image as a vector field and processes it directly, without losing color information.

The paper [5] is the basis on which our work is founded. Le Bihan and Sangwine present the SVD, the Eigenvalue Decomposition and the KLT for quaternion matrices applied to color images: they call them QSVD, QEVD and QKLT respectively. Their work also refers to many previous papers on the topic of quaternion matrices.

An application where quaternion representation is finding an active field of research is digital watermarking. The next subsection of this paper will review the most representative and recent works devoted to quaternion-based watermarking.

2.2. Quaternion-Based Watermarking

The first application of quaternions for robust digital watermarking was within the Fourier framework. In particular, [10] applies the QFT defined in [11] to perform robust image watermarking; given that the QFT depends on a parameter μ (which is a pure quaternion satisfying $\mu^2 = -1$), a study aimed at finding the best μ value to achieve both invisibility and robustness to attacks is described in that paper. In [12], the watermarking algorithm inserts a robust watermark into the scalar part of some selected QFT coefficients, and the detection stage deals with attacks through a least squares support vector machine applied to pseudo-Zernike moments of the scalar QFT matrix of the possibly attacked image.

An et al. [13] also used the QFT for devising a robust watermarking scheme. To be able to hide large payloads, i.e., the number of features allocated for watermark embedding, an improved quantization index (QIM) algorithm was proposed to compress the watermark bits. Due to the use of QIM, the scheme is able to extract the watermark without the use of the host image, which greatly increases its applicability. The simulations carried out prove that the watermarking algorithm based on QFT and the improved QIM with distortion compensation attains a good tradeoff between invisibility and robustness.

Later, Tsui et al. [14] developed a pair of watermarking algorithms working in the La*b* space. The first one applies the Spatio-Chromatic Discrete Fourier transform (SCDFT) to the a* and b* pixel values, then a binary watermark is inserted into the yellow-blue component, maximizing the watermark intensity and keeping the distortion below a Just Noticeable Difference level. The second one embeds a quaternion watermark into the QFT coefficients of the image, taking into account the Human Visual System (HVS) characteristics.

To overcome the limitation of the previous works which spread the watermark information over a limited number of RGB color channels, Chen et al. [15] employ a full 4D discrete QFT (QDFT) of the host color image. In their complete framework, which introduces three schemes, they provide the symmetry preconditions of the unit quaternions necessary for the QDFT in order to achieve the correct watermark extraction in the case of no-attacks. The experimental results show that the proposed framework offers a good performance in terms of capacity and robustness against attacks. However, the imposed preconditions for the unit pure quaternion affects the payload of the watermark, i.e., the number of features allocated for watermark embedding. Furthermore, the study lacks a theoretical analysis of the probability of false detection and a thorough comparison with the existing works.

In [16], Shao et al. have explored a joint robust watermarking and encryption technology based on the quaternion gyrator transform (QGT) [17] and double random phase encoding (DRPE). The main idea is to encrypt the watermark via DRPE and then to insert the encrypted bits into the mid-frequency coefficients of the QGT of the host image. It is important to note that the scheme requires some side information, related to the host image, in order to recover the watermark.

The Quaternion Polar Harmonic Transform (QPHT) has been used in [18] to devise a robust watermarking scheme in order to increase the security of the watermark information. The transform is a parameterized version of the linear canonical transform with the parameters belonging to the real special linear group defined as the set of 2×2 real matrices having the determinant equal to 1. Due to the large space where the correct parameters for the forward and backward transform are lying, the proposed scheme has a high level of security. Moreover, the scheme shows satisfactory performance in terms of robustness, capacity and imperceptibility of the watermark.

Yang et al. [19] also apply the quaternion algebra and Polar Harmonic Transform (PHT) to introduce an invariant color image watermarking scheme. The selection of PHT was motivated by its appealing properties compared to others moment-based transforms, e.g., (pseudo) Zernike moments: noise robustness, low computational complexity, better reconstruction accuracy and numerical stable solutions. An in-depth analysis of invariance properties (rotation, scaling and translation) of the QPHT moments is given in the paper along with the criterion used for the selection of the watermarking coefficients. More precisely, only the set of coefficients that offer the highest reconstruction accuracy are chosen by using the following relation

$$S = \left\{ M_{n,l}^{R} \text{ and } M_{n,l}^{L},\ l = 4m, m \in \mathbb{Z} \right\}, \tag{1}$$

where $M_{n,l}^{R}$ and $M_{n,l}^{L}$ denote the QPHT right and left coefficients respectively, n is the order and l is the repetition parameter. The encrypted watermark bits are inserted by adaptively modulating (via quantization) the selected coefficients. For instance, for the right coefficients the embedding rule is

$$\left|M'^{R}_{n,l}\right| = \begin{cases} 2\Delta \cdot \text{round}\left(\frac{\left|M_{n,l}^{R}\right|}{2\Delta}\right) + \frac{\Delta}{2},\ w_k = 1 \\ 2\Delta \cdot \text{round}\left(\frac{\left|M_{n,l}^{R}\right|}{2\Delta}\right) - \frac{\Delta}{2},\ w_k = 0 \end{cases} \quad (0 \le k < P \times Q)\ , \tag{2}$$

where $|\cdot|$ denotes the modulus operator, round$(\cdot)$ represents the round operator, Δ is the watermark strength factor, w_k is the bit of the watermark of size $P \times Q$. One of the drawbacks of the scheme is the inability to fully extract the watermark from the watermarked coefficient $\left|M'^{R}_{n,l}\right|$ since the QPHT coefficients can be obtained approximately for a digital image.

2.3. Image Authentication

The problem of image authentication has been also addressed by Al-Otum [20]. The paper proposes a semi-fragile watermarking scheme based on the DWT. The watermark, i.e., the authentication information, is implanted into the DWT coefficients of all image blocks of a color channel, randomly chosen. In order to better capture the characteristics of the image, a modified DWT (a reminiscent of the wavelet packet decomposition [21]) is used to compute the approximation of the horizontal, vertical and diagonal components. The method is semi-blind since it requires some auxiliary information (i.e., quantization thresholds are computed and passed to the detector) in order to extract the watermark, which limits its applicability. A security issue with this scheme is that for each block the authentication value, i.e., weighted mean values of the difference-color features, is computed only from the approximation horizontal and vertical components, ignoring the diagonal coefficients.

Besides detecting whether it has been tampered by common signal processing operations, the Lin et al. [22] scheme adds the recovery functionality of the affected image regions. To achieve these goals, they make use of several tools chosen to meet the requirements of watermarking schemes: lattice-based embedding into the DCT coefficients to lower the impact of the hidden data and a secret sharing approach to reconstruct the watermark with recovery capability.

In [23], a watermarking method is introduced which authenticates the compressed indexed representation of a color image. The authentication watermark is embedded into the LBSs of indexes of the compressed color images. To overcome the issue that arises when the modified index LSB coincides with the watermark bit, the scheme adopts an improved tamper verification procedure which consists of introducing interdependency relationships among pixels in each row or column.

In [24], the authors exploit the standard deviation information to devise an authentication method. Two sorts of information are embedded into the image: an authentication watermark and some image information that enables the recovery of tampered blocks. The two watermarks follow different insertion procedures. The authentication bits are just inserted into the LSBs of the image. To insert the recovery data, the scheme proceeds by firstly using the standard deviation to classify the image blocks into three classes. Afterwards, each block is prepared for embedding by mapping it to the DCT domain followed by quantization. Interestingly, the amount of information to be embedded in each block is adaptively modified and is determined by its class.

In the following section, we briefly introduce two algorithms for the authentication of images, which will be used to perform a comparison with our method (these algorithms do not make use of quaternion representations, but have very high Peak Signal-to-Noise Ratio (PSNR) and sensitivity).

A secure method for fragile image watermarking was introduced in [25]. The algorithm contrasts Vector Quantization attacks, may detect tampering and authenticates an image. The image is subdivided into a hierarchy of blocks. The LSBs of a block contain authentication information (a Message Authentication Code, MAC, or a signature) of the block itself and part of the MAC (or signature) of the upper layers in the hierarchy obtained by merging the block with other blocks (like in a quadtree decomposition). Thus, a block not in the lowest level of the hierarchy is authenticated by bits contained in the LSBs of the blocks at the lowest level composing it. This algorithm has a 100% tamper detection capability, but suffers a low PSNR due to the large quantity of data necessary to store a MAC or a signature.

MIMIC 9 is a modular framework developed for the fragile watermarking of color images. It embeds a watermark in a transformed secret space (defined by the Karhunen-Loève transform), using several different embedding techniques, such as LSB embedding and syndrome coding.

3. Quaternions

Quaternions are a representation of numbers in a hypercomplex space. A quaternion q is defined in a four-dimensional *vector space* $\mathbb{H}$ as $q = s\vec{1} + v_i\vec{i} + v_j\vec{j} + v_k\vec{k}$, where the basis of the vector space is composed by the vectors $\vec{1}, \vec{i}, \vec{j}, \vec{k}$. The number s is called the scalar part and $\vec{v} = v_i\vec{i} + v_j\vec{j} + v_k\vec{k}$ constitutes the vector part, so a quaternion is also written as $q = (s, \vec{v})$. The basis vectors have the property that

$$\vec{i}^2 = \vec{j}^2 = \vec{k}^2 = -\vec{1}, \tag{3}$$

which has several implications, like the non-commutativity of multiplication. The four operations are performed in the usual way of vector spaces, taking into account the previous property. When $\vec{v} = \vec{0} = (0, 0, 0)$ the quaternion is said *scalar*, whilst when $s = 0$ it is called *pure*.

The L_2 norm (or simply the norm) of a quaternion $q = (s, \vec{v})$ is $\| q \| = \sqrt{s^2 + v_i^2 + v_j^2 + v_k^2}$ and the L_1 norm is $\| q \|_1 = |s| + |v_i| + |v_j| + |v_k|$.

The multiplicative inverse of a quaternion q is computed as $q^{-1} = \overline{q}/\| q \|^2$ where $\overline{q} = (s, -\vec{v})$ is the complex conjugate of q. Due to the non-commutative property of multiplication, it is possible to divide a quaternion a by q in two possible ways, namely aq^{-1} and $q^{-1}a$.

A very good introduction to quaternions is [26]. In the following, we will use matrices having quaternion elements. An overview on this topic can be found in [27].

Quaternions and Color Images

Following the use of quaternions to represent and process color image pixels introduced in [28], a color pixel expressed in the RGB space by the triple (r, g, b) is represented by the pure quaternion $(0, r\vec{i} + g\vec{j} + b\vec{k})$, so every pixel is considered as an element of $\mathbb{H}$.

4. The Quaternion Karhunen-Loève Transform

The general definition of a linear transform is a mapping between different bases of vector spaces. Given a d-dimensional column vector x, a linear transform is defined by a $d \times d$ orthonormal matrix Φ (called kernel, whose rows form a vector basis) that maps x to y, i.e., the same vector expressed in a different basis, by means of the relation (forward transform) $y = \Phi x$. The vector x may be again obtained from y by means of the inverse transform $x = \Phi^{-1}y$, that given the orthonormality of Φ may be also written as $x = \Phi' y$ (where Φ' is the conjugate transpose of Φ). Differently from some common transforms which have fixed kernels for a fixed d, like Discrete Cosine, Fourier, Hadamard, and Haar transforms, the discrete Karhunen-Loève Transform, KLT, (also known as Hotelling Transform, Principal Component Analysis, or Eigenvector Transform) computes a kernel from a set of vectors.

The ability of the KLT is to orient the basis Φ along the d directions of maximum data expansion of the vectors used to compute the kernel.

The QKLT is based on the Quaternion Eigenvalue Decomposition (QED) [5].

An eigenvector associated to a matrix $\boldsymbol{A}$ is defined as a vector $\boldsymbol{x}$ which multiplied by the matrix results in a multiple of the vector itself, i.e., in general $\boldsymbol{A}\boldsymbol{x} = \lambda\boldsymbol{x} = \boldsymbol{x}\lambda$, where λ is the associated eigenvalue. But in $\mathbb{H}$ the product is not commutative, so two possible kinds of eigenvectors can be defined, namely left eigenvectors/eigenvalues for which $\boldsymbol{A}\boldsymbol{x}_L = \lambda_L\boldsymbol{x}_L$ and right eigenvectors/eigenvalues satisfying $\boldsymbol{A}\boldsymbol{x}_R = \boldsymbol{x}_R\lambda_R$. As stated in [5] and papers cited therein, left eigenvalues pose many theoretical problems, so in our work we will use right eigenvectors/eigenvalues only.

It is possible to compute the right eigenvalues and their associated eigenvectors of a quaternion matrix $\boldsymbol{A}$ with the decomposition $\boldsymbol{A} = \boldsymbol{V}\boldsymbol{E}\boldsymbol{V}'$ as stated in [5], where the columns of $\boldsymbol{V}$ contain the eigenvectors and $\boldsymbol{E}$ is a diagonal matrix containing the eigenvalues. If $\boldsymbol{A}$ is hermitian (i.e., $\forall a_{ij} \in \boldsymbol{A}$ at row i and column j, $a_{ij} = \overline{a_{ji}}$), as is the case with the covariance matrix we will compute on a color image, then the eigenvalues are real, i.e., $\lambda_R \in \mathbb{R}$ (in general, instead, $\lambda_R \in \mathbb{C}$).

Consider a set of column vectors $\boldsymbol{U} = \{\boldsymbol{x}_i\}$, compute the average vector, $\boldsymbol{m}_U = E[\boldsymbol{x}_i]$ where $E[\cdot]$ is the expected value operator, and successively the Hermitian covariance matrix $\boldsymbol{C}_U = E\left[(\boldsymbol{x}_i - \boldsymbol{m}_U)(\boldsymbol{x}_i - \boldsymbol{m}_U)'\right]$. Decomposing $\boldsymbol{C}_U$ as $\boldsymbol{C}_U = B\ \Gamma\ B'$ returns, as previously stated, the eigenvalues in the diagonal of Γ and the associated eigenvectors forming an *orthonormal basis* as columns of B. It is useful to give an ordering to the eigenvectors: sorting in non-increasing order of the eigenvalues' norm results in moving in the first positions the eigenvectors having, on average, more importance in the reconstruction of the vectors in $\boldsymbol{U}$. The KLT (or QKLT, if the vectors' components are quaternions) of a vector $\boldsymbol{z}$ (of size d) may then be written as $\boldsymbol{y} = B^T(\boldsymbol{z} - \boldsymbol{m}_U)$, where the transpose operator ^{T} moves the eigenvectors in rows: the d components of $\boldsymbol{y}$ are called coefficients of the transform and the position in $\boldsymbol{y}$ is called *order* of the coefficient. Obviously, the inverse QKLT is computed as $\boldsymbol{z} = \left(B^T\right)^{-1}\boldsymbol{y} + \boldsymbol{m}_U$. A more extensive discussion of the QKLT is presented in [5] and references cited therein.

5. Genetic Algorithms

A GA is a method of computation inspired by the evolution of living beings.

When the parameters of a problem may be encoded in a data structure (called individual) and a function exists to evaluate the quality of an individual in representing a solution to the problem, then a GA may be employed. The GA explores the space of the possible outcomes evolving a population of individuals (initially randomly created, as in real evolution) evaluating them through a *fitness function*, and reproducing the best ones to converge to an individual that represents a (local) optimal solution.

The GA executes a cycle for a maximum number of times (each iteration is called *epoch*) or until a viable solution is found, according to the following high level Algorithm 1:

The population is evolved by first reproducing the individuals and then picking the ones that will be part of the population for the next epoch.

Reproduction operates by selecting pairs of individuals (we made this with tournament selection, in which two pairs are chosen, and the individual with the best fitness in each pair is selected) and applying with probability p_c a crossover operator, which exchanges random subsets of genes in corresponding positions. The resulting individuals have a probability p_m to have a mutation of one of its genes (this aims to create potentially better individuals avoiding to fall into local optima).

After reproduction has taken place, all the individuals are evaluated according to the fitness function to create the new population for the next epoch: strategies like tournament selection, total or partial replacement may be applied in this phase. Also, if an individual has a fitness below a pre-defined THRESHOLD then the cycle is terminated and the individual given as output (we assume that the smaller the fitness the better the individual).

Algorithm 1: GENETIC EVOLUTION

INITIALIZE POPULATION
solution = ϕ
$e = 0$
best fitness = THRESHOLD
while *best fitness* $\geq$ THRESHOLD **and** $e <$ MAX_EPOCHS
    REPRODUCE POPULATION
    EVALUATE FITNESS OF EVERY INDIVIDUAL
    IF *best fitness* < THRESHOLD **THEN**
        solution = individual having best fitness
    ELSE $e = e + 1$
    UPDATE POPULATION
RETURN *solution*

6. The QKLT-FW Watermarking Algorithm

This section describes the various components of the watermarking algorithm. We call this algorithm Quaternion Karhunen-Loève Transform Fragile Watermarking (QKLT-FW).

The input to the algorithm are the host color image I_h to be watermarked, and a secret key, in the form of a color image I_k. The images, both in bitmap format, are divided into contiguous non-overlapping sub-images (or blocks) of size $n \times n$: if the dimensions of I_k are not a multiple of n then the remaining part ($<n$) is not considered, whilst for simplicity we assume that for I_h of size $N \times M$ both N and M are multiples of n leading to $N_B = N \times M/n^2$ blocks. The output of the algorithm is a fragile watermarked image I_f.

The idea of the whole method is to hide a key-and-host-image dependent binary watermark into a set of secret features (QKLT coefficients) defined by the key image, considering the *color pixels as pure quaternions*. The host image is divided into sub-images (of size $n \times n$) and a portion of the watermark is embedded in each of them (see Figure 1). The alteration of a sub-image of I_f will, with high probability (we will discuss this point in a following section), modify the part of the watermark embedded in it, allowing the detection and localization of the attack by simple comparison of the expected and extracted watermarks.

In the following, we describe the steps to be performed to watermark an image with QKLT-FW.

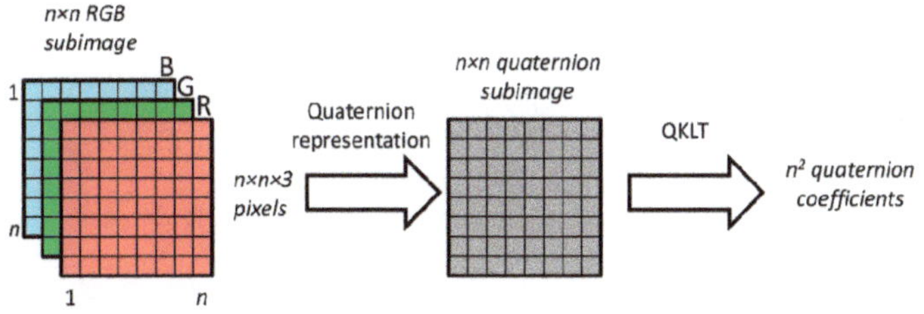

Figure 1. Quaternion representation of a sub-image and Quaternion Karhunen-Loève Transform (QKLT) coefficients derivation.

6.1. QKLT-FW Basis Generation

This step is performed by a unit that, receiving as input a key image I_k, divides it into non-overlapping blocks of size $n \times n$ and returns a QKLT orthonormal basis and an average sub-image as previously described: obviously the basis is composed of n^2 basis vectors each composed of n^2 quaternions, and the average sub-image is a vector of n^2 quaternions.

This computation may be performed only once for a set of images to be watermarked using the same key. Due to the dependence of the watermark on the host image, in principle a large amount of images may be watermarked with the same key without any possible leak on the computed secret basis.

6.2. QKLT-FW Integrity Data Generation

The secret bit string W used as watermark is made dependent on the host image to avoid transplantation and cut-and-paste attacks. It is obvious that this step must be executed for every host image to be watermarked with a particular key. Also, W should depend on the key (image), otherwise the bit string will not be secret (this is not strictly necessary, given the use of a secret space of embedding, but improves the whole security avoiding a possible search of embedded known bit strings); thus $W = S(I_h, I_k)$.

We implemented S (that may, and must, be public) as a procedure that:

- selects a set of pixels K_p in I_k in pre-defined positions;
- uses the values of these pixels to address a set of pixels H_p in I_h;
- applies the values of the pixels in H_p to address a set of pixels in I_k which in turn are used as seed of a cryptographic hash function (c.h.f.) like SHA-3;
- generates W by iteratively applying the c.h.f. until the required length is reached.

The pixels in the set H_p are not modified by the embedding algorithm because if only one of them is altered, a different watermark W_A will be computed during verification, leading to a completely altered image. Indeed, this is the result obtained from an attack that modifies a pixel in H_p: the localization property is lost but not the alteration detection.

If cropping is considered a possible attack (e.g., in case the protected images may be of any size), it is necessary to make the watermark dependent also on the image size (i.e., concatenating the height and the width of the image to the seed of the c.h.f.). A cropped (or enlarged) image will produce a different watermark during verification, thus many blocks will be flagged as forged: the localization will be lost but the attack will still be detected (effectively, cropping changes the relative position of a block with regard to borders).

6.3. QKLT-FW Embedding

To watermark a host image I_h composed of N_B sub-images, the algorithm inserts s watermark bits in every block of $n \times n$ color pixels (we call this payload s bits-per-block, or s bpb): thus, the previous step *Integrity data generation* will build a string W of size $s \cdot N_B$ bits.

To perform the insertion, different methods may be used. In the MIMIC framework [4], various embedding techniques were presented, but we briefly discuss only the two that are used here in conjunction with the QKLT:

- Bit Collect Module (BCM): s QKLT coefficients are selected to store each one a bit of the s watermark bits;
- Syndrome Coding Module (SCM): the s watermark bits per block are recorded as the syndrome of a word of r bits; these r bits are stored in r QKLT coefficients. Many possible codes may be used to perform syndrome coding, e.g., Hamming, Hadamard, Golay, BCH. See [4] for a deeper discussion on this topic.

In the present implementation, we chose to store one bit in one QKLT coefficient, having chosen a-priori the orders of these coefficients. From previous studies, we found that the order does not particularly influence the performance of the whole algorithm, so we presently use contiguous coefficients starting from the third (a key dependent choice is also an option, but we feel this a little improvement in security against an increased implementation complexity). A (quaternion) coefficient c is considered to carry the bit value b in position p (fixed a-priori) computed as

$$b = odd(int(2^{-p} \parallel c \parallel)), \tag{4}$$

where $\parallel c \parallel$ is the norm of the coefficient, *int* is a function that truncates a real number to its integer part, and *odd* is a function that returns 1 if its argument is odd and 0 otherwise.

In general, a sub-image does not already contain the required watermark string of s bits: the duty of the GA is to compute a modification Ξ of the sub-image such that it stores the correct bit string. The modification Ξ is, in the present case, a vector of n^2 components specifying the alterations to be applied to the RGB values of the n^2 pixels: we found that for an 8 bit-per-channel image, it is sufficient to have the set of possible alterations as small as $\{-3, -2, -1, 0, 1, 2, 3\}$, leading, as we will show, to a very good PSNR. Thus, the GA evolves a population of individuals each composed by n^2 genes: the absolute value of the alteration indexes the channel (i.e., 1 is red, 2 is green and 3 is blue) whilst the sign specifies if the pixel must be incremented or decremented by 1 (0 meaning no modification to the pixel). The fitness function of the GA takes into account two main properties of the resulting sub-image: the presence of the watermark and the PSNR with regard to the original. The GA runs for a maximum amount of epochs or until a viable solution is found.

When all the blocks have been processed by the GA, then the watermark has been embedded into I_h producing I_f.

We may summarize a high level behavior of the GA as:

- $I_f = I_h$
- For every sub-image Σ of I_f
  a. have a population of individuals representing modifications Ξ_i
  b. apply, in turn, every modification Ξ_i to Σ obtaining Σ_i
  c. compute the QKLT of every Σ_i and extract the bits according to (4)
  d. if, for some Σ_i, the watermark is stored and the PSNR is high (i.e., above a threshold) then save Σ_i in place of Σ and proceed to the next sub-image
  e. otherwise evolve the population Ξ_i and go to step b.
- Output I_f

6.4. QKLT-FW Extraction

This step uses the key image I_k and the watermarked (and possibly altered) image I_f. From I_k, the QKLT basis and the average sub-image are derived. Then, the QKLT is performed on the N_B blocks of I_f, selecting the chosen coefficients and extracting s bits from every block (using BCM or SCM and Equation (4)). The extracted watermark W_E is the concatenation of the $s \cdot N_B$ bits.

6.5. QKLT-FW Verification

From both I_k and I_f the watermark bit string, W is computed as shown in the *QKLT-FW Integrity data generation* step: this string is compared with the one extracted in the *QKLT-FW Extraction* phase W_E. Differing bits in homologous positions mean an alteration in the corresponding sub-image, signaling that an attack to the integrity of I_f has been performed.

6.6. Public and Secret

As the specific GA algorithm used and its parameters are instrumental for embedding only, they are not required by the verifier and are not part of the secret embedded, thus knowledge of the GA configuration would not compromise the security of the method and so its parameters may be left public.

The use of QKLT, the block size, the order and number of coefficients used and the bit embedding position may also be left public, but they give a hint in brute force attacks: anyway the space of all possible basis images is so large, for not naive (i.e., small) n, that the attack is unfeasible. The indexes used to address the pixels in I_k to create K_p may be public, and the suggestion is to keep the size of the set H_p small to reduce the probability that an attack alters the pixels in this set, leading to a loss of localization.

Finally, the key image *must* be kept secret: a compromised key image invalidates all the images authenticated with it.

7. Experimental Methodology

The parameters we used to evaluate the performance of the algorithms are the PSNR, the Structural Similarity index (SSIM) and the sensitivity.

For a d bit-per-pixel channel image, the PSNR of a watermarked image I_f with regard to the host image I_h is defined in the quaternion framework, considering the quaternion $m = (0,\ (2^d-1)\vec{i} + (2^d-1)\vec{j} + (2^d-1)\vec{k})$ that represents the peak pixel value, as:

$$\mathrm{PSNR} = 10\log_{10}\frac{\parallel m \parallel^2}{\mathrm{m.s.e.}} = 10\log_{10}\frac{\parallel m \parallel^2}{\frac{1}{NM}\sum_{z=1}^{NM}\parallel {I_h}^{(z)} - {I_f}^{(z)} \parallel^2}, \tag{5}$$

where m.s.e. is the mean squared error between the host and the watermarked images, and ${I_h}^{(z)}$ and ${I_f}^{(z)}$ are their z-th pixel quaternion representation.

The SSIM defined in [29] measures the difference between two images taking into account the characteristics of the Human Visual System. Its values range in the interval $[-1,\ 1]$, with a value of 1 expressing that two images are identical. As it resulted to be greater than 0.998 in all the experiments, we do not report the SSIM value explicitly for each setting in the result tables.

Sensitivity of level D refers to the percentage of detected altered image blocks when a single pixel of that block is modified by $+D$ or $-D$ intensity levels in a single channel. To compute the sensitivity of level D we initialize to 0 two counters TOTBLOCKS and DETECTED, and considering all the watermarked images of our experiments (as we will see, 500 images) for every image the respective watermark is generated and the following nested cycles are performed:

```
FOR EVERY BLOCK IN THE IMAGE
  FOR EVERY PIXEL IN THE BLOCK
    TOTBLOCKS = TOTBLOCKS + 1;
    ALTER THE PIXEL ADDING D AND, IF THE MODIFICATION IS POSSIBLE (That is, no overflow takes
    place), CHECK THE BLOCK USING THE VERIFICATION PROCEDURE TO TEST IF THE ATTACK IS
    DETECTED, IN WHICH CASE DO DETECTED = DETECTED + 1;
    ALTER THE PIXEL ADDING −D AND, IF THE MODIFICATION IS POSSIBLE (That is, no overflow takes
    place), CHECK THE BLOCK USING THE VERIFICATION PROCEDURE TO TEST IF THE ATTACK IS
    DETECTED,IN WHICH CASE DO DETECTED = DETECTED + 1;
```

When all the images have been examined, the ratio DETECTED/TOTBLOCKS represents the sensitivity of level D.

In this paper, we report the results for sensitivity levels 1 and 2, in order to show that QKLT-FW is very sensible to the smallest possible pixel alterations.

It should be pointed out that this kind of test is deeper that any image processing or compression algorithm that may be used to attack an image because any such algorithm, if it performs an alteration to a pixel, will be at least 1 intensity level: therefore, it is obvious that the detection of image processing or compression attacks can only lead to better performance than the worst cases examined by us.

The GA parameters were set as the following: a population size of 100 individuals, a crossover probability of 0.9 and a mutation probability of 0.04; the termination condition was 2000 generations total or the best solution did not change for 10 generations.

Firstly, we show the performance of the proposed algorithm in terms of PSNR, embedding time (on a Linux workstation equipped with 4GB RAM and an Intel(R) Xeon(R) E5410 2.33GHz processor) and sensitivity, in order to support our claims (b), (c) and (d) stated in the introduction. The set of images was composed of 500 images selected from the databases [30,31]; the images were cropped to 256×256 pixels to cut the computation times.

Table 1 reports the averages of PSNR, execution times and sensitivity for some system parameter settings (payload, embedding method, syndrome coding) of QKLT-FW. Moreover, the insertion position p (see Equation (4)) was fixed to 0.

Table 1. Results for some settings of Quaternion Karhunen-Loève Transform Fragile Watermarking (QKLT-FW) (the best performances with regard to the parameter are highlighted in boldface).

Payload (bpb)	Insertion Module	PSNR (dB)	Time (s)	Sensitivity ± 1 (%)	Sensitivity ± 2 (%)
8	BCM	61.97 ± 0.1	22.65 ± 1.67	67.29 ± 8.22	88.61 ± 2.82
12	SCM (Golay {24,12,8})	62.23 ± 0.11	121.5 ± 8.77	90.53 ± 2.91	99.25 ± 0.23
12	BCM	62.25 ± 0.18	34.7 ± 2.4	93.07 ± 1.05	99.78 ± 0.06
16	BCM	59.44 ± 0.1	87.68 ± 18.06	82.75 ± 7.07	97.36 ± 1.11

We also performed a test using smaller blocks of size 6×6, embedding 6 bpb using BCM as Insertion module: the PSNR resulted in 62.3961 ± 0.23853 dB with a computation time of 45.8534 ± 19.3611 s. As it may be seen, the quality is very high with a slightly increased computation time due to the augmented number of blocks: this is because the overhead is due to the computation of the genetic algorithm and the reduced dimension of the block does not completely compensate for this. This is only an example of the flexibility of the proposed algorithm, as the size and shape of the blocks can be set according to the localization resolutions, payload and running times required by the application.

As a visual example of the results of the watermarking and verification processes (i.e., what the naive user perceives), the Appendix A reports the F16 watermarked image with blocks of size 8×8, an attack to it and how the algorithm detects the tampered region(s).

Then, we compared the performance with those resulting from running the algorithms [25], implemented for color images, and [4] on the same set of images (as MIMIC was already compared with others in [4] and resulted to be qualitatively better). Table 2 shows the comparison among these algorithms. In the case of [25], the intrinsic properties of the algorithms forced the specific values of bpb and block size: on the contrary, both MIMIC and QKLT-FW revealed a better flexibility allowing more combinations of block sizes and payloads. Note that the sensitivity (of any level) of [25] is 100%, thanks to the use of MACs for the protection of the blocks. Due to the MAC size, [25] requires embedding a larger number of authentication bits reducing the PSNR.

Table 2. Comparison among different embedding algorithms.

Algorithm	Payload (bpb)	Block Size (pixel)	PSNR (dB)	Time (s)	Sensitivity ± 2 (%)
[25]	213.125	256	56.71	~2	100
MIMIC SRM [4]	12	64	62.7	37.32 ± 1.96	93.43
MIMIC BCM192 [4]	12	64	58.47	72.97 ± 0.83	96.49
QKLT-FW	12	64	62.25	34.70 ± 2.4	99.79

As can be seen in Table 2, QKLT-FW exhibits an improved performance over the baseline systems considered. Indeed, with an *equivalent* watermarked image *quality*, the quaternion approach *improves* with regard to *running time* and *sensitivity*. It is worth pointing out that the quaternion framework produces an integrated information with regard to the color components (or multi-channel components), allowing the application of all the properties of quaternions for color image processing, in particular the QKLT.

8. Conclusions

In summary, we emphasize that with a simple data type modification from real value to quaternion, the proposed system shows a better performance for detecting and localizing the content alterations. As already noted in [4], [25] has the advantages of a sensitivity of 100% and of a predictable computing time, but its main drawbacks are a lower PSNR and the space required to store the authentication data, that implies the use of large blocks needed to save the authentication information, reducing the localization capability.

The proposed algorithm based on QKLT has thus the following properties and advantages:

- high PSNR and high SSIM, resulting in very high quality of the watermarked images, both objective and subjective (see Table 1);
- high sensitivity to modifications because even single pixel modifications of two intensity levels may be detected in more than 99.7% of the cases (see Table 2): this makes the probability that any real attack goes undetected close to zero in practice;
- flexible and good localization capability, as shown working on blocks of different sizes, namely 6×6 and 8×8 color pixels, and different payloads;
- easily integrated into the MIMIC framework as a new Watermark Distilling Unit, improving the running times of previously developed algorithms in the same framework (as can be seen in Table 2).

It should be noted that, in some cases, the GA may not converge to a solution due to the intrinsic stochastic approach that embeds the watermark in every image block; it is possible to cope with this problem by running the GA multiple times on the blocks reporting a convergence problem, also reducing the tightness of some constraints (e.g., on the possible modification the GA may perform on the pixel values).

As for MIMIC [4], the security of the method is based on the secrecy of the image from which the KLT basis is derived: from that image a secret embedding space is derived, so the transform coefficients cannot be determined, in particular their less significant bits. Moreover, the watermark string is dependent both on the key image, on the host image, and on the host image size: this prevents copy-and-paste attacks, transplantation attacks, VQ attacks, cropping and embedding attacks. Note that in the MIMIC framework, a trivial substitution attack is always possible: changing a block with a random one will go undetected 1 every 2^s attempts, if s bpb are embedded. But, in an image with U blocks, the probability of not detecting any modification is $1/2^{Us}$: this is a very small number, even for an image with a small number of blocks. We stress the fact that the use of quaternions as color pixel representation opens up the possibility of applying the presented approach to color images with alpha channel (i.e., four-dimensional pixels), as an integrated approach. This is one of our future research directions.

Author Contributions: All authors equally contributed to the design and implementation of the described algorithms and experiments, as well as to write and proofread the paper.

Conflicts of Interest: The authors declare no conflict of interest.

Appendix A

In this appendix we report, as example, the result of watermarking one of the image used in the experiments: Figure A1 shows the watermarked image (we do not report the original image because, given the very high PSNR no differences with Figure A1 may be appreciated by the human eye), whilst Figure A2 presents the watermarked image altered by a tamper on the right part of the sign (the number 1100 has been modified, by copying and pasting two areas, to 1010).

In Figure A3, the output of the verification procedure on the image in Figure A2 is presented: the forged area is correctly evidenced by marking the blocks which contain at least one modified pixel.

Figure A1. Watermarked color image, publicly available from the McGill Calibrated Color Image Database [30] (http://tabby.vision.mcgill.ca/html/welcome.html), PSNR = 67.02 dB (with zoom on detail).

Figure A2. Tampered image (with zoom on tampered detail).

Figure A3. Verified image, with (nineteen) tampered blocks evidenced as crossed areas.

References

1. Cox, I.J.; Miller, M.L.; Bloom, J.A.; Fridrich, J.; Kalker, T. *Digital Watermarking and Steganography*, 2nd ed.; Morgan Kaufmann Publishers Inc.: San Francisco, CA, USA, 2008.
2. Botta, M.; Cavagnino, D.; Pomponiu, V. Fragile Watermarking using Karhunen-Loève transform: The KLT-F approach. *Soft Comput.* **2015**, *19*, 1905–1919. [CrossRef]

3. Aslantas, V.; Ozer, S.; Ozturk, S. Improving the performance of DCT-based fragile watermarking using intelligent optimization algorithms. *Opt. Commun.* **2009**, *282*, 2806–2817. [CrossRef]
4. Botta, M.; Cavagnino, D.; Pomponiu, V. A Modular Framework for Color Image Watermarking. *Signal Process.* **2016**, *119*, 102–114. [CrossRef]
5. Le Bihan, N.; Sangwine, S.J. Quaternion principal component analysis of color images. In Proceedings of the 2003 International Conference on Image Processing, Barcelona, Spain, 14–17 September 2013; Volume 1, pp. 809–812. [CrossRef]
6. Sangwine, S.J. Fourier transforms of colour images using quaternion, or hypercomplex numbers. *Electron. Lett.* **1996**, *32*, 1979–1980. [CrossRef]
7. Ell, T.A.; Sangwine, S.J. Hypercomplex Fourier transforms of color images. *IEEE Trans. Image Process.* **2007**, *16*, 22–35. [CrossRef] [PubMed]
8. Le Bihan, N.; Mars, J. Singular value decomposition of quaternion matrices: A new tool for vector-sensor signal processing. *Signal Process.* **2004**, *84*, 1177–1199. [CrossRef]
9. Angulo, J. Geometric algebra colour image representations and derived total orderings for morphological operators—Part I: Colour quaternions. *J. Vis. Commun. Image Represent.* **2010**, *21*, 33–48. [CrossRef]
10. Bas, P.; Le Bihan, N.; Chassery, J.-M. Color image watermarking using quaternion Fourier transform. In Proceedings of the 2003 IEEE International Conference on Acoustics, Speech, and Signal Processing, Hong Kong, China, 6–10 April 2003; Volume 3, pp. 521–524.
11. Sangwine, S.J.; Ell, T.A. Hypercomplex Fourier transforms of color images. In Proceedings of the ICIP, Thessaloniki, Greece, 7–10 October 2001; Volume 1, pp. 137–140.
12. Wang, X.-Y.; Wang, C.-P.; Yang, H.-Y.; Niu, P.-P. A robust blind color image watermarking in quaternion Fourier transform domain. *J. Syst. Softw.* **2013**, *86*, 255–277. [CrossRef]
13. An, M.; Wang, W.; Zhao, Z. Digital watermarking algorithm research of color images based on quaternion Fourier transform. In Proceedings of the SPIE 8917: Multispectral Image Acquisition, Processing, and Analysis, Wuhan, China, 26 October 2013; pp. 1–7.
14. Tsui, T.K.; Zhang, X.-P.; Androutsos, D. Color Image Watermarking Using Multidimensional Fourier Transforms. *IEEE Trans. Inf. Forensics Secur.* **2008**, *3*, 16–28. [CrossRef]
15. Chen, B.; Coatrieux, G.; Chen, G.; Sun, X.; Coatrieux, J.L.; Shu, H. Full 4-D quaternion discrete Fourier transform based watermarking for color images. *Digit. Signal Process.* **2014**, *28*, 106–119. [CrossRef]
16. Shao, Z.; Duan, Y.; Coatrieux, G.; Wu, J.; Meng, J.; Shu, H. Combining double random phase encoding for color image watermarking in quaternion gyrator domain. *Opt. Commun.* **2015**, *343*, 56–65. [CrossRef]
17. Shao, Z.; Wu, J.; Coatrieux, J.L.; Coatrieux, G.; Shu, H. Quaternion gyrator transform and its application to color image encryption. In Proceedings of the ICIP, Melbourne, Australia, 15–18 September 2013; pp. 4579–4582.
18. Qi, M.; Li, B.-Z.; Sun, H. Image watermarking using polar harmonic transform with parameters in SL(2,R). *Signal Process. Image Commun.* **2015**, *31*, 161–173. [CrossRef]
19. Yang, H.-Y.; Wang, X.-Y.; Niu, P.-P.; Wang, A.-L. Robust Color Image Watermarking Using Geometric Invariant Quaternion Polar Harmonic Transform. *ACM Trans. Multimed. Comput. Commun. Appl.* **2015**, *11*, 40:1–40:26. [CrossRef]
20. Al-Otum, H. Color image authentication using a zone-corrected error-monitoring quantization-based watermarking technique. *Opt. Eng.* **2016**, *55*, 083103. [CrossRef]
21. Akansu, A.N.; Haddad, R.A. *Multiresolution Signal Decomposition: Transforms, Subbands, and Wavelets*; Academic Press: Cambridge, MA, USA, 1992.
22. Lin, S.; Lin, J. Authentication and recovery of an image by sharing and lattice-embedding. *J. Electron. Imaging* **2010**, *19*, 043008. [CrossRef]
23. Lo, C.; Hu, Y.; Chen, W.; Chang, I. Probability-based image authentication scheme for indexed color images. *J. Electron. Imaging* **2014**, *23*, 033003. [CrossRef]
24. Wang, X.; Zhang, D.; Guo, X. Authentication and recovery of images using standard deviation. *J. Electron. Imaging* **2013**, *22*, 033012. [CrossRef]
25. Celik, M.U.; Sharma, G.; Saber, E.; Tekalp, A.M. Hierarchical watermarking for secure image authentication with localization. *IEEE Trans. Image Process.* **2002**, *11*, 585–595. [CrossRef] [PubMed]
26. Vince, J. *Quaternions for Computer Graphics*; Springer: Berlin, Germany, 2011.
27. Zhang, F. Quaternions and matrices of quaternions. *Linear Algebra Its Appl.* **1997**, *251*, 21–57. [CrossRef]

28. Pei, S.-C.; Cheng, C.-M. A novel block truncation coding of color images by using quaternion-moment preserving principle. In Proceedings of the IEEE International Symposium on Circuits and Systems, Atlanta, GA, USA, 23–26 May 1996; Volume 2, pp. 684–687.
29. Wang, Z.; Bovik, A.C.; Sheikh, H.R.; Simoncelli, E.P. Image quality assessment: From error visibility to structural similarity. *IEEE Trans. Image Process.* **2004**, *13*, 600–612. [CrossRef]
30. Olmos, A.; Kingdom, F.A.A. A biologically inspired algorithm for the recovery of shading and reflectance images. *Perception* **2004**, *33*, 1463–1473. [CrossRef] [PubMed]
31. Fred Kingdom's Laboratory, McGill Vision Research. Available online: http://tabby.vision.mcgill.ca/ (accessed on 18 December 2014).

Journal of
Imaging

MDPI

Article

Histogram-Based Color Transfer for Image Stitching†

Qi-Chong Tian and Laurent D. Cohen *

Université Paris-Dauphine, PSL Research University, CNRS, UMR 7534, CEREMADE, 75016 Paris, France; tian@ceremade.dauphine.fr

* Correspondence: cohen@ceremade.dauphine.fr; Tel.: +33-1-4405-4678

† This paper is an extended version of our paper published in the 6th International Conference on Image Processing Theory, Tools and Applications (IPTA'16), Oulu, Finland, 12–15 December 2016.

Received: 5 July 2017; Accepted: 6 September 2017; Published: 9 September 2017

Abstract: Color inconsistency often exists between the images to be stitched and will reduce the visual quality of the stitching results. Color transfer plays an important role in image stitching. This kind of technique can produce corrected images which are color consistent. This paper presents a color transfer approach via histogram specification and global mapping. The proposed algorithm can make images share the same color style and obtain color consistency. There are four main steps in this algorithm. Firstly, overlapping regions between a reference image and a test image are obtained. Secondly, an exact histogram specification is conducted for the overlapping region in the test image using the histogram of the overlapping region in the reference image. Thirdly, a global mapping function is obtained by minimizing color differences with an iterative method. Lastly, the global mapping function is applied to the whole test image for producing a color-corrected image. Both the synthetic dataset and real dataset are tested. The experiments demonstrate that the proposed algorithm outperforms the compared methods both quantitatively and qualitatively.

Keywords: color transfer; color correction; image stitching; histogram specification; global mapping curve

---

1. Introduction

Image stitching [1] is the technique for producing a panorama large-size image from multiple small-size images. Due to the differences in imaging devices, camera parameter settings or illumination conditions, these multiple images are usually color inconsistent. This will affect visual results of image stitching. Thus, color transfer plays an important role in image stitching. It can maintain the color consistency and make the panorama be more natural than the results without color transfer.

Color transfer is also known as color correction, color mapping or color alignment in the literature [2–7]. This kind of technique is aimed to transfer the color style of a reference image to a test image. It can make these images to be color consistent. One example is shown in Figure 1, from which we can obviously see the effectiveness of color transfer in image stitching.

Pitie et al. [8,9] proposed an automated color mapping method using color distribution transfer. There are two parts in their algorithm. The first part is to obtain a one-to-one color mapping using three-dimensional probability density function transfer, which is iterative, nonlinear and convergent. The second part is to reduce grain noise artifacts via a post-processing algorithm, which adjusts the gradient field of the corrected image to match the test image. Fecker et al. [10] proposed a color correction algorithm using cumulative histogram matching. They used basic histogram matching algorithm for the luminance component and chrominance components. Then, the first and last active bin values of cumulative histograms are modified to satisfy the monotonic constraint, which can avoid possible visual artifacts. Nikolova et al. [11,12] proposed a fast exact histogram specification algorithm, which can be applied to color transfer. This approach relies on an ordering algorithm, which is based on a specialized variational method [13]. They used a fast fixed-point algorithm to minimize the functions and obtain color corrected images.

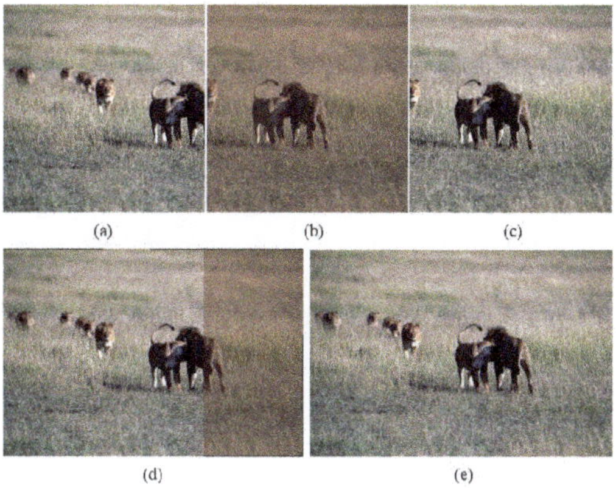

Figure 1. An example of color transfer in image stitching. (**a**) reference image; (**b**) test image; (**c**) color transfer for the test image using the reference color style; (**d**) stitching without color transfer; (**e**) stitching with color transfer. Image Source: courtesy of the authors and databases referred on [2,14].

Compared to the previous approaches described above, we combine the ideas of histogram specification and global mapping to produce a color transfer function, which can extend well the color mapping from the overlapping region to the entire image. The main advantage of our method is the color transfer ability for two images having small overlapping regions. The experiments also show that the proposed algorithm outperforms other methods in terms of objective evaluation and subjective evaluation.

This paper is an extended version of our previous work [15]. Compared with the conference paper [15], more related work are introduced, more comparisons and discussions are included in this paper. The rest of this paper is organized as follows. The related work is summarized in Section 2. The detailed proposed color transfer algorithm is presented in Section 3. The experiments and the result analysis are shown in Section 4. The discussion and conclusion are given in Section 5.

2. Related Work

Image stitching approaches can combine multiple small-size images together to produce a large-size panorama image. Generally speaking, image alignment and color transfer are the two important challenging tasks in image stitching, which has received a lot of attention recently [1,16–20]. Different kinds of image alignment methods or different color transfer algorithms can construct different approaches for image stitching. Even though color transfer method is the main topic studied in this paper, we also introduce the image alignment algorithms to make this research be comprehensive and be understood easily. A brief review of the methods for image alignment and color transfer is presented below.

2.1. Image Alignment

Motion models describe the mathematical relationships between the pixel coordinates in one image and the pixel coordinates in the other image. There are four main kinds of motion models in image stitching, including 2D translations, 3D translations, cylindrical and spherical coordinates, and lens distortions. For a specific application, a corresponding motion model needs to be defined first. Then, the parameters in the motion model can be estimated using corresponding algorithms. At last, the considered images can be aligned rightly to create a panorama image. We summarize two kinds of alignment algorithms, including pixel-based alignment and feature-based alignment.

2.1.1. Pixel-Based Alignment

The pixel-based alignment methods are to shift or warp the images relative to other images and to compare the corresponding pixels. Generally speaking, an error metric is firstly defined to compare the difference between the considered images. Then, a suitable search algorithm is applied to obtain the optimal parameters in the motion model. The detailed techniques and the comprehensive description are available in [1]. A simple description of this method is given below.

Given an image $I_0(x_i)$, the goal is to obtain where it is located in the other image $I_1(x_i)$. The simplest solution is to compute the minimum of the sum of squared difference function:

$$E(u) = \sum_i (I_1(x_i + u) - I_0(x_i))^2 = \sum_i e_i^2, \tag{1}$$

where u is the displacement vector, $e_i = I_1(x_i + u) - I_0(x_i)$ is the residual error. To solve this minimization problem, the search algorithms will be adopted. The simplest method is the full search technique. For speeding up the computation, coarse-to-fine techniques based on image pyramids are often used in the practical applications.

2.1.2. Feature-Based Alignment

The feature-based alignment methods are to extract distinctive features (interesting points) from each image and to match every feature. Then, the geometric transformation between the considered images is estimated. The most popular feature extraction method is the scale-invariant feature detection [21]. The most widely used solution for feature matching is the indexing schemes based on finding nearest neighbors in high-dimension spaces. For estimating the geometric transformation, a usual method is to use least squares to minimize the sum of squared residuals by

$$E_{LS} = \sum_i ||r_i||^2 = \sum_i ||\tilde{x}_i'(x_i; p) - \hat{x}_i'||^2, \tag{2}$$

where $\hat{x}_i'$ is the detected feature point location corresponding to point x_i in other images, $\tilde{x}_i'$ is the estimated location, and p is the estimated motion parameter. Equation (2) assumes all feature points are matched with the same accuracy, which does not work well in the real application. Thus, the weighted least square is often used to obtain more robust results via

$$E_{WLS} = \sum_i \sigma_i^{-2} ||r_i||^2, \tag{3}$$

where σ_i^2 is a variance estimate.

2.2. Color Transfer

The color transfer problem is well reviewed in [2,5]. A brief introduction is summarized below.

2.2.1. Geometry-Based Color Transfer

Geometric-based color transfer methods compute the color mapping functions using the corresponding feature points in multiple images. Feature detection algorithms are adopted to obtain the interesting points. Scale-Invariant Feature Transform (SIFT) [21] and Speeded-Up Robust Feature (SURF) [22] are the two most widely used methods for feature detection. After obtaining the features of each image, the correspondences between the considered images are matched using the RANdom SAmple Consensus algorithm (RANSAC), which can remove the outliers efficiently to improve the matching accuracy. Then, the correspondences are used to build a color transfer function via minimizing the color difference between the corresponding feature points. Finally, this transfer function is applied to the target image to produce the color transferred image.

2.2.2. Statistics-Based Color Transfer

When the feature detection and matching are not available, the geometry-based color transfer can not work. In this situation, the statistical correlation [23] between the reference image and the test image is used to create the color mapping function, which can transfer the color style of the reference image to the test image and enforce the considered images to share the same color style. Reinhard et al. [24] proposed a simple and traditional statistics-based algorithm to transfer colors between two images, which was also extended by many researchers. Papadakis et al. [25] proposed a variational model for color image histogram transfer, which used the energy functional minimization to finish the goal of transferring the image color style and maintaining the image geometry. Hristova et al. [26] presented a style-aware robust color transfer method, which was based on the style feature clustering and the local chromatic adaptation transform.

2.2.3. User-Guided Color Transfer

When the feature matching information and the statistical information of the considered images are both difficult to be obtained, it is essential to adopt user-guided methods to create the correspondences and use them to build the color transfer mapping function. The transfer function between images can be obtained from a set of strokes [27], which are user-defined by painting on the considered images. Then, the transfer function can be computed via different minimization approaches. The other kind of method is the color swatch based algorithm [28], which is more related to the construction of the correspondences between the considered images. The color mapping function is obtained from swatched regions in one image and can be applied to the corresponding regions in the other image.

3. The Proposed Approach

This paper proposes a method of color transfer in image stitching using histogram specification and global mapping. Generally speaking, there are four steps in this algorithm. Firstly, there are two given images to be stitched. The image with good visual quality is defined as the reference image, and the other is defined as the test image. Overlapping regions between these two images are obtained using a feature-based matching method. Secondly, histogram specification is conducted for the overlapping regions. Thirdly, using corresponding pixels in the overlapping region, which are original pixels and the pixels after histogram specification, the mapping function is computed with an iterative method for minimizing color differences. At last, the whole color transferred image is produced by applying the mapping function to the entire test image.

3.1. The Notations and the Algorithm Framework

$\boldsymbol{R}$ is a reference image,
$\boldsymbol{T}$ is a test image,
$\boldsymbol{R\_O}$ is the overlapping region in the reference image,
$\boldsymbol{T\_O}$ is the overlapping region in the test image,
$\boldsymbol{T\_O\_HS}$ is the result of histogram specification for $\boldsymbol{T\_O}$,
(i, j) is the location of pixels in images,
k is the pixel values, $\mathrm{k} \in [0, 1, \ldots, 255]$ for 8-bit images,
$\varepsilon(k) := \{(i,j) \in T\_O \mid T\_O(i,j) = k\}$,
$\boldsymbol{Map}$ is a color mapping function,
$\boldsymbol{T\_O\_Map}$ is the result of color transfer for $\boldsymbol{T\_O}$ using the color mapping function,
$\boldsymbol{Diff}$ is pixel differences between two images,
PSNR is the peak signal-to-noise ratio between two images.
The algorithm framework is described in Figure 2.

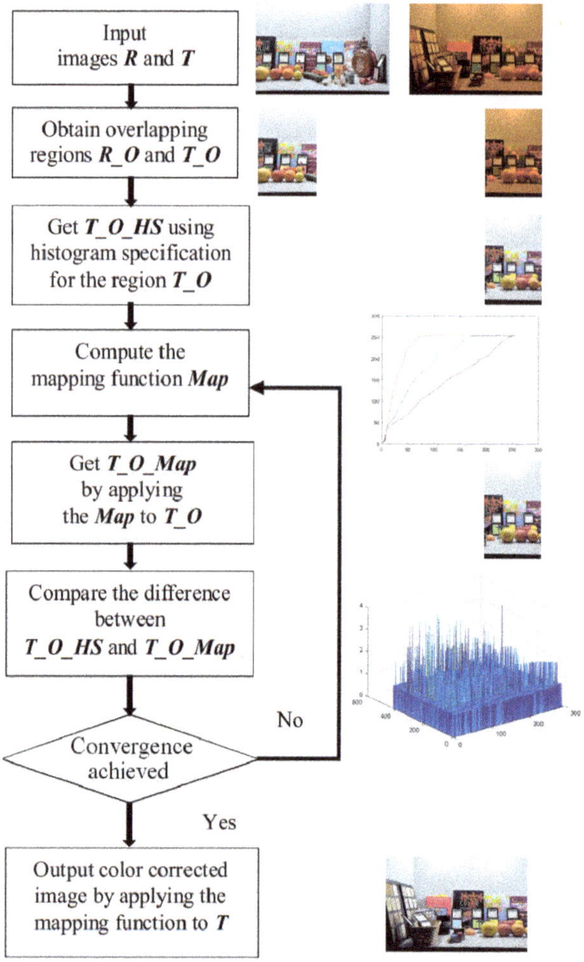

Figure 2. The framework of the proposed algorithm. Image Source: courtesy of the authors and databases referred on [29].

3.2. *The Detailed Description of This Algorithm*

In this section, we will describe the proposed algorithm in detail.

3.2.1. Obtain Overlapping Regions between Two Images

In the application of image stitching, there are overlapping regions between input images. Due to little changes of scenes, differences of image capture angles, differences of focal lengths and other factors, the corresponding overlapping regions are not exactly pixel-to-pixel. Firstly, we find matching points between the reference image and the test image, using the scale-and-rotation-invariant feature descriptor (SURF) [22]. Then, the geometric transformation will be estimated from the corresponding points. In our implementation, the projective transformation is applied. After that, these images can be transformed and placed to the same panorama [1]. At last, we obtain overlapping regions using the image correspondence location information. This part is described in Algorithm 1.

Algorithm 1 Obtain overlapping regions between two images.

1: Input two images $\boldsymbol{R}$ and $\boldsymbol{T}$, then compute the feature point correspondences $\boldsymbol{R}_i \leftrightarrow \boldsymbol{T}_i$ using SURF, $\boldsymbol{i = 1, 2, ..., N}$, where $\boldsymbol{N}$ is the number of feature point correspondences.

2: Estimate the geometric transform $\boldsymbol{tforms}$ using the correspondences, the following term is minimized:

$$\boldsymbol{min \sum_{i=1}^{N} (R_i - tform(T_i))^2}.$$

3: Warp these two images and put in the same panorama using the geometric transform $\boldsymbol{tforms}$, define two matrixes $\boldsymbol{M_1}$ and $\boldsymbol{M_2}$ to store position information.

4: Obtain overlapping regions using the image correspondence location information described in $\boldsymbol{M_1}$ and $\boldsymbol{M_2}$.

3.2.2. Histogram Specification for the Overlapping Region

In this step, we will make exact histogram specification for the overlapping region in the test image to match the histogram of the overlapping region in the reference image. The histogram is calculated as follows:

$$\boldsymbol{Hist(k)} = \frac{\boldsymbol{1}}{m \times n} \sum_{i=1}^{m} \sum_{j=1}^{n} \delta[k, T(i,j)], \tag{4}$$

where

$$\delta[a, b] = \begin{cases} 1, & \text{if } a = b, \\ 0, & \text{otherwise.} \end{cases}$$

T is an image, $\boldsymbol{k}$ are pixel values, $\boldsymbol{k} \in [0, 1, ..., 255]$ for 8-bit images, $\boldsymbol{m}$ and $\boldsymbol{n}$ are the height and width of the image, and $\boldsymbol{i}$ and $\boldsymbol{j}$ are the columns and rows of pixels.

Histogram specification is also known as histogram matching, which is aimed to transform an input image to an output image fitting a specific histogram. We adopt an algorithm in [11] to perform the histogram specification in overlapping regions between the reference image and the test image. The detailed algorithm is described in Algorithm 2.

Algorithm 2 Histogram specification for the overlapping region.

1: Input: $\boldsymbol{T\_O}$ is the overlapping region in the test image, $\boldsymbol{hist}$ is the histogram of $\boldsymbol{R\_O}$, $\boldsymbol{u^{(0)} = T\_O}$, $\boldsymbol{\alpha = 0.05, \beta = 0.1}$, iteration number $\boldsymbol{S = 5}$, $\boldsymbol{c_0 = 0}$.

2: For $\boldsymbol{s = 1, ..., S}$, compute

$$\boldsymbol{u^{(s)} = T\_O - \eta^{-1}(\beta \nabla^T \eta(\nabla u^{(s-1)}))},$$

where $\boldsymbol{\nabla}$ is the gradient operator, $\boldsymbol{\nabla^T}$ is the transposition of $\boldsymbol{\nabla}$, $\boldsymbol{\eta^{-1}(x) = \frac{\alpha x}{1-|x|}, \eta(x) = \frac{x}{\alpha+|x|}}$.

3: Order the values in $\boldsymbol{\Pi_N}$ according to the corresponding ascending entries of $\boldsymbol{u^{(S)}}$, where $\boldsymbol{\Pi_N := \{1, ..., N\}}$ denote the index set of pixels in $\boldsymbol{T\_O}$.

4: For $\boldsymbol{k = 0, 1, ..., 255}$,

set $\boldsymbol{c_{(k+1)} = c_{(k)} + hist_{(k)}}$ and $\boldsymbol{T\_O\_HS[c_{(k)} + 1] = ... = T\_O\_HS[c_{(k+1)}] = k}$.

3.2.3. Compute the Color Mapping Function

In this step, we will get the color mapping function from corresponding pixels in T_O and T_O_HS. This operation is conducted for the three color channels, respectively.

For each color channel, a mapping function is computed as follows:

$$Map(k) = \left\lfloor \left(\left(\min_{c} \sum_{(i,j)\in\varepsilon(k)} \left(T\_O\_HS(i,j) - c \right)^2 \right) + 0.5 \right) \right\rfloor, \tag{5}$$

where $k \in [0, 1, ..., 255]$ for 8-bits images, $\lfloor (x) \rfloor$ is the nearest integer of x towards minus infinity, $\varepsilon(k) := \{(i,j) \in T\_O \mid T\_O(i,j) = k\}$. The nearest integer of c is the mapping value corresponding to k. In the minimization problem of Equation (5), the value of c is usually not an integer. Thus, we use the nearest integer as the corresponding mapping value of k.

During the estimation of a color mapping function, we embed some constraint conditions like the related methods [3,30]. Firstly, the mapping function must be monotonic. Secondly, some function values may be obtained by interpolation methods, due to some pixel values k not existing in the overlapping regions. In our implementation, the simple linear interpolation is used. The detailed algorithm is described in Algorithm 3.

Algorithm 3 Compute the color mapping function.

1: Input: T_O is the overlapping region in the test image, T_O_HS is the result of histogram specification for T_O. The following steps will be conducted for the three color channels, respectively.

2: For $k = 0, 1, ..., 255$, minimize the function:

$$Map(k) = \left\lfloor \left(\left(\min_{c} \sum_{(i,j)\in\varepsilon(k)} \left(T\_O\_HS(i,j) - c \right)^2 \right) + 0.5 \right) \right\rfloor.$$

3: For some value of k, the set $\varepsilon(k)$ is the empty set. Then, the corresponding k can not be computed in the above step and will be obtained using interpolation methods.

3.2.4. Minimize Color Differences Using an Iterative Method

Firstly, color transfer is conducted in the overlapping region T_O by the color mapping function obtained at the previous step. The result is denoted as T_O_Map. Secondly, pixel value differences $Diff$, and the PSNR between T_O_HS and T_O_Map is computed. Thirdly, the pixels (i,j) will be removed from $\varepsilon(k) := \{(i,j) \in T\_O \mid T\_O(i,j) = k\}$, when $Diff(i,j)$ is larger than the preset threshold **Thd_Diff**, since this kind of pixel is considered to be outliers. Finally, a new color mapping function can be obtained by the algorithm described in Algorithm 3.

Repeat these processes until reaching the preset iteration times or PSNR increase is smaller than the preset threshold **Thd_PSNR**. After these iterations, the final mapping function is applied to the whole test image. Then, the corrected image shares the same color style with the reference image. In other words, the two images are color consistent, which are suitable for image stitching. The detailed algorithm is described in Algorithm 4.

Algorithm 4 Minimize color differences using an iterative method.

1: Input: $\boldsymbol{T\_O}$ is the overlapping region in the test image, $\boldsymbol{Map}$ is the color mapping function obtained in **Algorithm 3**, $\varepsilon(k) := \{(i,j) \in T\_O \,|\, T\_O(i,j) = k\}$, maximal iteration number S, **Thd_Diff** is a threshold value, **Thd_PSNR** is a threshold value.

2: Obtain T_O_Map by applying Map to T_O, using

$$T\_O\_Map(i,j) = Map(T\_O(i,j)).$$

3: Compute pixel-to-pixel differences by

$$Diff(i,j) = |T\_O\_Map(i,j) - T\_O\_HS(i,j)|.$$

4: Remove pixels (i,j) from $\varepsilon(k)$, when $Diff(i,j)$ is larger than the preset threshold **Thd_Diff**.

5: Compute the PSNR increase for T_O_Map.

6: With the new sets $\varepsilon(k)$, repeat **Algorithm** 3 and steps 2 to 5 in **Algorithm** 4 until reaching the maximal iteration number or PSNR increase is smaller than the threshold **Thd_PSNR**.

4. Experiments

4.1. Test Dataset and Evaluation Metrics

Both synthetic image pairs and real image pairs are selected to compose the test dataset. Test images in this dataset are chosen from [2,3,14,29]. The synthetic data includes 40 reference/test image pairs. Each pair is from the same image, but with different color style. The image with good visual quality is assigned as a reference image, and the other is assigned as a test image. The real data includes 35 reference/test image pairs. These image pairs are taken under different capture conditions, including different exposures, different illuminations, different imaging devices or different capture time. For each pair, the image of good quality is assigned as a reference image and the other as a test image.

Anbarjafari [31] proposed an objective no-reference measure for illumination assessment. Xu and Mulligan [2] proposed an evaluation method for color correction in image stitching, which has been adopted in our evaluation. This method includes two components: color similarity between a corrected image G and a reference image R, and structure similarity between a corrected image G and a test image T.

The **C**olor **S**imilarity $CS(G,R)$ is defined as $CS(G,R) = PSNR(G\_O, R\_O)$. PSNR is the **P**eak **S**ignal-to-**N**oise **R**atio [32] and G_O,R_O are the overlapped regions of G and R, respectively. The higher value of $CS(G,R)$ indicates the more similar color style between the corrected image and the reference image. The definition of $PSNR$ is given by

$$\begin{aligned} PSNR(A,B) &= 10 \times log_{10}(\frac{L^2}{MSE(A,B)}), \\ MSE(A,B) &= \frac{1}{m \times n} \sum_{i=1}^{m} \sum_{j=1}^{n} (A(i,j) - B(i,j))^2, \end{aligned} \tag{6}$$

where A and B are the considered images, $L = 255$ for 8-bit images, and m and n are the height and width of the considered images.

The structure similarity $SSIM(G,T)$ is the **S**tructural **SIM**ilarity index, which is defined as a combination of luminance, contrast and structure components [33]. The higher value of $SSIM(G,T)$

indicates the more similar structure between the corrected image and the test image. The definition of $SSIM$ is described by

$$SSIM(A,B) = \frac{1}{N}\sum_{i=1}^{N} SSIM(a_i,b_i), \quad (7)$$

where N is the number of local windows for an image, and a_i, b_i are the image blocks at the ith local window of the image A and B, respectively. The detailed computation of $SSIM(a_i,b_i)$ is described by

$$SSIM(a,b) = [l(a,b)]^{\alpha} \times [c(a,b)]^{\beta} \times [s(a,b)]^{\gamma}, \quad (8)$$

where $l(a,b) = \frac{2\mu_a\mu_b + C_1}{\mu_a^2+\mu_b^2 + C_1}, c(a,b) = \frac{2\sigma_a\sigma_b + C_2}{\sigma_a^2+\sigma_b^2 + C_2}, s(a,b) = \frac{\sigma_{ab} + C_3}{\sigma_a\sigma_b + C_3}$, μ_a and μ_b are the mean luminance values of the windows a and b, respectively, σ_a and σ_b are the standard variance of the windows a and b, respectively, σ_{ab} is the auto-covariance between the windows a and b, C_1, C_2, C_3 are small constants to avoid divide-by- zero error, and α, β, γ are constants controlling the weight among the three components. The default settings recommended in [33] are: $C_1 = (0.01L)^2, C_2 = (0.03L)^2$, $C_3 = \frac{C_2}{2}, L = 255, \alpha = \beta = \gamma = 1$.

In the following parts, we compare our algorithm with the methods proposed in [9–11]. These methods transfer the color style of the whole reference image to the whole test image. The source codes of Pitie's and Nikolova's methods are downloaded from their homepages. The source code of Fecker's method is obtained from [2].

4.2. Experiments on Synthetic Image Pairs

Each synthetic image pair from [2,14,34,35] describes the same scene (exactly pixel-to-pixel) with different color styles. Our algorithm is applied to color correction in image stitching, so we cropped these image pairs to have various overlapping percentages, which simulates the situation in image stitching. Then, color transfer methods are applied to the corresponding image pairs that have different overlapping percentages. In the following experiments, we cropped each image pair with four different overlapping percentages (10%, 30%, 60% and 80%), respectively. Thus, we have 40 × 4 = 160 synthetic pairs to make numerical experiments. As shown in the Table 1, our algorithm outperforms other methods in terms of color similarity and structure similarity.

Table 1. Comparison for synthetic dataset (average of 40 image pairs for each overlapping percentage). **CS** is the **C**olor **S**imilarity index, **SSIM** is the **S**tructural **SIM**ilarity index.

Overlapping Percentage	CS (dB)				SSIM			
	Pitie	*Fecker*	*Nikolova*	*Proposed*	*Pitie*	*Fecker*	*Nikolova*	*Proposed*
10%	18.21	18.34	18.39	**22.03**	0.7924	0.8033	0.8165	**0.8834**
30%	20.16	20.28	20.31	**24.11**	0.8101	0.8181	0.8299	**0.8867**
60%	21.93	21.83	22.02	**24.19**	0.8417	0.8461	0.8545	**0.8853**
80%	23.39	23.24	23.43	**24.31**	0.8662	0.8674	0.8721	**0.8857**

From the experimental results of these algorithms, we can also make a conclusion that our algorithm obtains the better visual quality of correction results even though the overlapping percentage is very small. The ability of color transfer for image pairs having narrow overlapping regions is very important in the application of image stitching. This advantage can make our color correction algorithm more suitable for image stitching. In Table 1, we can also observe that the proposed method is not significantly better than other algorithms when the overlapping percentage is very large. For example, when the overlapping percentage is 80%, the difference between the proposed method and Nikolova's algorithm [11] is very small. Since we adopted Nikolova's algorithm for transferring the color style in the overlapping region, the proposed method is almost the same as Nikolova's algorithm when the overlapping percentage is close to 100%.

Some visual comparisons are shown in Figures 3–6. In Figure 3, the overlapping regions include the information describing the sky, the pyramid and the head of the camel. The red rectangles indicate the transferred color has some distance from the reference color style in the reference image. The yellow rectangle indicates the transferred color by the proposed method is almost the same as the reference color style. We can also observe easily that our algorithm transfers color information more accurately than other algorithms. For more accurate comparison, we show the histograms of the overlapping regions in Figure 4. The histograms of the overlapping regions in the reference image and in the test image are totally different. The histograms of the overlapping regions after color transfer algorithms are closer to the reference. In addition, the results by the proposed method are the closest one, which indicates the proposed method outperforms other algorithms.

Figure 3. Comparison for the synthetic image pair. Image Source: courtesy of the authors and databases referred on [2,14].

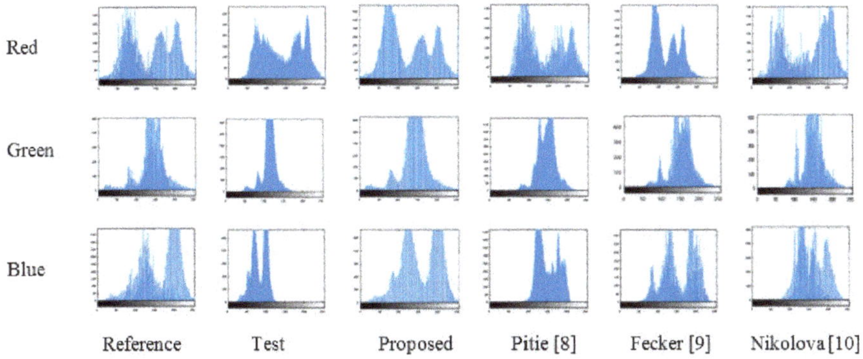

Figure 4. Histogram comparisons for overlapping regions in Figure 3. The first column shows the histograms (three color channels, respectively) of overlapping regions in the reference image, the second column shows the corresponding histograms in the test image, the third column shows the corresponding histograms of overlapping regions after the proposed method, the fourth column shows Pitie's result, the fifth column shows Fecker's result, and the last column shows Nikolova's result.

In Figure 5, the red rectangles show disadvantages of other algorithms, which have transferred the green color to the body of the sheep. The yellow rectangle indicates the advantage of our algorithm, which has transferred the right color to the sheep body. In Figure 6, the rectangles describe the airplane body that exists in the overlapping region. The red rectangles show disadvantages of other algorithms that transferred the inconsistent color to the airplane body. The yellow rectangle indicates that the proposed method transfers the consistent color to the airplane body.

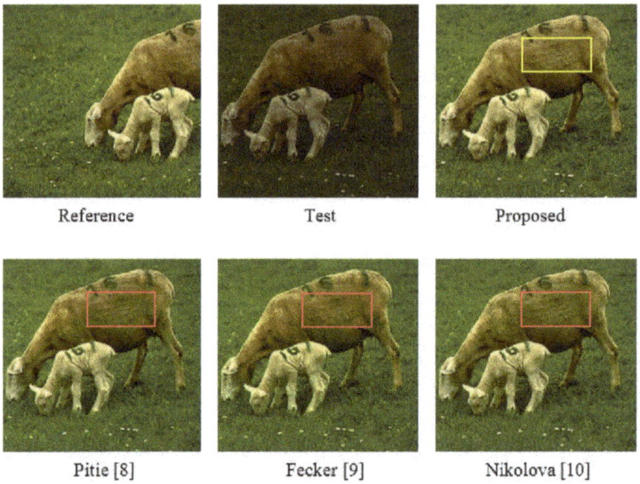

Figure 5. Comparison for the synthetic image pair. Image Source: courtesy of the authors and databases referred on [2,14].

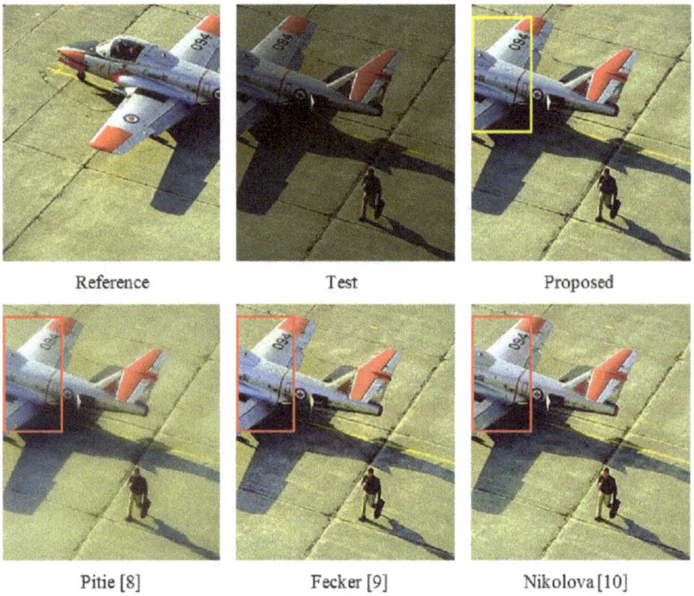

Figure 6. Comparison for the synthetic image pair. Image Source: courtesy of the authors and databases referred on [2,14].

4.3. Experiments on Real Image Pairs

In the experiments above, we make the comparisons using synthetic image pairs, which have exactly the same overlapping regions. However, overlapping regions are not usually exactly the same (not pixel-to-pixel) in the real application of image stitching. Thus, we make some experiments for real image pairs.

Objective comparisons are given in Table 2, which indicates that our algorithm outperforms other methods in terms of color similarity and structure similarity. Subjective visual comparisons are also presented in Figures 7–10. In Figure 7, the red rectangles show disadvantages of other algorithms, which have transferred the green color to the tree body and the windows. The yellow rectangles indicate the advantage of our algorithm, which transfers the right color to the mentioned regions. The histogram comparisons for the overlapping regions are shown in Figure 8, which indicates the proposed method outperforms other algorithms. More results and the comparisons are given in Figures 9 and 10.

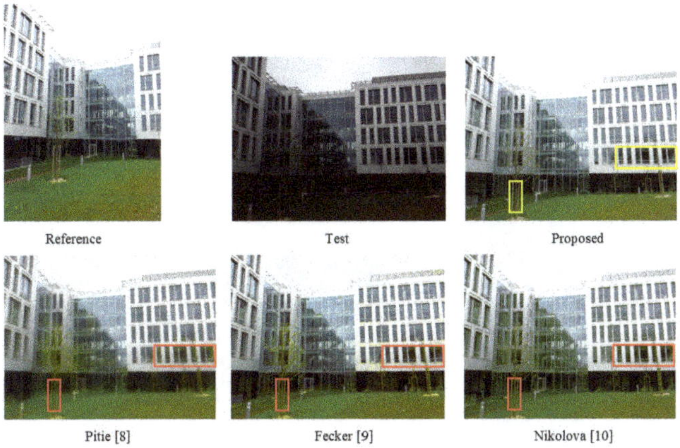

Figure 7. Comparison for the real image pair. Image Source: courtesy of the authors and databases referred on [29].

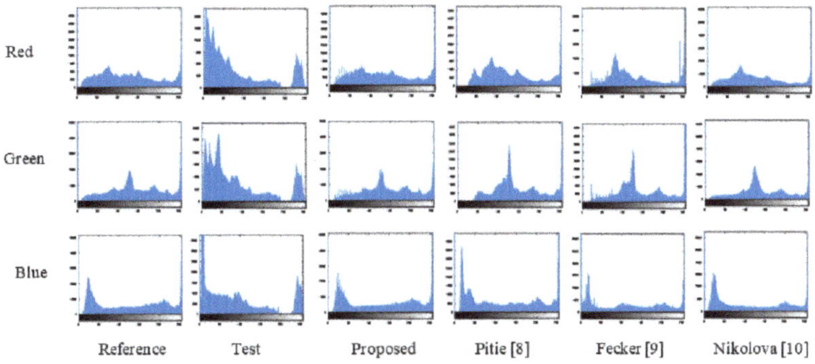

Figure 8. Histogram comparisons for overlapping regions in Figure 7. The first column shows the histograms of overlapping regions in the reference image, the second column shows the corresponding histograms in the test image, the third column shows the corresponding histograms of overlapping regions after the proposed method, the fourth column shows Pitie's result, the fifth column shows Fecker's result, and the last column shows Nikolova's result.

Table 2. Comparison for real image pairs (Average of 35 pairs).

	Pitie	*Fecker*	*Nikolova*	*Proposed*
CS(dB)	18.98	19.04	19.12	**21.19**
SSIM	0.8162	0.8334	0.8255	**0.8531**

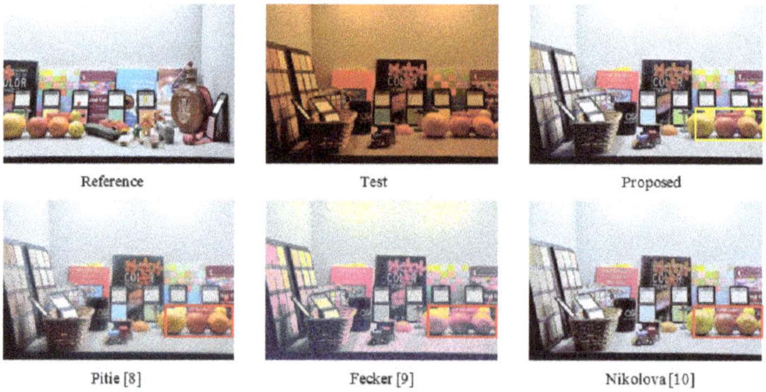

Figure 9. Comparison for the real image pair. Image Source: courtesy of the authors and databases referred on [29].

Figure 10. Comparison for the real image pair. Image Source: courtesy of the authors and databases referred on [3].

5. Discussion

In this paper, we have proposed an efficient color transfer method for image stitching, which combines the ideas of histogram specification and global mapping. The main contribution of the proposed method is using original pixels and the corresponding pixels after histogram specification to compute a global mapping function with an iteration method, which can effectively minimize color differences between a reference image and a test image. The color mapping function can spread well the color style from the overlapping region to the whole image. The experiments also demonstrate the advantages of our algorithm in terms of objective evaluation and subjective evaluation.

As our work relies on the exact histogram specification, bad results of histogram specification will decrease the visual quality of our results. Even though the problem of histogram specification has received considerable attention and has been well studied during recent years, some future work can be conducted to improve the results of this kind of algorithm.

In the detailed description of the proposed algorithm, we have shown that our method is building the color mapping functions using the global information and without using the local neighbor information. In future work, we will consider the information of local patches to construct the color mapping functions, which may be more accurate to transfer colors. Another problem is that the mapping function is computed for each color channel. This simple processing does not consider the relation of the three color channels, and this may produce some color artifacts. In our future work, we try to obtain the color mapping function considering the relation of the three color channels. The minimization is completed with an iteration framework, and the termination conditions include computing PSNR. These operations need high computation, so a fast minimization method will also be considered in the future work.

Acknowledgments: We would like to sincerely thank Hasan Sheikh Faridul, Youngbae Hwang, and Wei Xu for sharing the test images and permitting us to use the images in this paper. We greatly thank Charless Fowlkes for sharing the BSDS300 dataset for research purposes.

Author Contributions: Qi-Chong Tian designed the algorithm presented in this article, conducted the numerical experiments, and wrote the paper. Laurent D. Cohen proposed the idea of this research, analyzed the results, and revised the whole article. Qi-Chong Tian is a Ph.D. student supervised by Laurent D. Cohen.

Conflicts of Interest: The authors declare no conflict of interest.

References

1. Szeliski, R. Image Alignment and Stitching: A Tutorial. *Found. Trends Comput. Graph. Vis.* **2006**, *2*, 1–104.
2. Xu, W.; Mulligan, J. Performance evaluation of color correction approaches for automatic multi-view image and video stitching. In Proceedings of the IEEE Conference on Computer Vision and Pattern Recognition (CVPR'10), San Francisco, CA, USA, 13–18 June 2010; pp. 263–270.
3. Hwang, Y.; Lee, J.Y.; Kweon, I.S.; Kim, S.J. Color transfer using probabilistic moving least squares. In Proceedings of the IEEE Conference on Computer Vision and Pattern Recognition (CVPR'14), Columbus, OH, USA, 23–28 June 2014; pp. 3342–3349.
4. Faridul, H.; Stauder, J.; Kervec, J.; Trémeau, A. Approximate cross channel color mapping from sparse color correspondences. In Proceedings of the IEEE International Conference on Computer Vision (ICCV'13)—Workshop in Color and Photometry in Computer Vision (CPCV'13), Sydney, Australia, 8 December 2013; pp. 860–867.
5. Faridul, H.S.; Pouli, T.; Chamaret, C.; Stauder, J.; Reinhard, E.; Kuzovkin, D.; Tremeau, A. Colour Mapping: A Review of Recent Methods, Extensions and Applications. *Comput. Graph. Forum* **2016**, *35*, 59–88.
6. Fitschen, J.H.; Nikolova, M.; Pierre, F.; Steidl, G. A variational model for color assignment. In Proceedings of the International Conference on Scale Space and Variational Methods in Computer Vision (SSVM'15), Lege Cap Ferret, France, 31 May–4 June 2015; Volume LNCS 9087, pp. 437–448.
7. Moulon, P.; Duisit, B.; Monasse, P. Global multiple-view color consistency. In Proceedings of the European Conference on Visual Media Production (CVMP'13), London, UK, 6–7 November 2013.
8. Pitie, F.; Kokaram, A.C.; Dahyot, R. *N*-dimensional probability density function transfer and its application to color transfer. In Proceedings of the IEEE International Conference on Computer Vision (ICCV'05), Beijing, China, 17–21 October 2005; Volume 2, pp. 1434–1439.
9. Pitié, F.; Kokaram, A.C.; Dahyot, R. Automated colour grading using colour distribution transfer. *Comput. Vis. Image Underst.* **2007**, *107*, 123–137.
10. Fecker, U.; Barkowsky, M.; Kaup, A. Histogram-based prefiltering for luminance and chrominance compensation of multiview video. *IEEE Trans. Circuits Syst. Video Technol.* **2008**, *18*, 1258–1267.
11. Nikolova, M.; Steidl, G. Fast ordering algorithm for exact histogram specification. *IEEE Trans. Image Process.* **2014**, *23*, 5274–5283.
12. Nikolova, M.; Steidl, G. Fast hue and range preserving histogram specification: Theory and new algorithms for color image enhancement. *IEEE Trans. Image Process.* **2014**, *23*, 4087–4100.
13. Nikolova, M.; Wen, Y.W.; Chan, R. Exact histogram specification for digital images using a variational approach. *J. Math. Imaging Vis.* **2013**, *46*, 309–325.

14. Martin, D.; Fowlkes, C.; Tal, D.; Malik, J. A database of human segmented natural images and its application to evaluating segmentation algorithms and measuring ecological statistics. In Proceedings of the Eighth IEEE International Conference on Computer Vision (ICCV'01), Vancouver, BC, Canada, 7–14 July 2001; Volume 2, pp. 416–423.
15. Tian, Q.C.; Cohen, L.D. Color correction in image stitching using histogram specification and global mapping. In Proceedings of the The 6th International Conference on Image Processing Theory Tools and Applications (IPTA'16), Oulu, Finland, 12–15 December 2016; pp. 1–6.
16. Brown, M.; Lowe, D.G. Automatic panoramic image stitching using invariant features. *Int. J. Comput. Vis.* **2007**, *74*, 59–73.
17. Xiong, Y.; Pulli, K. Fast panorama stitching for high-quality panoramic images on mobile phones. *IEEE Trans. Consum. Electron.* **2010**, *56*, doi:10.1109/TCE.2010.5505931.
18. Wang, W.; Ng, M.K. A variational method for multiple-image blending. *IEEE Trans. Image Process.* **2012**, *21*, 1809–1822.
19. Shan, Q.; Curless, B.; Furukawa, Y.; Hernandez, C.; Seitz, S.M. Photo Uncrop. In Proceedings of the 13th European Conference on Computer Vision (ECCV'14), Zurich, Switzerland, 6–12 September 2014; pp. 16–31.
20. Lin, C.C.; Pankanti, S.U.; Natesan Ramamurthy, K.; Aravkin, A.Y. Adaptive as-natural-as-possible image stitching. In Proceedings of the IEEE Conference on Computer Vision and Pattern Recognition (CVPR'15), Boston, MA, USA, 7–12 June 2015; pp. 1155–1163.
21. Lowe, D.G. Distinctive image features from scale-invariant keypoints. *Int. J. Comput. Vis.* **2004**, *60*, 91–110.
22. Bay, H.; Ess, A.; Tuytelaars, T.; Van Gool, L. Speeded-up robust features (SURF). *Comput. Vis. Image Underst.* **2008**, *110*, 346–359.
23. Provenzi, E. *Variational Models for Color Image Processing in the RGB Space Inspired by Human Vision*; Habilitation à Diriger des Recherches; ED 386: École doctorale de sciences mathématiques de Paris centre, UPMC, France, 2016.
24. Reinhard, E.; Adhikhmin, M.; Gooch, B.; Shirley, P. Color transfer between images. *IEEE Comput. Graph. Appl.* **2001**, *21*, 34–41.
25. Papadakis, N.; Provenzi, E.; Caselles, V. A variational model for histogram transfer of color images. *IEEE Trans. Image Process.* **2011**, *20*, 1682–1695.
26. Hristova, H.; Le Meur, O.; Cozot, R.; Bouatouch, K. Style-aware robust color transfer. In Proceedings of the workshop on Computational Aesthetics. Eurographics Association, Istanbul, Turkey, 20–22 June 2015; pp. 67–77.
27. Wen, C.L.; Hsieh, C.H.; Chen, B.Y.; Ouhyoung, M. *Example-Based Multiple Local Color Transfer by Strokes*; Computer Graphics Forum; Wiley Online Library: Hoboken, NJ, USA, 2008; Volume 27, pp. 1765–1772.
28. Welsh, T.; Ashikhmin, M.; Mueller, K. *Transferring Color to Greyscale Images*; ACM Transactions on Graphics (TOG); ACM: New York, NY, USA, 2002; Voume 21, pp. 277–280.
29. Faridul, H.S.; Stauder, J.; Trémeau, A. Illumination and device invariant image stitching. In Proceedings of the IEEE International Conference on Image Processing (ICIP'14), Paris, France, 27–30 October 2014; pp. 56–60.
30. Tai, Y.W.; Jia, J.; Tang, C.K. Local color transfer via probabilistic segmentation by expectation-maximization. In Proceedings of the IEEE Conference on Computer Vision and Pattern Recognition (CVPR'05), 20–25 June 2005; Volume 1, pp. 747–754.
31. Anbarjafari, G. An Objective No-Reference Measure of Illumination Assessment. *Meas. Sci. Rev.* **2015**, *15*, 319–322.
32. Maitre, H. *From Photon to Pixel: The Digital Camera Handbook*; Wiley Online Library: Hoboken, NJ, USA, 2015.
33. Wang, Z.; Bovik, A.C.; Sheikh, H.R.; Simoncelli, E.P. Image quality assessment: from error visibility to structural similarity. *IEEE Trans. Image Process.* **2004**, *13*, 600–612.
34. Color Correction Images. Available online: https://www.researchgate.net/publication/282652076_color_correction_images (accessed on 2 September 2017).
35. The Berkeley Segmentation Dataset and Benchmark. Available online: https://www2.eecs.berkeley.edu/Research/Projects/CS/vision/bsds/ (accessed on 2 September 2017).

Journal of
Imaging

MDPI

Article

Robust Parameter Design of Derivative Optimization Methods for Image Acquisition Using a Color Mixer†

HyungTae Kim [1,*], KyeongYong Cho [2], Jongseok Kim[1], KyungChan Jin [1] and SeungTaek Kim [1]

1 Smart Manufacturing Technology Group, KITECH, 89 Yangdae-Giro RD., CheonAn 31056, ChungNam, Korea; jongseok@kitech.re.kr (J.K.); kcjin@kitech.re.kr (K.J.); stkim@kitech.re.kr (S.K.)

2 UTRC, KAIST, 23, GuSung, YouSung, DaeJeon 305-701, Korea; yong00@kaist.ac.kr

* Correspondence: htkim@kitech.re.kr; Tel.: +82-41-589-8478

† This paper is an extended version of the paper published in Kim, HyungTae, KyeongYong Cho, SeungTaek Kim, Jongseok Kim, KyungChan Jin, SungHo Lee. "Rapid Automatic Lighting Control of a Mixed Light Source for Image Acquisition using Derivative Optimum Search Methods." In *MATEC Web of Conferences*, Volume 32, EDP Sciences, 2015.

Received: 27 May 2017; Accepted: 15 July 2017; Published: 21 July 2017

Abstract: A tuning method was proposed for automatic lighting (auto-lighting) algorithms derived from the steepest descent and conjugate gradient methods. The auto-lighting algorithms maximize the image quality of industrial machine vision by adjusting multiple-color light emitting diodes (LEDs)—usually called color mixers. Searching for the driving condition for achieving maximum sharpness influences image quality. In most inspection systems, a single-color light source is used, and an equal step search (ESS) is employed to determine the maximum image quality. However, in the case of multiple color LEDs, the number of iterations becomes large, which is time-consuming. Hence, the steepest descent (STD) and conjugate gradient methods (CJG) were applied to reduce the searching time for achieving maximum image quality. The relationship between lighting and image quality is multi-dimensional, non-linear, and difficult to describe using mathematical equations. Hence, the Taguchi method is actually the only method that can determine the parameters of auto-lighting algorithms. The algorithm parameters were determined using orthogonal arrays, and the candidate parameters were selected by increasing the sharpness and decreasing the iterations of the algorithm, which were dependent on the searching time. The contribution of parameters was investigated using ANOVA. After conducting retests using the selected parameters, the image quality was almost the same as that in the best-case parameters with a smaller number of iterations.

Keywords: derivative optimization; light control; multi-color source; RGB mixer; robust parameter design; Taguchi method

1. Introduction

The quality of images acquired from an industrial machine vision system determines the performance of the inspection process during manufacturing [1]. Image-based inspection using machine vision is currently widespread, and the image quality is critical in automatic optical inspection [2]. The image quality is affected by focusing, which is usually automatized, and illumination, which is still a manual process. Illumination in machine vision has many factors, such as intensity, peak wavelength, bandwidth, light shape, irradiation angle, distance, uniformity, diffusion, and reflection. The active control factors in machine vision are intensity and peak wavelength, though other factors are usually invariable. Although image quality is sensitive to intensity and peak wavelength, the optimal combination of these factors may be varied by the material of the target object [3]. Because the active control factors are currently manually changed, it is considerably labor intensive to adjust the illumination condition of the inspection machines in cases of the initial setup or product change. However, the light intensity of a light-emitting diode (LED) can be easily adjusted by varying the electric current. A few articles have been written about auto-lighting by controlling the intensity of a single-color light source [4–6]. A single-color lighting based on fuzzy control logic is applied to a robot manipulator [7]. The light intensity from a single-color source is mostly determined using in an equal step search (ESS), which varies the intensity from minimum to maximum in small intervals.

Color mixers synthesize various colors of light from multiple LEDs. The LEDs are arranged in an optical direction in a back plane, and an optical mixer is attached to a light output [8]. The color is varied using a combination of light intensities, which can be adjusted using electric currents. Optical collimators are the most popular device to combine the lights from the multiple LEDs [9–11]. These studies aim to achieve exact color generation, uniformity in a target plane, and thermal stability. They do not focus on the image quality. Optimal illumination can increase the color contrast in machine vision [12], hence spectral approaches in bio-medical imaging [13,14].

When color mixers are applied to machine vision, the best color and intensity must be found manually. Because automatic search is applied using the ESS, the searching time is long, which is caused by the vast number of light combinations. Thus, we have been studying fast optimization between color and image quality in industrial machine vision [15–17]. Because the above-mentioned studies were based on non-differential optimum methods, they were stably convergent, but required multiple calls of a cost function for iterations, leading to a longer processing time. Derivative optimum search methods are well-known, simple, and easy to implement [18]. The derivative optimum methods are less stable and more oscillatory, but usually faster [18,19]. In this study, arbitrary N color sources and image quality were considered for steepest descent (STD) and conjugate gradient (CJG). The optimum methods are composed of functions, variables, and coefficients which are difficult to determine for the inspection process. Algorithm parameters also affect the performance of image processing methods [20], and they can be determined using optimum methods. Thus, a tuning step is necessary to select the value of the coefficients when applying the methods to inspection machines. The relation between the LED inputs and the image quality is complex, difficult to describe, and is actually a black box function. The coefficients are sensitive to convergence, number of iterations, and oscillation, but the function is unknown. The Taguchi method is one of the most popular methods for determining the optimal process parameters with a minimum number of experiments when the system is unknown, complex, and non-linear. The contribution of process parameters can be investigated using ANOVA, and many cases have been proposed in machining processes [21,22]. The Taguchi method for robust parameter design was applied to tune the auto-lighting algorithm for achieving the fastest search time and best image quality in the case of a mixed-color source.

2. Derivative Optimum for Image Quality

2.1. Index for Image Quality

The conventional inspection system for color lighting comprises a mixed light source, industrial camera, framegrabber, controller, and light control board. Figure 1 shows a conceptual diagram of the color mixer and machine vision system. The color mixer generates a mixed light and emits it toward a target object, and the camera acquires a digital image, which is a type of response to the mixed light. The digital image is analyzed to study the image properties (e.g., image quality) and to determine the intensity levels of the LEDs in the color mixer. The intensity levels are converted into voltage level using a digital-to-analog converter (DAC) board. The electric current to drive the LEDs is generated using a current driver according to the voltage level. The color mixer and the machine vision form a feedback loop.

The image quality must be evaluated to use optimum methods. There are various image indices proposed in many papers; these are evaluated using pixel operations [23,24]. For instance, the brightness, $\bar{I}$, is calculated from the conception of the average grey level of an $m \times n$ pixel image.

$$\bar{I} = \frac{1}{mn}\sum_{x}^{m}\sum_{y}^{n} I(x,y) \tag{1}$$

where the grey level of pixels is $I(x, y)$ and the size of the image is $m \times n$.

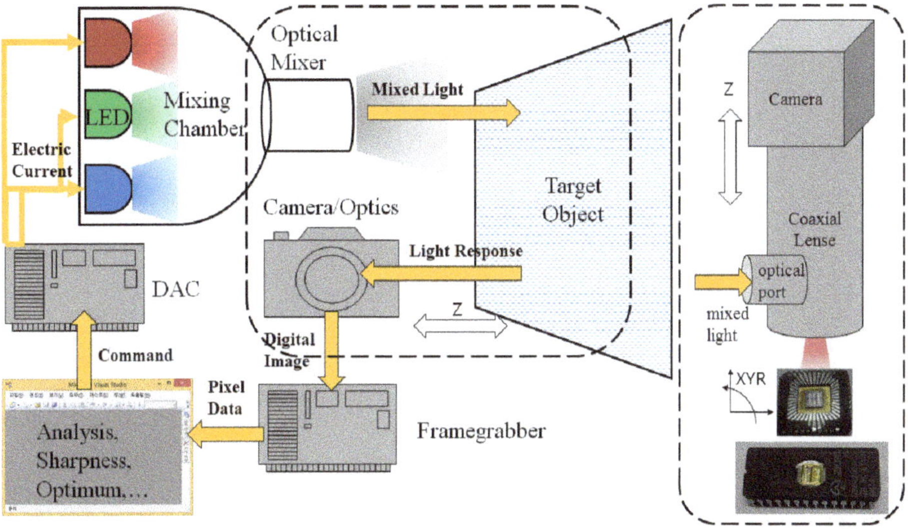

Figure 1. System Diagram for color mixing and automatic lighting.

Image quality is one of the image indices, and is usually estimated using sharpness. Sharpness actually indicates the deviation and difference of grey levels among pixels. There are dozens of definitions for sharpness, and standard deviation is widely used as sharpness in machine vision [25]. Thus, sharpness σ can be written as follows.

$$\sigma^2 = \frac{1}{mn}\sum_{x}^{m}\sum_{y}^{n} (I(x,y) - \bar{I}) \tag{2}$$

Industrial machine vision usually functions in a dark room so that the image acquired by a camera completely depends on the lighting system. The color mixer employed in this study uses multiple color LEDs having individual electric inputs. Because the inputs are all adjusted using the voltage level, the color mixer has a voltage input vector for N LEDs as follows.

$$V = (v_1, v_2, v_3, \ldots, v_N) \tag{3}$$

As presented in section I, the relationship between the LED inputs and the image quality involve electric, spectral, and optical responses. This procedure cannot easily be described using a mathematical model, and the relationship from (1) to (3) is a black box function which can be denoted as an arbitrary function f, which is a cost function in this study.

$$\sigma = f(V) \tag{4}$$

The best sharpness can be obtained by adjusting V. However, V is an unknown vector. The maximum sharpness can be found using optimum methods, but negative sharpness ρ must be defined because optimum methods are designed for finding the minimum. Hence, negative sharpness is a cost function.

$$\rho = -\sigma = -f(V) \tag{5}$$

The optimum methods have a general form of problem definition using a cost function as follows [17]:

$$\min_{\mathrm{V}} \ \rho = -f(V) \ \text{ for } \forall \mathrm{V} \tag{6}$$

2.2. Derivative Optimum Methods

The steepest descent and conjugate gradient methods are representative of the derivative optimum methods, which involve the differential operation of a cost function written as a gradient.

$$\nabla\rho = \left(\frac{\partial\rho}{\partial v_1}, \frac{\partial\rho}{\partial v_2}, \frac{\partial\rho}{\partial v_3}, \ldots, \frac{\partial\rho}{\partial v_N}\right) \tag{7}$$

The STD iterates the equations until it finds a local minimum; a symbol k is necessary to show the current iteration. The STD updates current inputs ${}^{k}V$ by adding a negative gradient to the current inputs. α is originally determined at $\partial\rho(\alpha)/\partial\alpha = 0$ in STD [18], however it is difficult to obtain using an experimental apparatus. In this study, the α is assumed to be a constant, α.

$${}^{k+1}V = {}^{k}V - \alpha\nabla({}^{k}\rho) = {}^{k}V - \alpha({}^{k}\xi) \tag{8}$$

The CJG has the same method of updating the current inputs. However, the difference lies in calculating the index of the updates ξ.

$${}^{k+1}V = {}^{k}V - \alpha({}^{k}\xi) \tag{9}$$

$${}^{k}\xi = -\nabla({}^{k}\rho) + \left|\frac{\nabla({}^{k}\rho)}{\nabla({}^{k-1}\rho)}\right|^2 {}^{(k-1)}\xi \tag{10}$$

${}^{k}\xi$ usually has an unpredictably large or small value, which causes divergence or oscillation near the optimum. Consequently, the following boundary conditions are given before updating the current inputs.

$$\alpha\,{}^{k}\xi = \begin{cases} -\eta\tau & {}^{k}\xi < -\tau \\ -\eta({}^{k}\xi) & -\tau <{}^{k}\xi < \tau \\ \eta\tau & {}^{k}\xi > \tau \end{cases} \tag{11}$$

where η is the convergence coefficient for a limited range and τ is the threshold. The updating of inputs and the acquisition of sharpness are iterated until the gradient becomes smaller than the terminal condition ϵ_1, which indicates that auto-lighting finds the maximum sharpness and the best image quality.

$$|{}^{k}\xi| < \epsilon_1 \tag{12}$$

where ϵ_1 is an infinitesimal value for the terminal condition.

The cost function is acquired using hardware, and the terminal condition considers differential values. The values are discrete and sensitive to noises; hence, an additional terminal condition, ϵ_2, is applied as follows:

$$|{}^{k}\rho - {}^{k-1}\rho| < \epsilon_2 \tag{13}$$

3. Robust Parameter Design

3.1. System for Experiment

The sharpness and derivative methods were applied to a test system which was constructed in our previous study [6]. The test system was composed of a 4 M pixel camera (SVS-4021, SVS-VISTEK, Seefeld, Germany), a coaxial lens (COAX, Edmund Optics, Barrington, NJ, USA), a framegrabber (SOL6M, Matrox, Dorval, QC, Canada), a multi-channel DAC board (NI-6722, NI, Austin, TX, USA), and an RGB mixing light source. Commercial integrated circuits (ICs) of EPROMs were used as sample targets A (EP910JC35, ALTERA, San Jose, CA, USA) and B (Z86E3012KSES, ZILOG, Milpitas, CA, USA), as shown in Figure 2. The camera and the ICs were fixed on Z and XYR axes, respectively. The coaxial lens was attached to the camera, and faced the ICs. Optical fiber from the RGB source was connected to the coaxial lens and illuminated the ICs. Images of the ICs were acquired and transferred into a PC through a CAMERALINK port on the framegrabber. Operating software was constructed using a development tool (Visual Studio 2008, Microsoft, Redmond, WA, USA) and vision library (MIL 8.0, Matrox, Dorval, QC, Canada). Location of the ICs in an image was adjusted using XYR axes after focusing was performed using the Z axis. The inputs of the RGB source were connected to the DAC board. The light color and intensity were adjusted through the board. The STD and CJG for optimum light condition were implemented into the software.

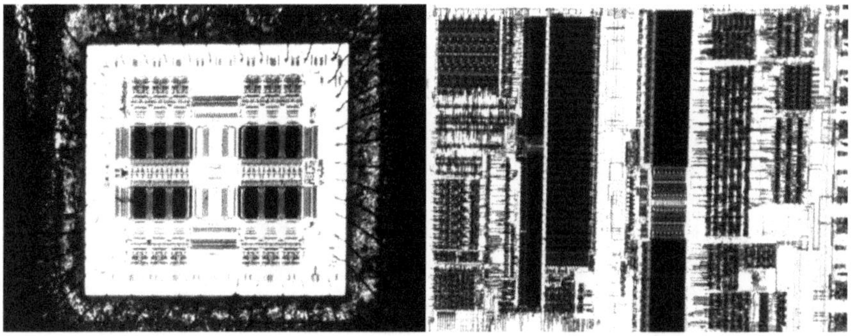

(a) (b)

Figure 2. Target patterns acquired by maximum sharpness: (**a**) Pattern A; (**b**) Pattern B.

3.2. Taguchi Method

The Taguchi method is commonly used to tune the algorithm parameters and optical design in machine vision [26–28]. A neural network is a massive and complex numerical model, and derivative optimal methods are frequently applied to its training parameters [29,30]. Taguchi method is useful to find the learning parameters of neural network and increase learning efficiency in machine vision system [31]. Considering the non-linear, multi-dimensional, and black box function systems in this study, we expected that the Taguchi method could be useful in tuning the auto-lighting algorithm. The performance of the algorithm was largely evaluated using the minimum number of iterations and the maximum sharpness. Hence, "the smaller the better" concept was applied in case of the number of iterations and "the larger the better" concept was applied in the case of the sharpness while calculating the signal-to-noise (SN) ratio. Those SN ratios can be obtained using the following equations [32,33]:

$$SN = -10log\left(\frac{1}{w}\sum_{j=1}^{w}\frac{1}{u_j^2}\right) \tag{14}$$

$$SN = -10log\left(\frac{1}{w}\sum_{j=1}^{w} u_j^2\right) \tag{15}$$

where u_j is the performance index (e.g., sharpness and iteration), and w is the number of experiments.

3.3. Experiment Design

The selected parameters were initial voltages of red, green, and blue (RGB) LEDs, $V = (v_{R0}, v_{G0}, v_{B0})$, the convergence constant η, and the threshold τ. Because the maximum sharpness is usually formed in low-voltage regions under a single-light condition, the range of the initial voltage was less than half of the full voltage. The ranges of η and τ were between 0.0 and 1.0. These five factors were chosen as control factors. Because all the ranges are divided into five intervals, the level was set at 5. Therefore, the $L_{25}(5^5)$ model is organized using five control factors and five levels, as shown in Table 1. The combination of the experiment is 25, which is quite a small value considering the multiple color sources and the algorithm parameters. Two sample targets were used for the experiments, as proposed.

Table 1. Control factors and levels for derivative optimum methods.

Factors	Code	Level				
		1	2	3	4	5
V_{R0}: Initial V_R	A	0.5	1.0	1.5	2.0	2.5
V_{G0}: Initial V_G	B	0.5	1.0	1.5	2.0	2.5
V_{B0}: Initial V_B	C	0.5	1.0	1.5	2.0	2.5
τ: Threshold	D	0.2	0.4	0.6	0.8	1.0
η: Convergence Constant	E	0.2	0.4	0.6	0.8	1.0

4. Results

The maximum sharpness found using the ESS was $\sigma_{max} = 392.76$ at $V = (0,\ 0,\ 1.2)$ for Pattern A, and $\sigma_{max} = 358.87$ at $V = (1.0,\ 0,\ 0)$ for Pattern B. The total step number of combinations for RGB was $50^3 = 125,000$. The $L_{25}(5^5)$ orthogonal arrays for steepest descent and conjugate gradient methods were constructed as shown in Tables 2 and 3. σ_{max}, k_{max}, V_R, V_G and V_B were the optimal statuses found by the steepest descent method by using the selected parameters. Some combinations showed almost the same sharpness as that of the exact solution, some combinations reached the maximum after several steps, and some cases failed to converge. These facts show that parameter selection for a derivative optimum is important because of stability. The SN ratios were calculated using MINITAB for mathematical operations of Taguchi analysis. Figures 3–6 are the results of Taguchi analysis and show the trend of the control factors. The variation in the sharpness was very small, whereas the variation in the number of the iteration was larger, which implied that the parameters were sensitive to iteration.

Table 2. Orthogonal array of steepest descent method for Patterns A and B.

Run #	Control Factors					Pattern A					Pattern B				
	A	B	C	D	E	σ_{max}	k_{max}	V_R	V_G	V_B	σ_{max}	k_{max}	V_R	V_G	V_B
1	1	1	1	1	1	389.43	117	0.98	0.00	0.41	353.88	153	0.53	0.30	0.00
2	1	2	2	2	2	390.50	255	0.00	0.00	1.25	-	-	-	-	-
3	1	3	3	3	3	-	-	-	-	-	340.47	3	0.00	0.42	0.42
4	1	4	4	4	4	382.4	2	0.00	0.72	0.72	-	-	-	-	-
5	1	5	5	5	5	390.43	152	0.00	0.00	1.30	317.63	2	0.00	0.53	0.82
6	2	1	2	3	4	386.09	189	0.52	0.02	0.52	344.02	1	0.52	0.02	0.52
7	2	2	3	4	5	387.22	1	0.20	0.20	0.70	337.22	1	0.20	0.20	0.70
8	2	3	4	5	1	390.40	123	0.00	0.00	1.26	333.35	6	0.00	0.30	0.80
9	2	4	5	1	2	390.36	49	0.00	0.00	1.27	335.65	22	0.00	0.24	0.74
10	2	5	1	2	3	384.64	6	0.00	1.06	0.00	346.93	108	0.00	0.58	0.00
11	3	1	3	5	2	389.01	109	0.75	0.02	0.57	-	-	-	-	-
12	3	2	4	1	3	388.87	263	0.78	0.00	0.51	340.46	11	0.42	0.00	0.68
13	3	3	5	2	4	390.36	164	0.00	0.00	1.29	324.30	107	0.00	0.00	0.90
14	3	4	1	3	5	385.88	3	0.41	0.70	0.00	331.85	2	0.30	0.80	0.00
15	3	5	2	4	1	389.21	237	0.99	0.00	0.38	346.91	192	0.00	0.58	0.00
16	4	1	4	2	5	387.71	3	0.80	0.00	0.80	338.46	18	0.40	0.00	0.40
17	4	2	5	3	1	389.18	206	1.04	0.00	0.36	338.25	15	0.30	0.00	0.70
18	4	3	1	4	2	-	-	-	-	-	354.62	4	0.72	0.22	0.00
19	4	4	2	5	3	382.32	2	0.80	0.80	0.00	303.61	2	0.82	0.80	0.00
20	4	5	3	1	4	384.95	49	0.24	0.74	0.00	349.91	151	0.24	0.42	0.00
21	5	1	5	4	3	387.90	4	0.58	0.00	0.60	341.62	4	0.58	0.00	0.58
22	5	2	1	5	4	386.36	2	1.28	0.00	0.00	358.14	2	0.90	0.00	0.00
23	5	3	2	1	5	386.78	6	1.30	0.30	0.00	358.26	82	0.90	0.00	0.00
24	5	4	3	2	1	389.05	191	1.08	0.00	0.32	355.32	23	0.66	0.16	0.00
25	5	5	4	3	2	386.50	8	0.58	0.58	0.08	350.51	137	0.34	0.34	0.00

Table 3. Orthogonal array of conjugate gradient method for Patterns A and B.

Run #	Control Factors					Pattern A					Pattern B				
	A	B	C	D	E	σ_{max}	k_{max}	V_R	V_G	V_B	σ_{max}	k_{max}	V_R	V_G	V_B
1	1	1	1	1	1	389.28	59	0.99	0.00	0.43	358.49	143	0.95	0.00	0.00
2	1	2	2	2	2	390.58	41	0.00	0.00	1.25	343.00	24	0.48	0.16	0.43
3	1	3	3	3	3	390.43	139	0.00	0.00	1.20	340.21	3	0.00	0.42	0.42
4	1	4	4	4	4	382.42	2	0.00	0.72	0.72	354.05	9	0.80	0.00	0.00
5	1	5	5	5	5	384.66	2	0.00	0.57	0.78	316.46	2	0.00	0.50	0.88
6	2	1	2	3	4	386.07	1	0.52	0.02	0.52	343.92	1	0.52	0.02	0.52
7	2	2	3	4	5	387.16	1	0.20	0.20	0.70	337.14	1	0.20	0.20	0.70
8	2	3	4	5	1	390.40	50	0.00	0.00	1.30	358.43	168	0.98	0.00	0.00
9	2	4	5	1	2	390.54	28	0.00	0.00	1.27	338.89	29	0.24	0.16	0.66
10	2	5	1	2	3	388.64	178	0.74	0.02	0.71	350.01	21	0.48	0.41	0.00
11	3	1	3	5	2	390.42	24	0.00	0.00	1.28	331.95	2	0.70	0.00	0.70
12	3	2	4	1	3	390.50	144	0.00	0.00	1.25	352.28	16	0.58	0.24	0.08
13	3	3	5	2	4	390.37	17	0.00	0.00	1.31	324.42	15	0.00	0.00	0.90
14	3	4	1	3	5	388.16	5	1.48	0.00	0.00	347.42	33	0.00	0.60	0.00
15	3	5	2	4	1	390.39	284	0.00	0.00	1.29	348.20	26	0.11	0.63	0.00
16	4	1	4	2	5	387.61	3	0.80	0.00	0.80	338.48	124	0.40	0.00	0.40
17	4	2	5	3	1	390.43	273	0.00	0.00	1.24	342.81	38	0.28	0.28	0.43
18	4	3	1	4	2	390.43	264	0.00	0.00	1.25	354.42	4	0.72	0.22	0.00
19	4	4	2	5	3	387.57	7	0.95	0.60	0.00	304.10	2	0.80	0.80	0.00
20	4	5	3	1	4	387.84	39	.00	0.27	0.91	349.67	13	0.24	0.42	0.00
21	5	1	5	4	3	387.92	5	0.74	0.08	0.70	338.57	4	0.58	0.00	0.65
22	5	2	1	5	4	384.69	1	1.70	0.20	0.00	358.02	2	0.90	0.00	0.00
23	5	3	2	1	5	387.46	8	0.90	0.30	0.20	357.96	8	0.90	0.00	0.00
24	5	4	3	2	1	389.12	80	0.93	0.00	0.42	358.45	118	0.95	0.00	0.00
25	5	5	4	3	2	386.35	8	0.58	0.58	0.08	350.36	9	0.34	0.34	0.00

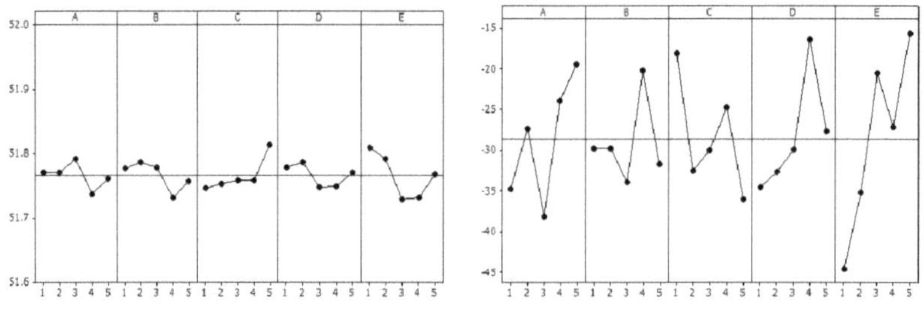

(a) (b)

Figure 3. Signal-to-noise (SN) ratios of control factors for Pattern A in the case of steepest descent method: (**a**) Sharpness; (**b**) Iterations.

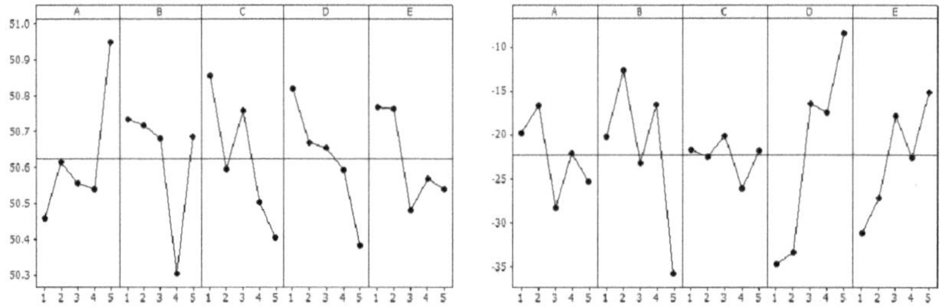

(a) (b)

Figure 4. SN ratios of control factors for Pattern B in the case of steepest descent method: (**a**) Sharpness; (**b**) Iterations.

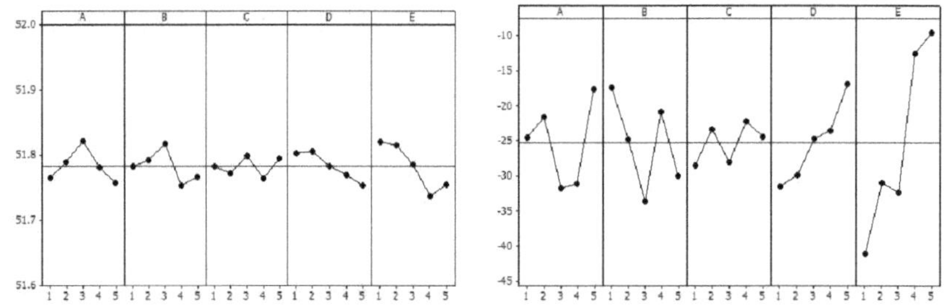

(a) (b)

Figure 5. SN ratios of control factors for Pattern A in the case of conjugate gradient method: (**a**) Sharpness; (**b**) Iterations.

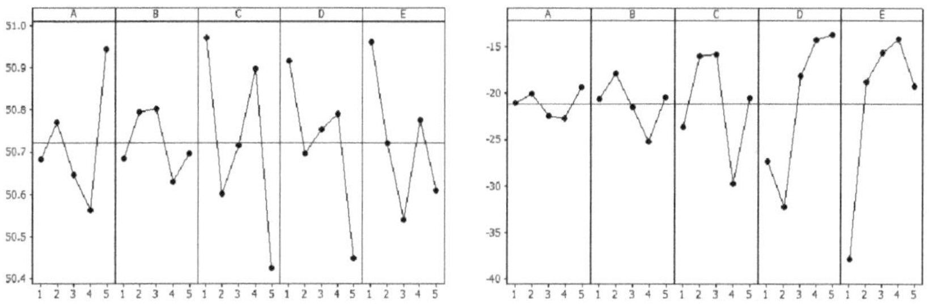

Figure 6. SN ratios of control factors for Pattern B in the case of conjugate gradient method: (**a**) Sharpness; (**b**) Iterations.

However, the trends of sharpness and the number of iterations were inverse. Sharpness is more important than the number of iterations because the inspection in a manufacturing process must be accurate. Hence, we chose the initial voltage in the sharpness, and τ and η in the number of the iteration. Retest combinations of STD were determined considering figures such as $A_3B_2C_5D_2E_5$ and $A_5B_1C_1D_1E_1$ for Patterns A and B, respectively. $A_3B_3C_3D_5E_5$, and $A_5B_3C_1D_5E_4$ were selected for Patterns A and B in case of CJG. The retest results using $A_3B_2C_5D_2E_5$ were $\sigma_{max} = 390.07$, $V = (0.00, 0.00, 1.09)$, and 19 iterations. The combination of $A_5B_1C_1D_1E_1$ was $\sigma_{max} = 357.97$, $V = (1.02, 0.00, 0.00)$, and 37 iterations. The retest results using $A_3B_3C_3D_5E_5$ were $\sigma_{max} = 383.73$, $V = (0.31, 0.30, 0.3)$, and 16 iterations. The value of this point was 2% lower than the ESS, and the coordinate is far from the ESS. This indicates a different local minimum compared with the ESS results. However, when the terminal condition is tightly given, a result similar to the ESS can be obtained with 74 iterations. The retest results using $A_5B_3C_1D_5E_4$ were $\sigma_{max} = 357.09$, $V = (1.02, 0.02, 0.00)$, and 37 iterations. Contributions of the parameters in the STD were evaluated using ANOVA, as shown in Tables 4 and 5. The results of ANOVA were obtained using general linear model in MINITAB. The η was the most significant factor for Pattern A, but initial point was significant for Pattern B. Tables 6 and 7 show contributions of the parameters in the CJD. η was the most significant factor for the sharpness and the iteration. However, initial point was significant for the sharpness, and the iteration was more significant for the iteration. Hence, convergence constant, η, is the most important and the initial point is the second to find optimum of color lighting. τ was a minor factor in the experiments.

Table 4. ANOVA of Pattern A for contribution of steepest descent method.

Control Factors			σ_{max}			I		
Source	**Parameter**	**DF**	**SS**	**MS**	**Contribution (%)**	**SS**	**MS**	**Contribution (%)**
A	Initial V_R	4	19.548	4.887	14.2	71,168	17,792	24.5
B	Initial V_G	4	18.847	4.712	13.7	36,910	9228	12.7
C	Initial V_B	4	34.290	8.572	23.0	65,697	16,424	22.6
D	τ	4	17.188	4.297	12.5	43,406	10,851	14.9
E	η	4	41.656	10.414	30.3	71,069	17,767	24.5
Error		2	5.806	2.903	4.2	2171	1085	0.7
Total		22	137.335			290,421		

Table 5. ANOVA of Pattern B for contribution of steepest descent method.

Control Factors			σ_{max}			l		
Source	Parameter	DF	SS	MS	Contribution (%)	SS	MS	Contribution (%)
A	Initial V_R	4	1094.64	273.66	24.0	4027	1007	4.5
B	Initial V_G	4	916.54	229.13	20.1	40,860	10,215	45.6
C	Initial V_B	4	1019.66	254.91	22.4	2807	702	3.1
D	τ	4	715.99	179	15.7	17,557	4389	19.6
E	η	4	516.07	129.02	11.3	10,919	2730	12.2
Error		4	291.90	72.97	6.4	13,524	3381	15.1
Total		24	4554.80			89,694		

Table 6. ANOVA of Pattern A for contribution of conjugate gradient method.

Control Factors			σ_{max}			l		
Source	Parameter	DF	SS	MS	Contribution (%)	SS	MS	Contribution (%)
A	Initial V_R	4	24.802	6.2	18.5	30,195	7549	14.8
B	Initial V_G	4	23.749	5.937	17.7	34,288	8572	16.9
C	Initial V_B	4	8.295	2.074	6.2	9756	2439	4.8
D	τ	4	19.159	4.8	14.3	24,720	6180	12.1
E	η	4	51.27	12.818	38.2	72,863	18,216	35.8
Error		4	7.054	1.763	5.3	31,656	7914	15.6
Total		24	134.329			203,478		

Table 7. ANOVA of Pattern B for contribution of conjugate gradient method.

Control Factors			σ_{max}			l		
Source	Parameter	DF	SS	MS	Contribution (%)	SS	MS	Contribution (%)
A	Initial V_R	4	637.8	159.5	14.3	1884.4	471.1	3.3
B	Initial V_G	4	161.4	40.3	3.6	5903.6	1475.9	10.3
C	Initial V_B	4	1491.5	372.9	33.4	8974.8	2243.7	15.6
D	τ	4	840.4	210.1	18.8	8401.6	2100.4	14.6
E	η	4	794.9	198.7	17.8	29,353.6	7338.4	51.1
Error		4	537.4	134.3	12.0	2888	722	5.0
Total		24	4463.4			57,406		

Figures 7 and 8 show the convergence of maximum sharpness by employing the STD and the CJD methods. In the figures, V_R, V_G, and V_B are mapped virtually in Cartesian coordinates. The starting point is shown in blue, and the color is varied into others during iteration. The terminal point is marked with red. The paths shaped smooth curve lines compared to direct and non-differentiation optimum search methods showing discrete pattern. The starting points of individual pattern determined using Taguchi method were different, but they approached the same point.

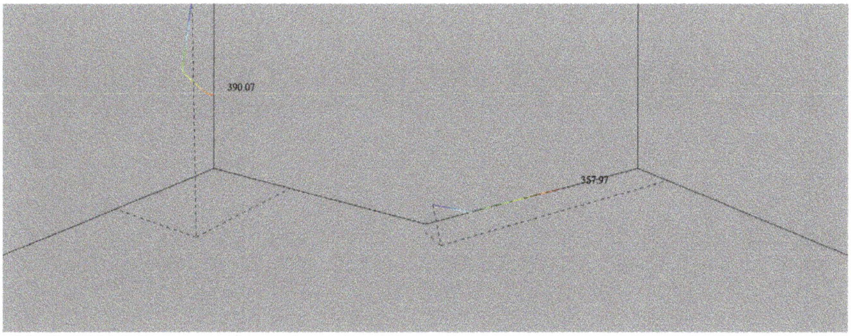

(a) (b)

Figure 7. Search path formed by steepest descent method using Patterns (**a**) A and (**b**) B.

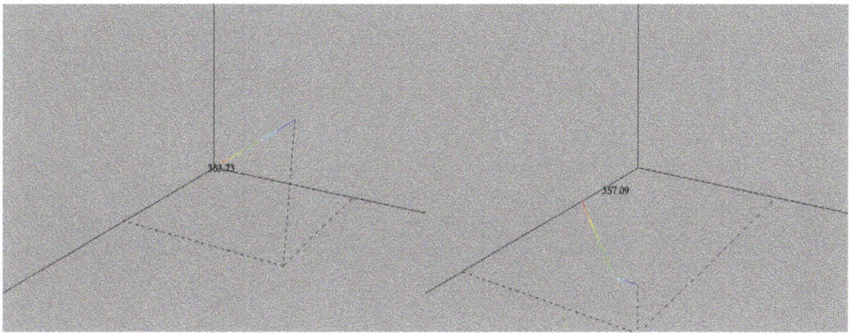

(a) (b)

Figure 8. Search path formed by conjugate gradient method using Patterns (**a**) A and (**b**) B.

The sharpnesses in the results were almost the same as that observed in the best-case parameters. However, the number of iterations was relatively small compared to the average number of iterations—even the numbers using ESS. One result had almost the same sharpnesses as that of the exact solution using different voltage. The retest results show that the Taguchi method provides useful parameters with a small number of experiments. Although the maximum sharpness value determined by the proposed methods was a little lower than that determined by ESS, the number of iterations was much smaller. Therefore, the proposed auto-lighting algorithm can reduce the number of iterations, while the image quality remains almost the same. Furthermore, the Taguchi method can reduce laborious tasks and the setup time for the inspection process in manufacturing.

5. Conclusions

A tuning method was proposed for the auto-lighting algorithm using the Taguchi method. The algorithm maximizes the image quality by adjusting multiple light sources in the shortest time, thus providing a function called auto-lighting. The image quality is defined as sharpness—the standard deviation of the grey level in pixels of an inspected image. The best image quality was found using two differential optimum methods—STD and CJG. The image quality was represented using sharpness, and the minimum of the negative sharpness was found using the steepest descent and conjugate gradient methods. These methods are modified for auto-lighting algorithms.

The Taguchi method was applied to determine the algorithm parameters, such as initial voltage, convergence constant, and threshold. The $L_{25}(5^5)$ orthogonal array was constructed considering five control factors and five levels of the parameter ranges. The SN ratio of the sharpness was calculated using "the larger the better", and that of the number of iterations was calculated using "the smaller the better". The desired combinations were determined after the Taguchi analysis using the orthogonal array. A retest was conducted by using the desired combination, and the results showed that the Taguchi method provides useful parameter values, and the performance is almost equal to that of the best-case parameters. The Taguchi method will be useful in reducing tasks and the time required to set up the inspection process in manufacturing.

Acknowledgments: We would like to acknowledge the financial support from the R & D Program of MOTIE (Ministry of Trade, Industry and Energy) and KEIT (Korea Evaluation Institute of Industrial Technology) of Republic of Korea (Development of Vision Block supporting the combined type I/O of extensible and flexible structure, 10063314). The authors are grateful to AM Technology (http://www.amtechnology.co.kr) for supplying RGB mixable color sources.

Author Contributions: All of the authors contributed extensively to this work presented in the paper. HyungTae Kim conceived the optimum algorithm, conducted the experiments and wrote the paper. KyeongYong Cho conducted the experiment using the Taguchi method. KyungChan Jin designed the experimental apparatus and Jongseok Kim constructed the apparatus. SeungTaek Kim derived color coordinates and analyzed the data.

Conflicts of Interest: The authors declare no conflict of interest.

Abbreviations

STD	Steepest descent method
CJG	Conjugate gradient method
LED	Light emitting diode
RGB	Red, green and blue
sRGB	Standard red, green and blue
ESS	Equal step search
TAE	Trial-and-error
SN	Signal-to-noise
$I(x, y)$	Grey level of an image pixel
$\bar{I}$	Brightness, average grey level of an image
k	Current iteration
m	Horizontal pixel number of an image
N	Number of voltage inputs for a color mixer
n	Vertical pixel number of an image
u	the performance index
V	Vector of voltage inputs for a color mixer
v	Individual voltage input for an LED
w	the number of experiments
x	Horizontal coordinate of an image
y	Vertical coordinate of an image
α	Convergence coefficient
ϵ	Terminal condition
η	Convergence coefficient for limited range
ρ	Negative sharpness, cost function
σ	Sharpness, image quality
τ	Threshold
ξ	Index of update for conjugate gradient method

References

1. Gruber, F.; Wollmann, P.; Schumm, B.; Grahlert, W.; Kaskel, S. Quality Control of Slot-Die Coated Aluminum Oxide Layers for Battery Applications Using Hyperspectral Imaging. *J. Imaging* **2016**, *2*, 12.
2. Neogi, N.; Mohanta, D.K.; Dutta, P.K. Review of vision-based steel surface inspection systems. *EURASIP J. Image Video Process.* **2014**, *50*, 1–19.

3. Arecchi, A.V.; Messadi, T.; Koshel, R.J. *Field Guide to Illimination*; SPIE Press: Bellingham, WA, USA, 2007; pp. 110–115.
4. Pfeifer, T.; Wiegers, L. Reliable tool wear monitoring by optimized image and illumination control in machine vision. *Measurement* **2010**, *28*, 209–218.
5. Jani, U.; Reijo, T. Setting up task-optimal illumination automatically for inspection purposes *Proc. SPIE* **2007**, *6503*, 65030K.
6. Kim, T.H.; Kim, S.T.; Cho, Y.J. Quick and Efficient Light Control for Conventional AOI Systems. *Int. J. Precis. Eng. Manuf.* **2015**, *16*, 247–254.
7. Chen, S.Y.; Zhang, J.W.; Zhang, H.X.; Kwok, N.M.; Li, Y.F. Intelligent Lighting Control for Vision-Based Robotic Manipulation. *IEEE Trans. Ind. Electron.* **2012**, *59*, 3254–3263.
8. Victoriano, P.M.A.; Amaral, T.G.; Dias, O.P. Automatic Optical Inspection for Surface Mounting Devices with IPC-A-610D compliance. In Proceedings of the 2011 International Conference on Power Engineering, Energy and Electrical Drives (POWERENG), Malaga, Spain, 11–13 May 2011; pp. 1–7.
9. Muthu, S.; Gaines, J. Red, Green and Blue LED-based White Light Source: Implementation Challenges and Control Design. In Proceedings of the 2003 38th IAS Annual Meeting, Conference Record of the Industry Applications Conference, Salt Lake City, UT, USA, 12–16 October 2003; Volume 1, pp. 515–522.
10. Esparza, D.; Moreno, I. Color patterns in a tapered lightpipe with RGB LEDs. *Proc. SPIE* **2010**, *7786*, 77860I.
11. Van Gorkom, R.P.; van AS, M.A.; Verbeek, G.M.; Hoelen, C.G.A.; Alferink, R.G.; Mutsaers, C.A.; Cooijmans, H. Etendue conserved color mixing. *Proc. SPIE* **2007**, *6670*, 66700E.
12. Zhu, Z.M.; Qu, X.H.; Liang, H.Y.; Jia, G.X. Effect of color illumination on color contrast in color vision application. *Proc. SPIE Opt. Metrol. Insp. Ind. Appl.* **2010**, *7855*, 785510.
13. Park, J.I.; Lee, M.H.; Grossberg, M.D.; Nayar, S.K. Multispectral Imaging Using Multiplexed Illumination. In Proceedings of the IEEE 11th International Conference on Computer Vision (ICCV 2007), Rio de Janeiro, Brazil, 14–20 October 2007; pp. 1–8.
14. Lee, M.H.; Seo, D.K.; Seo, B.K.; Park, J.I. Optimal Illumination Spectrum for Endoscope. In Proceedings of the 2011 17th Korea-Japan Joint Workshop on Frontiers of Computer Vision (FCV), Ulsan, Korea, 9–11 February 2011; pp. 1–6.
15. Kim, T.H.; Kim, S.T.; Cho, Y.J. An Optical Mixer and RGB Control for Fine Images using Grey Scale Distribution. *Int. J. Optomech.* **2012**, *6*, 213–225.
16. Kim, T.H.; Kim, S.T.; Kim, J.S. Mixed-color illumination and quick optimum search for machine vision. *Int. J. Optomech.* **2013**, *7*, 208–222.
17. Kim, T.H.; Cho, K.Y.; Jin, K.C.; Yoon, J.S.; Cho, Y.J. Mixing and Simplex Search for Optimal Illumination in Machine Vision. *Int. J. Optomech.* **2014**, *8*, 206–217.
18. Arora, J.S. *Introduction to Optimum Design*, 2nd ed.; Academic Press: San Diego, CA, USA, 2004; pp. 433–465.
19. Kim, T.H.; Cho, K.Y.; Kim, S.T.; Kim, J.S.; Jin, K.C.; Lee, S.H. Rapid Automatic Lighting Control of a Mixed Light Source for Image Acquisition using Derivative Optimum Search Methods. In Proceedings of the International Symposium of Optomechatronics Technology (ISOT 2015), Neuchâtel, Switzerland, 14–16 October 2015; p. 07004.
20. Dey, N.; Ashour, A.S.; Beagum, S.; Pistola, D.S.; Gospodinov, M.; Gospodinova, E.P.; Tavares, J.M.R.S. Parameter Optimization for Local Polynomial Approximation based Intersection Confidence Interval Filter Using Genetic Algorithm: An Application for Brain MRI Image De-Noising. *J. Imaging* **2015**, *1*, 60–84.
21. Lazarevic, D.; Madic, M.; Jankovic, P.; Lazarevic, A. Cutting Parameters Optimization for Surface Roughness in Turning Operation of Polyethylene (PE) Using Taguchi Method. *Tribol. Ind.* **2012**, *34*, 68–73.
22. Verma, J.; Agrawal, P.; Bajpai, L. Turning Parameter Optimization for Surface Roughness of Astm A242 Type-1 Alloys Steel by Taguchi Method. *Int. J. Adv. Eng. Technol.* **2012**, *3*, 255–261.
23. Firestone, L.; Cook, K.; Culp, K.; Talsania, N.; Preston, K., Jr. Comparision of Autofocus Methods for Automated Microscopy. *Cytometry* **1991**, *12*, 195–206.
24. Bueno-Ibarra, M.A.; Alvarez-Borrego, J.; Acho, L.; Chavez-Sanchez, M.C. Fast autofocus algorithm for automated microscopes. *Opt. Eng.* **2005**, *44*, 063601.
25. Sun, Y.; Duthaler, S.; Nelson, B.J.; Autofocusing in computer microscopy: Selecting the optimal focus algorithm. *Microsc. Res. Tech.* **2004**, *65*, 139–149.
26. Muruganantham, C.; Jawahar, N.; Ramamoorthy, B.; Giridhar, D. Optimal settings for vision camera calibration. *Int. J. Manuf. Technol.* **2010**, *42*, 736–748.

27. Li, M.; Milor, L.; Yu, W. Developement of Optimum Annular Illumination: A Lithography-TCAD Approach. In Proceedings of the Advanced Semiconductor Manufacturing Conference and Workshop (IEEE/SEMI), Cambridge, MA, USA, 10–12 September 1997; pp. 317–321.
28. Kim, T.H.; Cho, K.Y.; Kim, S.T.; Kim, J.S. Optimal RGB Light-Mixing for Image Acquisition Using Random Search and Robust Parameter Design. In Proceedings of the 16th International Workshop on Combinatorial Image Analysis, Brno, Czech Republic, 28–30 May 2014; Volume 8466, pp. 171–185.
29. Zahlay, F.D.; Rao, K.S.R.; Baloch, T.M. Autoreclosure in Extra High Voltage Lines using Taguchi's Method and Optimized Neural Network. In Proceedings of the 2008 Electric Power Conference (EPEC), Vancouver, BC, Canada, 6–7 October 2008; Volume 2, pp. 151–155.
30. Sugiono; Wu, M.H.; Oraifige, I. Employ the Taguchi Method to Optimize BPNN's Architectures in Car Body Design System. *Am. J. Comput. Appl. Math.* **2012**, 2, 140–151.
31. Su, T.L.; Chen, H.W.; Hong, G.B.; Ma, C.M. Automatic Inspection System for Defects Classification of Stretch Kintted Fabrics. In Proceedings of the 2010 International Conference on Wavelet Analysis and Pattern Recognition (ICWAPR), Qingdao, China, 11–14 July 2010; pp. 125–129.
32. Wu, Y.; Wu, A. Quality engineering and experimental design. In *Taguchi Methods for Robust Design*; The American Society of Mechanical Engineers: Fairfield, NJ, USA, 2000; pp. 3–16.
33. Yoo, W.S.; Jin, Q.Q.; Chung, Y.B. A Study on the Optimization for the Blasting Process of Glass by Taguchi Method. *J. Soc. Korea Ind. Syst. Eng.* **2007**, *30*, 8–14.

Journal of
Imaging

MDPI

Article

Improving CNN-Based Texture Classification by Color Balancing

Simone Bianco [1,*], Claudio Cusano [2], Paolo Napoletano [1] and Raimondo Schettini [1]

1 Department of Computer Science, Systems and Communications, University of Milano-Bicocca, 20126 Milan, Italy; paolo.napoletano@disco.unimib.it (P.N.); schettini@disco.unimib.it (R.S.)

2 Department of Electrical, Computer and Biomedical Engineering, University of Pavia, 27100 Pavia, Italy; claudio.cusano@unipv.it

* Correspondence: bianco@disco.unimib.it; Tel.: +39-02-6448-7827

Received: 29 June 2017; Accepted: 21 July 2017; Published: 27 July 2017

Abstract: Texture classification has a long history in computer vision. In the last decade, the strong affirmation of deep learning techniques in general, and of convolutional neural networks (CNN) in particular, has allowed for a drastic improvement in the accuracy of texture recognition systems. However, their performance may be dampened by the fact that texture images are often characterized by color distributions that are unusual with respect to those seen by the networks during their training. In this paper we will show how suitable color balancing models allow for a significant improvement in the accuracy in recognizing textures for many CNN architectures. The feasibility of our approach is demonstrated by the experimental results obtained on the RawFooT dataset, which includes texture images acquired under several different lighting conditions.

Keywords: convolutional neural networks; color balancing; deep learning; texture classification; color constancy; color characterization

1. Introduction

Convolutional neural networks (CNNs) represent the state of the art for many image classification problems [1–3]. They are trained for a specific task by exploiting a large set of images representing the application domain. During the training and the test stages, it is common practice to preprocess the input images by centering their color distribution around the mean color computed on the training set. However, when test images have been taken under acquisition conditions unseen during training, or with a different imaging device, this simple preprocessing may not be enough (see the example reported in Figure 1 and the work by Chen et al. [4]).

The most common approach to deal with variable acquisition conditions consists of applying a color constancy algorithm [5], while to obtain device-independent color description a color characterization procedure is applied [6]. A standard color-balancing model is therefore composed of two modules: the first discounts the illuminant color, while the second maps the image colors from the device-dependent RGB space into a standard device-independent color space. More effective pipelines have been proposed [7,8] that deal with the cross-talks between the two processing modules.

In this paper we systematically investigate different color-balancing models in the context of CNN-based texture classification under varying illumination conditions. To this end, we performed our experiments on the RawFooT texture database [9] which includes images of textures acquired under a large number of controlled combinations of illumination color, direction and intensity.

Concerning CNNs, when the training set is not big enough, an alternative to the full training procedure consists of adapting an already trained network to a new classification task by retraining only a small subset of parameters [10]. Another possibility is to use a pretrained network as a feature extractor for another classification method (nearest neighbor, for instance). In particular, it is common to use networks trained for the ILSVRC contest [11]. The ILSVRC training set includes over one million images taken from the web to represent 1000 different concepts. The acquisition conditions of training images are not controlled, but we may safely assume that they have been processed by digital processing pipelines that mapped them into the standard sRGB color space. We will investigate how different color-balancing models permit adapting images from the RawFooT dataset in such a way that they can be more reliably classified by several pretrained networks.

The rest of the paper is organized as follows: Section 2 summarizes the state of the art in both texture classification and color-balancing; Section 3 presents the data and the methods used in this work; Section 4 describes the experimental setup and Section 5 reports and discusses the results of the experiments. Finally, Section 6 concludes the paper by highlighting its main outcomes and by outlining some directions for future research on this topic.

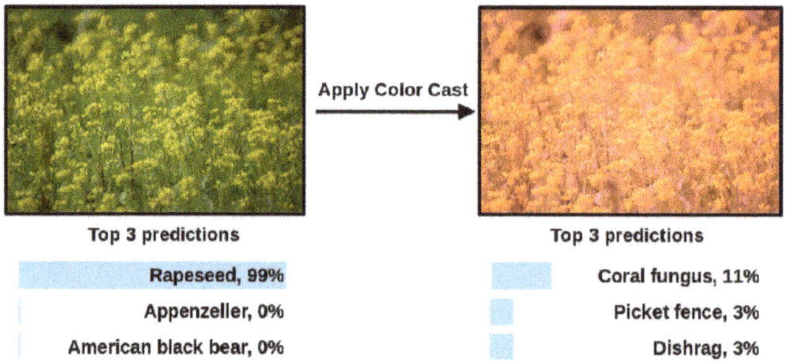

Figure 1. Example of correctly predicted image and mis-predicted image after a color cast is applied.

2. Related Works

2.1. Color Texture Classification under Varying Illumination Conditions

Most of the research efforts on the topic of color texture classification have been devoted to the definition of suitable descriptors able to capture the distinctive properties of the texture images while being invariant, or at least robust, with respect to some variations in the acquisition conditions, such as rotations and scalings of the image, changes in brightness, contrast, light color temperature, and so on [12].

Color and texture information can be combined in several ways. Palm categorized them into parallel (i.e., separate color and texture descriptors), sequential (in which color and texture analysis are consecutive steps of the processing pipeline) and integrative (texture descriptors computed on different color planes) approaches [13]. The effectiveness of several combinations of color and texture descriptors has been assessed by Mäenpää, and Pietikäinen [14], who showed how the descriptors in the state of the art performed poorly in the case of a variable color of the illuminant. Their findings have been more recently confirmed by Cusano et al. [9].

In order to successfully exploit color in texture classification the descriptors need to be invariant (or at least robust) with respect to changes in the illumination. For instance, Seifi et al. proposed characterizing color textures by analyzing the rank correlation between pixels located in the same neighborhood and by using a correlation measure which is related to the colors of the pixels, and is not sensitive to illumination changes [15]. Cusano et al. [16] proposed a descriptor that measures the local contrast: a property that is less sensitive than color itself to variations in the color of the illuminant. The same authors then enhanced their approach by introducing a novel color space where changes in illumination are even easier to deal with [17]. Other strategies for color texture recognition have been proposed by Drimbarean and Whelan who used Gabor filters and co-occcurrence matrices [18], and by Bianconi et al. who used ranklets and the discrete Fourier transform [19].

Recent works suggested that, in several application domains, carefully designed features can be replaced by features automatically learned from a large amount of data with methods based on deep learning [20]. Cimpoi et al., for instance, used Fisher Vectors to pool features computed by a CNN trained for object recognition [21]. Approaches based on CNNs have compared against combinations of traditional descriptors by Cusano et al. [22], who found that CNN-based features generally outperform the traditional handcrafted ones unless complex combinations are used.

2.2. Color Balancing

The aim of color constancy is to make sure that the recorded color of the objects in the scene does not change under different illumination conditions. Several computational color constancy algorithms have been proposed [5], each based on different assumptions. For example, the gray world algorithm [23] is based on the assumption that the average color in the image is gray and that the illuminant color can be estimated as the shift from gray of the averages in the image color channels. The white point algorithm [24] is based on the assumption that there is always a white patch in the scene and that the maximum values in each color channel are caused by the reflection of the illuminant on the white patch, and they can be thus used as the illuminant estimation. The gray edge algorithm [25] is based on the assumption that the average color of the edges is gray and that the illuminant color can be estimated as the shift from the gray of the averages of the edges in the image color channels. Gamut mapping assumes that for a given illuminant, one observes only a limited gamut of colors [26]. Learning-based methods also exist, such as Bayesian [27], CART-based [28], and CNN-based [29,30] approaches, among others.

The aim of color characterization of an imaging device is to find a mapping between its device-dependent and a device-independent color representation. The color characterization is performed by recording the sensor responses to a set of colors and the corresponding colorimetric values, and then finding the relationship between them. Numerous techniques in the state of the art have been proposed to find this relationship, ranging from empirical methods requiring the acquisition of a reference color target (e.g., a GretagMacbeth ColorChecker [31]) with known spectral reflectance [8], to methods needing the use of specific equipment such as monochromators [32]. In the following we will focus on empirical methods that are the most used in practice, since they do not need expensive laboratory hardware. Empirical device color characterization directly relates measured colorimetric data from a color target and the corresponding camera raw RGB data obtained by shooting the target itself under one or more controlled illuminants. Empirical methods can be divided into two classes: the methods belonging to the first class rely on model-based approaches, that solve a set of linear equations by means of pseudo-inverse approach [6] , constrained least squares [33], exploiting a non-maximum ignorance assumption [33,34], exploiting optimization to solve more meaningful objective functions [7,35,36], or lifting the problem into a higher dimensional polynomial space [37,38]. The second class instead contains methods that do not explicitly model the relationship between device-dependent and device-independent color representations such as three-dimensional lookup tables with interpolation and extrapolation [39], and neural networks [40,41].

3. Materials and Methods

3.1. RawFooT

The development of texture analysis methods heavily relies on suitably designed databases of texture images. In fact, many of them have been presented in the literature [42,43]. Texture databases are usually collected to emphasize specific properties of textures such as the sensitivity to the acquisition device, the robustness with respect to the lighting conditions, and the invariance to image rotation or scale, etc. The RawFooT database has been especially designed to investigate the performance of color texture classification methods under varying illumination conditions [9]. The database includes images of 68 different samples of raw foods, each one acquired under 46 different lighting conditions (for a total of $68 \times 46 = 3128$ acquisitions). Figure 2 shows an example for each class.

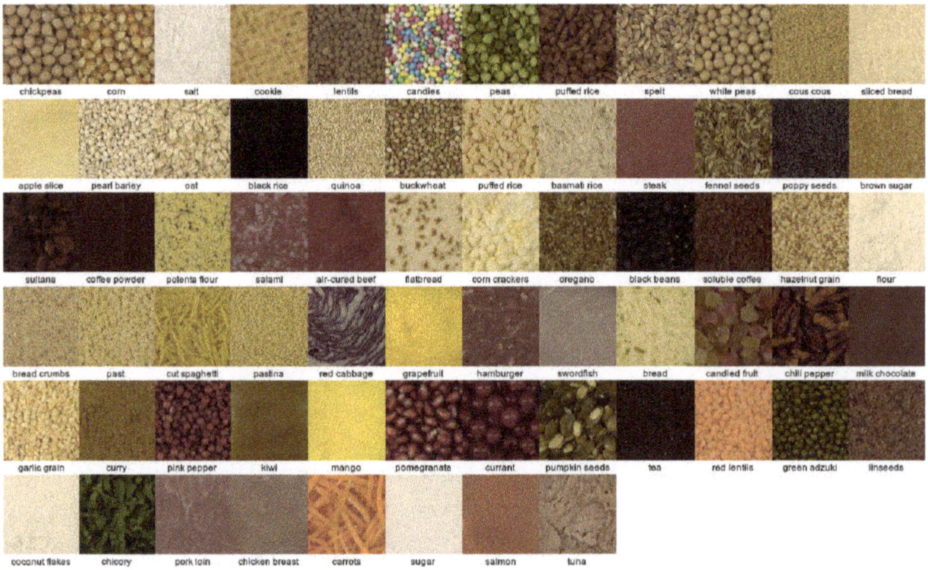

Figure 2. A sample for each of the 68 classes of textures composing the RawFooT database.

Images have been acquired with a Canon EOS 40D DSLR camera. The camera was placed 48 cm above the sample to be acquired, with the optical axis perpendicular to the surface of the sample. The lenses used had a focal length of 85 mm, and a camera aperture of f/11.3; each picture has been taken with four seconds of exposition time. For each 3944×2622 acquired image a square region of 800×800 pixels has been cropped in such a way that it contains only the surface of the texture sample without any element of the surrounding background. Note that, while the version of the RawFooT database that is publicly available includes a conversion of the images in the sRGB color space, in this work we use the raw format images that are thus encoded in the device-dependent RGB space.

To generate the 46 illumination conditions, two computer monitors have been used as light sources (two 22-inch Samsung SyncMaster LED monitors). The monitors were tilted by 45 degrees facing down towards the texture sample, as shown in Figure 3. By illuminating different regions of one or both monitors it was possible to set the direction of the light illuminating the sample. By changing the RGB values of the pixels it was also possible to control the intensity and the color of the light sources. To do so, both monitors have been preliminarily calibrated using a X-Rite i1 spectral colorimeter by setting their white point to D65.

With this setup it was possible to approximate a set of diverse illuminants. In particular, 12 illuminants have been simulated, corresponding to 12 daylight conditions differing in the color temperature. The CIE-xy chromaticities corresponding to a given temperature T have been obtained by applying the following equations [44]:

$$\begin{aligned} x &= a_0 + a_1 \frac{10^3}{T} + a_2 \frac{10^6}{T^2} + a_3 \frac{10^9}{T^3}, \\ y &= -3x^2 + 2.87x - 0.275, \end{aligned} \tag{1}$$

where $a_0 = 0.244063$, $a_1 = 0.09911$, $a_2 = 2.9678$, $a_3 = -4.6070$ if $4000\,\text{K} \leq T \leq 7000\,\text{K}$, and $a_0 = 0.23704$, $a_1 = 0.24748$, $a_2 = 1.9018$, $a_3 = -2.0064$ if $7000\,\text{K} < T \leq 25{,}000\,\text{K}$. The chromaticities were then converted in the monitor RGB space [45] with a scaling of the color channels in such a way that largest value was 255.

$$\begin{pmatrix} R' \\ G' \\ B' \end{pmatrix} = \begin{bmatrix} 3.2406 & -1.5372 & -0.4986 \\ -0.9689 & 1.8758 & 0.0415 \\ 0.0557 & -0.2040 & 1.0570 \end{bmatrix} \cdot \begin{pmatrix} x \\ y \\ 1-x-y \end{pmatrix} \times \frac{1}{y}, \tag{2}$$

$$\begin{pmatrix} R \\ G \\ B \end{pmatrix} = \frac{255}{\max\{R', G', B'\}} \times \begin{pmatrix} R' \\ G' \\ B' \end{pmatrix}. \tag{3}$$

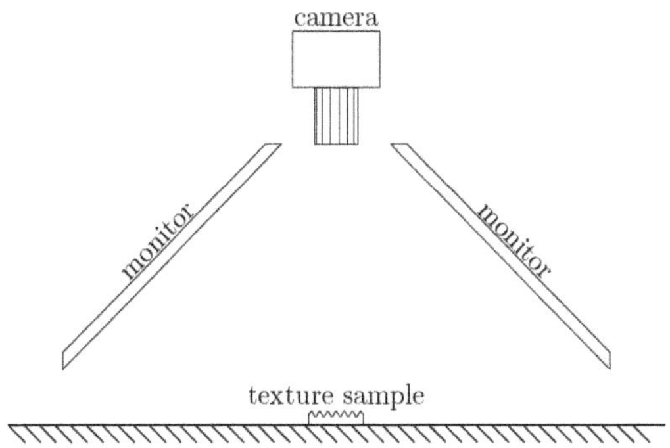

Figure 3. Scheme of the acquisition setup used to take the images in the RawFooT database.

The twelve daylight color temperatures that have been considered are: 4000 K, 4500 K, 5000 K, ..., 9500 K (we will refer to these as D40, D45, ..., D95).

Similarly, six illuminants corresponding to typical indoor light have been simulated. To do so, the CIE-xy chromaticities of six LED lamps (six variants of SOLERIQ®S by Osram) have been obtained from the data sheets provided by the manufacturer. Then, again the RGB values were computed and scaled to 255 in at least one of the three channels. These six illuminants are referred to as L27, L30, L40, L50, L57, and L65 in accordance with the corresponding color temperature.

Figure 4 shows, for one of the classes, the 46 acquisitions corresponding to the 46 different lighting conditions in the RawFooT database. These include:

- 4 acquisitions with a D65 illuminant of varying intensity (100%, 75%, 50%, 25% of the maximum);
- 9 acquisitions which were only a portion of one of the monitors lit to obtain directional light (approximately 24, 30, 36, 42, 48, 54, 60, 66 and 90 degrees between the direction of the incoming light and the normal to the texture sample);
- 12 acquisitions with both monitors entirely projecting simulated daylight (D4, ..., D95);
- 6 acquisitions with the monitor simulating artificial light (L27, ..., L65);
- 9 acquisitions with simultaneously change of both the direction and the color of light;
- 3 acquisitions with the two monitors simulating a different illuminant (L27+D65, L27+D95 and D65+D95);
- 3 acquisitions with both monitors projecting pure red, green and blue light.

Figure 4. Example of the 46 acquisitions included in the RawFooT database for each class (here the images show the acquisitions of the "rice" class).

In this work we are interested in particular in the effects of changes in the illuminant color. Therefore, we limited our analysis to the 12 illuminants simulating daylight conditions, and to the six simulating indoor illumination.

Beside the images of the 68 texture classes, the RawFooT database also includes a set of acquisitions of a color target (the Macbeth color checker [31]). Figure 5 shows these acquisitions for the 18 illuminants considered in this work.

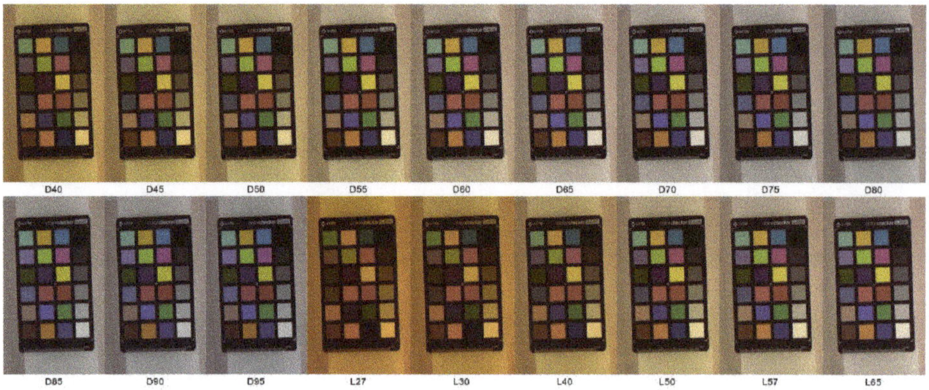

Figure 5. The Macbeth color target, acquired under the 18 lighting conditions considered in this work.

3.2. Color Balancing

An image acquired by a digital camera can be represented as a function ρ mainly dependent on three physical factors: the illuminant spectral power distribution $I(\lambda)$, the surface spectral reflectance $S(\lambda)$, and the sensor spectral sensitivities $C(\lambda)$. Using this notation, the sensor responses at the pixel with coordinates (x, y) can be described as:

$$\rho(x,y) = \int_{\omega} I(x,y,\lambda)S(x,y,\lambda)C(\lambda)d\lambda, \tag{4}$$

where ω is the wavelength range of the visible light spectrum, and ρ and $C(\lambda)$ are three-component vectors. Since the three sensor spectral sensitivities are usually more sensitive respectively to the low, medium and high wavelengths, the three-component vector of sensor responses $\rho = (\rho_1, \rho_2, \rho_3)$ is also referred to as the sensor or camera raw RGB triplet. In the following we adopt the convention that ρ triplets are represented by column vectors.

As previously said, the aim of color characterization is to derive the relationship between device-dependent and device-independent color representations for a given device. In this work, we employ an empirical, model-based characterization. The characterization model that transforms the i-th input device-dependent triplet ρ^{IN} into a device-independent triplet ρ^{OUT} can be compactly written as follows [46]:

$$\rho_i^{OUT} = \left(\alpha I M \rho_i^{IN}\right)^{\gamma}, \tag{5}$$

where α is an exposure correction gain, M is the color correction matrix, I is the illuminant correction matrix, and $(\cdot)^{\gamma}$ denotes an element-wise operation.

Traditionally [46], M is fixed for any illuminant that may occur, while α and I compensate for the illuminant power and color respectively, i.e.,

$$\rho_{i,j}^{OUT} = \left(\alpha_j I_j M \rho_{i,j}^{IN}\right)^{\gamma}. \tag{6}$$

The model can be thus conceptually split into two parts: the former compensates for the variations of the amount and color of the incoming light, while the latter performs the mapping from the device-dependent to the device-independent representation. In the standard model (Equation (6)) α_j is a single value, I_j is a diagonal matrix that performs the Von Kries correction [47], and M is a 3×3 matrix.

In this work, different characterization models have been investigated together with Equation (6) in order to assess how the different color characterization steps influence the texture recognition accuracy. The first tested model does not perform any kind of color characterization, i.e.,

$$\rho_{i,j}^{OUT} = \rho_{i,j}^{IN}. \tag{7}$$

The second model tested performs just the compensation for the illuminant color, i.e., it balances image colors as a color constancy algorithm would do:

$$\rho_{i,j}^{OUT} = \left(I_j \rho_{i,j}^{IN}\right)^{\gamma}. \tag{8}$$

The third model tested uses the complete color characterization model, but differently from the standard model given in Equation (6), it estimates a different color correction matrix M_j for each illuminant j. The illuminant is compensated for both its color and its intensity, but differently from the standard model, the illuminant color compensation matrix I_j for the j-th illuminant is estimated by using a different luminance gain $\alpha_{i,j}$ for each patch i:

$$\rho_{i,j}^{OUT} = \left(\alpha_{i,j} I_j M_j \rho_{i,j}^{IN}\right)^{\gamma}. \tag{9}$$

The fourth model tested is similar to the model described in Equation (9) but uses a larger color correction matrix M_j by polynomially expanding the device-dependent colors:

$$\rho_{i,j}^{OUT} = \left(\alpha_{i,j} I_j M_j T\left(\rho_{i,j}^{IN}\right)\right)^{\gamma}, \tag{10}$$

where $T(\cdot)$ is an operator that takes as input the triplet ρ and computes its polynomial expansion. Following [7], in this paper we use $T(\rho) = (\rho(1), \rho(2), \rho(3), \sqrt{\rho(1)\rho(2)}, \sqrt{\rho(1)\rho(3)}, \sqrt{\rho(2)\rho(3)})$, i.e., the rooted second degree polynomial [38].

Summarizing, we have experimented with five color-balancing models. They all take as input the device-dependent raw values and process them in different ways:

1. *device-raw*: it does not make any correction to the device-dependent raw values, leaving them unaltered from how they are recorded by the camera sensor;
2. *light-raw*: it performs the correction of the illuminant color, similarly to what is done by color constancy algorithms [5,30,48] and chromatic adaptation transforms [49,50]. The output color representation is still device-dependent, but with the discount of the effect of the illuminant color;
3. *dcraw-srgb*: it performs a full color characterization according to the standard color correction pipeline. The chosen characterization illuminant is the D65 standard illuminant, while the color mapping is linear and fixed for all illuminants that may occur. The correction is performed using the DCRaw software (available at http://www.cybercom.net/~dcoffin/dcraw/);
4. *linear-srgb*: it performs a full color characterization according to the standard color correction pipeline, but using different illumination color compensation and different linear color mapping for each illuminant;
5. *rooted-srgb*: it performs a full color characterization according to the standard color correction pipeline, but using a different illuminant color compensation and a different color mapping for each illuminant. The color mapping is no more linear but it is performed by polynomially expanding the device-dependent colors with a rooted second-degree polynomial.

The main properties of the color-balancing models tested are summarized in Table 1.

Table 1. Main characteristics of the tested color-balancing models. Regarding the color-balancing steps, the open circle denotes that the current step is not implemented in the given model, while the filled circle denotes its presence. Regarding the mapping properties, the dash denotes that the given model does not have this property.

Model Name	Color-Balancing Steps			Mapping Properties	
	Illum. Intensity Compensation	Illum. Color Compensation	Color Mapping	Mapping Type	Number of Mappings
Device-raw (Equation (7))	○	○	○	–	–
Light-raw (Equation (8))	○	●	○	–	–
Dcraw-srgb (Equation (6))	●	◐ fixed for D65	●	Linear	1
Linear-srgb (Equation (9))	●	●	●	Linear	1 for each illum.
Rooted-srgb (Equation (10))	●	●	●	Rooted 2nd-deg. poly.	1 for each illum.

All the correction matrices for the compensation of the variations of the amount and color of the illuminant and the color mapping are found using the set of acquisitions of the Macbeth color checker available in the RawFooT using the optimization framework described in [7,36]. An example of the effect of the different color characterization models on a sample texture class of the RawFooT database is reported in Figure 6.

(a)

Figure 6. *Cont.*

(c)

(d)

(e)

Figure 6. Example of the effect of the different color-balancing models on the "rice" texture class: device-raw (**a**); light-raw (**b**); dcraw-srgb (**c**); linear-srgb (**d**); and rooted-srgb (**e**).

4. Experimental Setup

Given an image, the experimental pipeline includes the following operations: (1) color balancing; (2) feature extraction; and (3) classification. All the evaluations have been performed on the RawFooT database.

4.1. RawFooT Database Setup

For each of the 68 classes we considered 16 patches obtained by dividing the original texture image, that is of size 800×800 pixels, in 16 non-overlapping squares of size 200×200 pixels. For each class we selected eight patches for training and eight for testing alternating them in a chessboard pattern. We form subsets of $68 \times (8 + 8) = 1088$ patches by taking the training and test patches from images taken under different lighting conditions.

In this way we defined several subsets, grouped in three texture classification tasks.

1. **Daylight temperature**: 132 subsets obtained by combining all the 12 daylight temperature variations. Each subset is composed of training and test patches with different light temperatures.
2. **LED temperature**: 30 subsets obtained by combining all the six LED temperature variations. Each subset is composed of training and test patches with different light temperatures.
3. **Daylight vs. LED**: 72 subsets obtained by combining 12 daylight temperatures with six LED temperatures.

4.2. Visual Descriptors

For the evaluation we select a number of descriptors from CNN-based approaches [51,52]. All feature vectors are L^2-normalized (each feature vector is divided by its L^2-norm.). These descriptors are obtained as the intermediate representations of deep convolutional neural networks originally trained for scene and object recognition. The networks are used to generate a visual descriptor by removing the final softmax nonlinearity and the last fully-connected layer. We select the most representative CNN architectures in the state of the art [53] by considering different accuracy/speed trade-offs. All the CNNs are trained on the ILSVRC-2012 dataset using the same protocol as in [1]. In particular we consider the following visual descriptors [10,54]:

- *BVLC AlexNet* (BVLC AlexNet): this is the AlexNet trained on ILSVRC 2012 [1].
- *Fast CNN* (Vgg F): it is similar to that presented in [1] with a reduced number of convolutional layers and the dense connectivity between convolutional layers. The last fully-connected layer is 4096-dimensional [51].
- *Medium CNN* (Vgg M): it is similar to the one presented in [55] with a reduced number of filters in the fourth convolutional layer. The last fully-connected layer is 4096-dimensional [51].
- *Medium CNN* (Vgg M-2048-1024-128): it has three modifications of the Vgg M network, with a lower-dimensional last fully-connected layer. In particular we use a feature vector of 2048, 1024 and 128 size [51].
- *Slow CNN* (Vgg S): it is similar to that presented in [56], with a reduced number of convolutional layers, fewer filters in layer five, and local response normalization. The last fully-connected layer is 4096-dimensional [51].
- *Vgg Very Deep 19 and 16 layers* (Vgg VeryDeep 16 and 19): the configuration of these networks has been achieved by increasing the depth to 16 and 19 layers, which results in a substantially deeper network than the previously ones [2].
- *ResNet 50* is a residual network. Residual learning frameworks are designed to ease the training of networks that are substantially deeper than those used previously. This network has 50 layers [52].

4.3. Texture Classification

In all the experiments we used the nearest neighbor classification strategy: given a patch in the test set, its distance with respect to all the training patches is computed. The prediction of the classifier is the class of the closest element in the training set. For this purpose, after some preliminary tests with several descriptors in which we evaluated the most common distance measures, we decided to use the L2-distance: $d(\mathbf{x}, \mathbf{y}) = \sqrt{\sum_{i=1}^{N}(x(i) - y(i))^2}$, where $\mathbf{x}$ and $\mathbf{y}$ are two feature vectors. All the experiments have been conducted under the *maximum ignorance* assumption, that is, no information about the lighting conditions of the test patches is available for the classification method and for the descriptors. Performance is reported as classification rate (i.e., the ratio between the number of correctly classified images and the number of test images). Note that more complex classification schemes (e.g., SVMs) would have been viable. We decided to adopt the simplest one in order to focus the evaluation on the features themselves and not on the classifier.

5. Results and Discussion

The effectiveness of each color-balancing model has been evaluated in terms of texture classification accuracy. Table 2 shows the average accuracy obtained on each classification task (*daylight temperature*, *LED temperature* and *daylight vs LED*) by each of the visual descriptors combined with each balancing model. Overall, the *rooted-srgb* and *linear-srgb* models achieve better performance than others models with a minimum improvement of about 1% and a maximum of about 9%. In particular the *rooted-srgb* model performs slightly better than *linear-srgb*. The improvements are more visible in Figure 7 that shows, for each visual descriptor, the comparison between all the balancing models.

Each bar represents the mean accuracy over all the classification tasks. ResNet-50 is the best-performing CNN-based visual descriptor with a classification accuracy of 99.52%, that is about 10% better than the poorest CNN-based visual descriptor. This result confirms the power of deep residual nets compared to sequential network architectures such as AlexNet, and VGG etc.

Table 2. Classification accuracy obtained by each visual descriptor combined with each model, the best result is reported in bold.

Features	Device-Raw	Light-Raw	Dcraw-Srgb	Linear-Srgb	Rooted-Srgb
VGG-F	87.81	90.09	93.23	**96.25**	95.83
VGG-M	91.26	92.69	94.71	95.85	**96.14**
VGG-S	90.36	92.64	93.54	**96.83**	96.65
VGG-M-2048	89.83	92.09	94.08	95.37	**96.15**
VGG-M-1024	88.34	90.92	93.74	94.31	**94.92**
VGG-M-128	82.52	85.99	87.35	90.17	**90.97**
AlexNet	84.65	87.16	93.34	93.58	**93.68**
VGG-VD-16	91.15	94.68	95.79	**98.23**	97.93
VGG-VD-19	92.22	94.87	95.38	**97.71**	97.51
ResNet-50	97.42	98.92	98.67	99.28	**99.52**

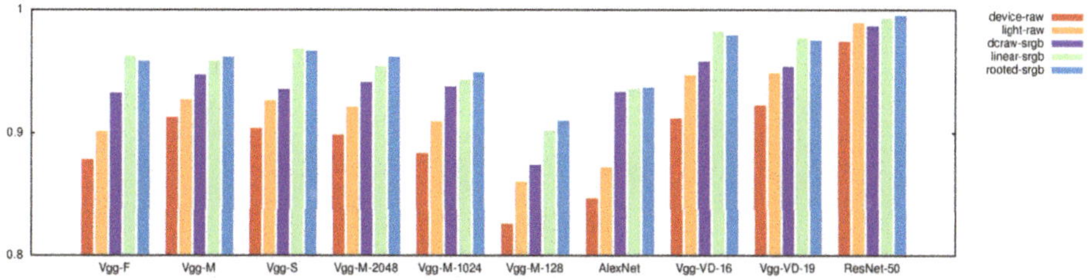

Figure 7. Classification accuracy obtained by each visual descriptor combined with each model.

To better show the usefulness of color-balancing models we focused on the *daylight temperature* classification task, where we have images taken under 12 daylight temperature variations from 4000 K to 9500 K with an increment of 500 K. To this end, Figure 8 shows the accuracy behavior (y-axis) with respect to the difference (ΔT measured in Kelvin degrees) of *daylight temperature* (x-axis) between the training and the test sets. The value $\Delta T = 0$ corresponds to no variations. Each graph shows, given a visual descriptor, the comparison between the accuracy behaviors of each single model. There is an evident drop in performance for all the networks when ΔT is large and no color-balancing is applied. The use of color balancing is able to make uniform the performance of all the networks independently of the difference in color temperature. The *dcraw-srgb* model represents the most similar conditions to those of the ILSVRC training images. This explains why this model obtained the best performance for low values of ΔT. However, since *dcraw-srgb* does not include any kind of color normalization for high values of ΔT we observe a severe loss in terms of classification accuracy. Both *linear-srgb* and *rooted-srgb* are able, instead, to normalize the images with respect to the color of the illumination, making all the plots in Figure 8 almost flat. The effectiveness of these two models also depends on the fact that they work in a color space similar to those used to train the CNNs. Between the linear and the rooted models, the latter performs slightly better, probably because its additional complexity increases the accuracy in balancing the images.

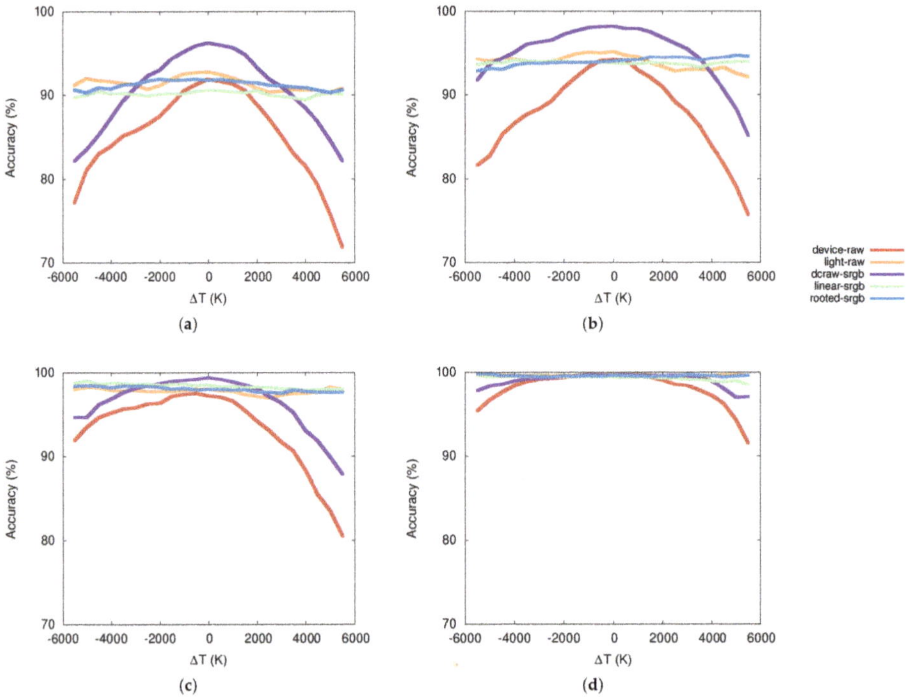

Figure 8. Accuracy behavior with respect to the difference (ΔT) of *daylight temperature* between the training and the test: (**a**) setsVGG-M-128; (**b**) AlexNet; (**c**) VGG-VD-16; (**d**) ResNet-50.

6. Conclusions

Recent trends in computer vision seem to suggest that convolutional neural networks are so flexible and powerful that they can substitute in toto traditional image processing/recognition pipelines. However, when it is not possible to train the network from scratch due to the lack of a suitable training set, the achievable results are suboptimal. In this work we have extensively and systematically evaluated the role of color balancing that includes color characterization as a preprocessing step in color texture classification in presence of variable illumination conditions. Our findings suggest that to really exploit CNNs, an integration with a carefully designed preprocessing procedure is a must. The effectiveness of color balancing, in particular of the color characterization that maps device-dependent RGB values into a device-independent color space, has not been completely proven since the RawFooT dataset has been acquired using a single camera. As future work we would like to extend the RawFooT dataset and our experimentation acquiring the dataset using cameras with different color transmittance filters. This new dataset will make more evident the need for accurate color characterization of the cameras.

Acknowledgments: The authors gratefully acknowledge the support of NVIDIA Corporation with the donation of the Tesla K40 GPU used for this research.

Author Contributions: All the four authors equally contributed to the design of the experiments and the analysis of the data; all the four authors contributed to the writing, proof reading and final approval of the manuscript.

Conflicts of Interest: The authors declare no conflict of interest.

References

1. Krizhevsky, A.; Sutskever, I.; Hinton, G.E. Imagenet classification with deep convolutional neural networks. In *Advances in Neural Information Processing Systems*; The MIT Press: Cambridge, MA, USA, 2012; pp. 1097–1105.
2. Simonyan, K.; Zisserman, A. Very deep convolutional networks for large-scale image recognition. *arXiv preprint arXiv:1409.1556* **2014**, *pp. 1–14*.
3. Zhou, B.; Lapedriza, A.; Xiao, J.; Torralba, A.; Oliva, A. Learning Deep Features for Scene Recognition using Places Database. In *Advances in Neural Information Processing Systems 27*; Neural Information Processing Systems (NIPS): Montreal, QC, Canada, 2014; pp. 487–495.
4. Chen, Y.H.; Chao, T.H.; Bai, S.Y.; Lin, Y.L.; Chen, W.C.; Hsu, W.H. Filter-invariant image classification on social media photos. In Proceedings of the 23rd ACM International Conference on Multimedia, Brisbane, Australia, 26–30 October 2015.
5. Gijsenij, A.; Gevers, T.; Van De Weijer, J. Computational color constancy: Survey and experiments. *IEEE Trans. Image Process.* **2011**, *20*, 2475–2489.
6. Barnard, K.; Funt, B. Camera characterization for color research. *Color Res. Appl.* **2002**, *27*, 152–163.
7. Bianco, S.; Schettini, R. Error-tolerant Color Rendering for Digital Cameras. *J. Math. Imaging Vis.* **2014**, *50*, 235–245.
8. Bianco, S.; Schettini, R.; Vanneschi, L. Empirical modeling for colorimetric characterization of digital cameras. In Proceedings of the 2009 16th IEEE International Conference on Image Processing (ICIP), Cairo, Egypt, 7–10 November 2009.
9. Cusano, C.; Napoletano, P.; Schettini, R. Evaluating color texture descriptors under large variations of controlled lighting conditions. *J. Opt. Soc. Am. A* **2016**, *33*, 17–30.
10. Razavian, A.S.; Azizpour, H.; Sullivan, J.; Carlsson, S. CNN features off-the-shelf: An astounding baseline for recognition. In Proceedings of the 2014 IEEE Conference on Computer Vision and Pattern Recognition Workshops (CVPRW), Columbus, OH, USA, 23–28 June 2014.
11. Russakovsky, O.; Deng, J.; Su, H.; Krause, J.; Satheesh, S.; Ma, S.; Huang, Z.; Karpathy, A.; Khosla, A.; Bernstein, M.; et al. ImageNet Large Scale Visual Recognition Challenge. *Int. J. Comput. Vis.* **2015**, *115*, 211–252.
12. Bianconi, F.; Harvey, R.; Southam, P.; Fernández, A. Theoretical and experimental comparison of different approaches for color texture classification. *J. Electron. Imaging* **2011**, *20*, 043006.
13. Palm, C. Color texture classification by integrative Co-occurrence matrices. *Pattern Recognit.* **2004**, *37*, 965–976.
14. Mäenpää, T.; Pietikäinen, M. Classification with color and texture: Jointly or separately? *Pattern Recognit.* **2004**, *37*, 1629–1640.
15. Seifi, M.; Song, X.; Muselet, D.; Tremeau, A. Color texture classification across illumination changes. In Proceedings of the Conference on Colour in Graphics, Imaging, and Vision, Joensuu, Finland, 14–17 June 2010.
16. Cusano, C.; Napoletano, P.; Schettini, R. Combining local binary patterns and local color contrast for texture classification under varying illumination. *J. Opt. Soc. Am. A* **2014**, *31*, 1453–1461.
17. Cusano, C.; Napoletano, P.; Schettini, R. Local Angular Patterns for Color Texture Classification. In *New Trends in Image Analysis and Processing – ICIAP 2015 Workshops*. Murino, V., Puppo, E., Sona, D., Cristani, M., Sansone, C., Eds.; Springer International Publishing: Cham, Switzerland, 2015; pp. 111–118.
18. Drimbarean, A.; Whelan, P. Experiments in colour texture analysis. *Pattern Recognit. Lett.* **2001**, *22*, 1161–1167.
19. Bianconi, F.; Fernández, A.; González, E.; Armesto, J. Robust color texture features based on ranklets and discrete Fourier transform. *J. Electron. Imaging* **2009**, *18*, 043012.
20. LeCun, Y.; Bengio, Y.; Hinton, G. Deep learning. *Nature* **2015**, *521*, 436–444.
21. Cimpoi, M.; Maji, S.; Vedaldi, A. Deep filter banks for texture recognition and segmentation. In Proceedings of the IEEE Conference on Computer Vision and Pattern Recognition, Boston, MA, USA, 7–12 June 2015.
22. Cusano, C.; Napoletano, P.; Schettini, R. Combining multiple features for color texture classification. *J. Electron. Imaging* **2016**, *25*, 061410.
23. Buchsbaum, G. A spatial processor model for object colour perception. *J. Frankl. Inst.* **1980**, *310*, 1–26.

24. Cardei, V.C.; Funt, B.; Barnard, K. White point estimation for uncalibrated images. In Proceedings of the Seventh Color Imaging Conference: Color Science, Systems, and Applications Putting It All Together, CIC 1999, Scottsdale, AZ, USA, 16–19 November 1999.
25. Van de Weijer, J.; Gevers, T.; Gijsenij, A. Edge-based color constancy. *IEEE Trans. Image Process.* **2007**, *16*, 2207–2214.
26. Forsyth, D.A. A novel algorithm for color constancy. *Int. J. Comput. Vis.* **1990**, *5*, 5–35.
27. Gehler, P.V.; Rother, C.; Blake, A.; Minka, T.; Sharp, T. Bayesian color constancy revisited. In Proceedings of the 2008 IEEE Conference on Computer Vision and Pattern Recognition, Anchorage, AK, USA, 23–28 June 2008.
28. Bianco, S.; Ciocca, G.; Cusano, C.; Schettini, R. Improving Color Constancy Using Indoor-Outdoor Image Classification. *IEEE Trans. Image Process.* **2008**, *17*, 2381–2392.
29. Bianco, S.; Cusano, C.; Schettini, R. Color Constancy Using CNNs. In Proceedings of the IEEE Conference on Computer Vision and Pattern Recognition Workshops (CVPRW), Boston, MA, USA, 7–12 June 2015.
30. Bianco, S.; Cusano, C.; Schettini, R. Single and Multiple Illuminant Estimation Using Convolutional Neural Networks. *IEEE Trans. Image Process.* **2017**, doi:10.1109/TIP.2017.2713044.
31. McCamy, C.S.; Marcus, H.; Davidson, J. A color-rendition chart. *J. App. Photog. Eng.* **1976**, *2*, 95–99.
32. ISO. *ISO/17321-1:2006: Graphic Technology and Photography – Colour Characterisation of Digital Still Cameras (DSCs) – Part 1: Stimuli, Metrology and Test Procedures*; ISO: Geneva, Switzerland, 2006.
33. Finlayson, G.D.; Drew, M.S. Constrained least-squares regression in color spaces. *J. Electron. Imaging* **1997**, *6*, 484–493.
34. Vrhel, M.J.; Trussell, H.J. Optimal scanning filters using spectral reflectance information. In *IS&T/SPIE's Symposium on Electronic Imaging: Science and Technology*. International Society for Optics and Photonics: San Jose, CA, USA; 1993; pp. 404–412.
35. Bianco, S.; Bruna, A.R.; Naccari, F.; Schettini, R. Color correction pipeline optimization for digital cameras. *J. Electron. Imaging* **2013**, *22*, 023014.
36. Bianco, S.; Bruna, A.; Naccari, F.; Schettini, R. Color space transformations for digital photography exploiting information about the illuminant estimation process. *J. Opt. Soc. Am. A* **2012**, *29*, 374–384.
37. Bianco, S.; Gasparini, F.; Schettini, R.; Vanneschi, L. Polynomial modeling and optimization for colorimetric characterization of scanners. *J. Electron. Imaging* **2008**, *17*, 043002.
38. Finlayson, G.D.; Mackiewicz, M.; Hurlbert, A. Root-polynomial colour correction. In Proceedings of the 19th Color and Imaging Conference, CIC 2011, San Jose, CA, USA, 7–11 November 2011.
39. Kang, H.R. *Computational Color Technology*; Spie Press: Bellingham, WA, USA, 2006.
40. Schettini, R.; Barolo, B.; Boldrin, E. Colorimetric calibration of color scanners by back-propagation. *Pattern Recognit. Lett.* **1995**, *16*, 1051–1056.
41. Kang, H.R.; Anderson, P.G. Neural network applications to the color scanner and printer calibrations. *J. Electron. Imaging* **1992**, *1*, 125–135.
42. Bianconi, F.; Fernández, A. An appendix to "Texture databases—A comprehensive survey". *Pattern Recognit. Lett.* **2014**, *45*, 33–38.
43. Hossain, S.; Serikawa, S. Texture databases—A comprehensive survey. *Pattern Recognit. Lett.* **2013**, *34*, 2007–2022.
44. Wyszecki, G.; Stiles, W.S. *Color Science*; Wiley: New York, NY, USA, 1982.
45. Anderson, M.; Motta, R.; Chandrasekar, S.; Stokes, M. Proposal for a standard default color space for the internet—sRGB. In Proceedings of the Color and imaging conference, Scottsdale, AZ, USA, 19–22 November 1996.
46. Ramanath, R.; Snyder, W.E.; Yoo, Y.; Drew, M.S. Color image processing pipeline. *IEEE Signal Process. Mag.* **2005**, *22*, 34–43.
47. Von Kries, J. Chromatic adaptation. In *Festschrift der Albrecht-Ludwigs-Universität*; Universität Freiburg im Breisgau: Freiburg im Breisgau, Germany, 1902; pp. 145–158.
48. Bianco, S.; Schettini, R. Computational color constancy. In Proceedings of the 2011 3rd European Workshop on Visual Information Processing (EUVIP), Paris, France, 4–6 July 2011.
49. Nayatani, Y.; Takahama, K.; Sobagaki, H.; Hashimoto, K. Color-appearance model and chromatic-adaptation transform. *Color Res. Appl.* **1990**, *15*, 210–221.

50. Bianco, S.; Schettini, R. Two new von Kries based chromatic adaptation transforms found by numerical optimization. *Color Res. Appl.* **2010**, *35*, 184–192.
51. Chatfield, K.; Simonyan, K.; Vedaldi, A.; Zisserman, A. Return of the devil in the details: Delving deep into convolutional nets. *arXiv preprint arXiv:1405.3531* **2014**, 1–11.
52. He, K.; Zhang, X.; Ren, S.; Sun, J. Deep residual learning for image recognition. In Proceedings of the IEEE Conference on Computer Vision and Pattern Recognition, Seattle, WA, USA, 27–30 June 2016.
53. Vedaldi, A.; Lenc, K. MatConvNet—Convolutional Neural Networks for MATLAB. *CoRR* **2014**. doi:10.1145/2733373.2807412.
54. Napoletano, P. Hand-Crafted vs Learned Descriptors for Color Texture Classification. In *International Workshop on Computational Color Imaging*. Springer: Berlin, Germany, 2017; pp. 259–271.
55. Zeiler, M.D.; Fergus, R. Visualizing and understanding convolutional networks. In *Computer Vision–ECCV 2014*; Springer: Berlin, Germany, 2014; pp. 818–833.
56. Sermanet, P.; Eigen, D.; Zhang, X.; Mathieu, M.; Fergus, R.; LeCun, Y. Overfeat: Integrated recognition, localization and detection using convolutional networks. *arXiv preprint arXiv:1312.6229* **2013** , 1–16.

Journal of
Imaging

MDPI

Article

Automatic Recognition of Speed Limits on Speed-Limit Signs by Using Machine Learning

Shigeharu Miyata

School of Engineering, Department of Robotics, Kindai University, Takayaumenobe 1, Higashihiroshima 739-2116, Japan; miyata@hiro.kindai.ac.jp; Tel.: +81-82-434-7000

Received: 30 May 2017; Accepted: 1 July 2017; Published: 5 July 2017

Abstract: This study describes a method for using a camera to automatically recognize the speed limits on speed-limit signs. This method consists of the following three processes: first (1) a method of detecting the speed-limit signs with a machine learning method utilizing the local binary pattern (LBP) feature quantities as information helpful for identification, then (2) an image processing method using Hue, Saturation and Value (HSV) color spaces for extracting the speed limit numbers on the identified speed-limit signs, and finally (3) a method for recognition of the extracted numbers using a neural network. The method of traffic sign recognition previously proposed by the author consisted of extracting geometric shapes from the sign and recognizing them based on their aspect ratios. This method cannot be used for the numbers on speed-limit signs because the numbers all have the same aspect ratios. In a study that proposed recognition of speed limit numbers using an Eigen space method, a method using only color information was used to detect speed-limit signs from images of scenery. Because this method used only color information for detection, precise color information settings and processing to exclude everything other than the signs are necessary in an environment where many colors similar to the speed-limit signs exist, and further study of the method for sign detection is needed. This study focuses on considering the following three points. (1) Make it possible to detect only the speed-limit sign in an image of scenery using a single process focusing on the local patterns of speed limit signs. (2) Make it possible to separate and extract the two-digit numbers on a speed-limit sign in cases when the two-digit numbers are incorrectly extracted as a single area due to the light environment. (3) Make it possible to identify the numbers using a neural network by focusing on three feature quantities. This study also used the proposed method with still images in order to validate it.

Keywords: pattern recognition; image processing; computer vision; traffic sign; machine learning; Local binary pattern; neural network

1. Introduction

As one part of research related to Intelligent Transport Systems (ITS), in Japan, much research has been conducted regarding Advanced cruise-assist Highway Systems (AHS), aiming to ensure the safety and smoothness of automobile driving [1–9]. AHS is composed of the following sections: information collection, support for driver operations, and fully automated driving. The means of information collection include the use of radar to detect obstacles in front, and camera images for detecting traffic signs and signals. Driving support systems using radar have already been commercialized and are installed in many vehicles. However, driving support systems using cameras remain in the research phase both for use on expressways and ordinary roads. Commercialization of these systems will require the following for accurate and stable provision of information about the vehicle surroundings to the driver: high processing speed, high recognition accuracy, detection of all detectable objects without omission, and robustness in response to changes in the surrounding environment.

It is said that when driving, a driver relies 80–90% on visual information for awareness of the surrounding conditions and for operating the vehicle. According to the results of investigations, large numbers of accidents involving multiple vehicles occur due to driver inattention which results in the driver not looking ahead while driving, distracted driving, or failure to stop at an intersection. Of such accidents, those that result in death are primarily due to excessive speed. Therefore, from the perspective of constructing a driver support system, it will be important to automatically detect the speed limit and provide a warning display or voice warning to the driver urging him/her to observe the posted speed limit when the vehicle is travelling too fast.

Traffic restriction signs such as "vehicle passage prohibited" and "speed limits" have a round external shape and are distinguished by a red ring around the outside edge. There have been many proposals regarding methods of using this distinguishing feature to extract traffic signs from an image [10–30]. Frejlichwski [10] proposed the Polar-Fourier greyscale descriptor, which applies the information about silhouette and intensity of an object, for extracting an object from a digital image. Gao et al. [11] proposed the novel local histogram feature representation to improve the traffic sign recognition performance. Sun et al. [12] proposed an effectiveness and efficient algorithm based on a relatively new artificial network, extreme learning machine in designing a feasible recognition approach. Liu et al. [13] proposed a new approach to tackle the traffic sign recognition problem by using the group sparse coding. Zaklouta et al. [14] proposed the real-time traffic sign recognition system using an efficient linear SVM with HOG features. Stallkamp et al. [15] compared the traffic sign recognition performance of humans to that of a state-of-art machine learning algorithm. Berger et al. [16] proposed a new approach for traffic sign recognition based on the virtual generalizing random access memory weightless neural networks (VG-RAM WNN). Gu et al. [18] proposed the dual-focal active camera system to obtain a high-resolution image of the traffic signs. Zaklouta et al. [19] proposed a real-time traffic sign recognition system under bad weather conditions and poor illumination. Ruta et al. [20] proposed novel image representation and discriminative feature selection algorithm in a three-stage framework involving detection, tracking and recognition. Chen et al. [22] proposed the reversible image watermarking approach that works on quantized Discrete Cosine Transform (DCT) coefficients. Zhang [23] proposed the method to pre-pure feature points by clustering to improve the efficiency of the feature matching. Zhang et al. [24] proposed a multiple description method based on fractal image coding in image transformation. Zin et al. [25] proposed a robust road sign recognition system under various illumination conditions to work for various types of circular and pentagonal road signs. Kohashi et al. [26] proposed the high speed and high accurate road sign recognition method which is realized by emitting near-infrared ray to respond to bad lighting environment. Yamauchi et al. [27] proposed the traffic sign recognition method by using string matching technique to match the signatures to template signatures. Matsuura, Uchimura et al. [28,30] proposed the algorithm for estimation of internal area in the circular road signs from color scene image. Yabuki et al. [29] proposed an active net for region detection and a shape extraction. Many recognition methods have been proposed using means such as template matching and improved template matching, or using applied genetic algorithms [30] and neural network [29]. A method using shape features [21], and an Eigen space method based on the KL transform [17] have also been proposed. These have generally delivered good results. However, before commercialization of these methods is possible, a number of problems arising during the actual processing will have to be resolved. These problems include the need to convert an extracted image of unknown size to an image that is approximately the same size as the template, the need to make corrections for rotation and displacement, and the large time that is required for these processing calculations.

The method for traffic sign recognition [21] previously proposed by the authors of this study was focused on extracting traffic signs in real time from a moving image as a step toward system commercialization. As a result, the recognition method proposed was a relatively simple method that was based on the geometric features of the signs and that was able to shorten the processing time. Because it extracted the geometric shape on the sign and then conducted recognition based on its aspect ratio, the method was not able to identify the numbers on a speed-limit sign, all of which have identical aspect ratios. However, in consideration of the fact that excessive speed is a major cause of fatal accidents, and because recognition of the speed-limit signs that play a major role in preventing fatal accidents and other serious accidents is more important, a speed-limit sign recognition method utilizing an Eigen space method based on the KL transform was proposed [17]. This method yielded a fast processing speed, was able to detect the targets without fail, and was robust in response to geometrical deformation of the sign image resulting from changes in the surrounding environment or the shortening distance between the sign and vehicle. However, because this method used only color information for detection of speed-limit signs, precise color information settings and processing to exclude everything other than the signs are necessary in an environment where many colors similar to the speed-limit signs exist, and further study of the method for sign detection is needed.

Aiming to resolve the problems with the method previously proposed, this study focuses its consideration on the following three points, and proposes a new method for recognition of the speed limits on speed-limit signs. (1) Make it possible to detect only the speed-limit sign in an image of scenery using a single process focusing on the local patterns of speed-limit signs. (2) Make it possible to separate and extract the two-digit numbers on a speed-limit sign in cases when the two-digit numbers are incorrectly extracted as a single area due to the light environment. (3) Make it possible to identify the numbers using a neural network without performing precise analysis related to the geometrical shapes of the numbers.

Section 2 describes a speed-limit sign detection method using the AdaBoost classifier based on image LBP feature quantities. Section 3 describes a method of reliably extracting individual speed numbers using an image in a HSV color space, and a method of classifying the indicated speed number image using a neural network that learned the probabilistic relationship between the feature vectors of the numbers and the classifications. Section 4 describes the discussion of this study and the remaining problem points.

2. Detection of Speed-Limit Signs

Figure 1 shows a landscape image which includes 40 km/h speed-limit signs installed above a road as seen looking forward from the driver's position during driving. Because one sign is at a distance in the image, it is too small and is difficult to recognize. However, there is another speed-limit sign in a position close to the center of the image. The two red shining objects in the center are traffic signals. The speed-limit sign is round in shape, with a red ring around the outer edge and blue numbers on a white background indicating the speed limit.

Figure 1. Typical landscape image that includes a speed-limit sign as seen during driving.

The image acquired from the camera uses the popular RGB. In order to obtain the hue information, the image is converted to the popular Hue, Saturation, Value (HSV) [31–33]. The resulting image is expressed as independent Hue, Saturation, and Value parameters, providing color information that is highly robust in terms of brightness. When the threshold value with respect to Hue H expressed at an angle of 0–2π is set to $-1/3\pi \leq H \leq 1/6\pi$, this yields the HSV image shown in Figure 2a. As shown in Figure 2b, when the image is converted to a binary image where only the regions where the Hue is within the set threshold value appear white and all other regions appear black, we can see that parts other than the sign have also been extracted.

(a) (b)

Figure 2. (**a**) Hue, Saturation and Value (HSV) image and (**b**) image of extracted $-1/3\pi \leq \mathrm{H} \leq 1/6\pi$ Hue regions when a threshold value is set for Hue (H) in order to extract the red ring at the outer edge of the sign.

Because regions other than the sign are included in the binary image as shown in Figure 2b, first at the labeling process for the extracted regions we remove the white regions that have small areas, and next we perform contour processing for the labeled regions. The roundness of the region is calculated from the area and peripheral length which are obtained as the contour information, and only the regions considered to be circular are extracted. The results are as shown in Figure 3.

Figure 3. Round traffic sign extracted based on the labeling process.

When the bounding box around the red circle is added based on the binary image of the extracted traffic sign (Figure 3), we obtain the coordinates of the top left and bottom right corners of the speed-limit sign. Based on these corners, we are able to extract only the part corresponding to the traffic sign from the landscape image (Figure 1). The extracted image is shown in Figure 4a. After converting this image to a HSV image (Figure 4b), and performing a binarization process by setting the threshold for the blue, the result is the extracted speed limit of 40 km/h, as can be seen in Figure 4c.

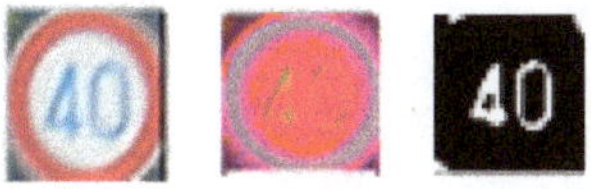

(a) (b) (c)

Figure 4. Extracted speed limit. (**a**) RGB image of Speed-limit sign extracted from the landscape image; (b) HSV image converted from RGB image (a); (**c**) Binarized image of image (**b**) by setting the threshold for the blue.

Outdoors, even when imaging the same scene, Hue, Saturation, and Value vary depending on the time, such as day or night, and when imaging the scene from different directions. As a result, it is necessary to use a processing method which considers these differences when extracting the signs. However, this requires making precise settings in advance to handle all of the various situations that may occur.

This study attempted to find a method of determining the speed-limit sign pattern (feature quantities) in the image by converting the image to a grayscale image so that the color differences are converted to differences in grayscale levels, instead of using the color information directly. The AdaBoost classifier [34] is used to detect the speed-limit signs based on feature quantities extracted from the image. The AdaBoost classifier is a machine learning algorithm that combines a number of simple classifiers (weak classifiers) with detection accuracy that is not particularly high to create a single advanced classifier (strong classifier).

Local binary pattern (LBP) feature quantities [35–37] are used as the feature quantities. LBP find the differences between the target pixel and the eight neighboring pixels, and assigns values of one if positive and zero if negative, and express the eight neighboring pixels as a binary number to produce a 256-grade grayscale image. These feature quantities have a number of characteristics, including the fact that they can be extracted locally from the image, are resistant to the effects from image lighting changes, can be used to identify the relationships with the surrounding pixels, and require little calculation cost.

200 images which contained speed-limit signs (positive images) and 1000 images which did not contain speed-limit signs (negative images) were used for training set. A testing set of 191 images was prepared separately from the training set. The positive images included images of a variety of conditions including images with clear sign colors, images with faded sign colors, and images of signs photographed at an angle. Examples of the positive and negative images are shown in Figures 5 and 6.

Figure 5. Examples of images containing speed-limit signs (positive images) in a variety of conditions.

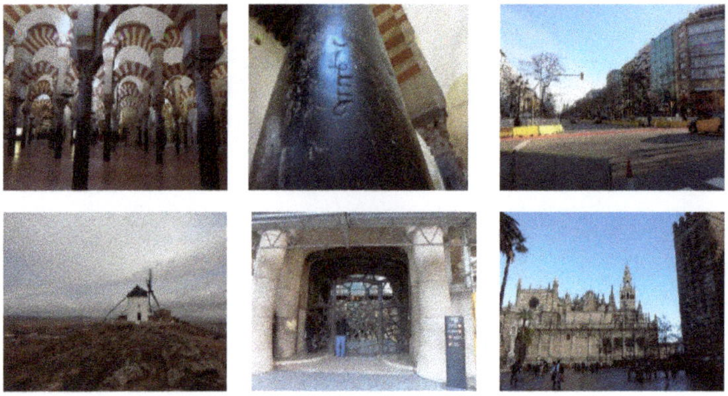

Figure 6. Examples of images not containing speed-limit signs (negative images).

Figure 7 shows the changes in the detection rate when the number of positive images was increased from 50 to 100, 150, and 200 for the same 1000 negative images. It shows that increasing the number of positive images increases the detection rate. Figure 8 shows the changes in the false detection rate (= (Number of detected areas that are not speed-limit signs)/Number of detected speed-limit signs)) when the number of negative images was increased from 100 to 7000. It shows that although the false detection rate dropped to 0 (%) with 1000 and 3000 negative images, a small number of false detections occurred when the number of negative images was 5000 or more. It is believed that using too many negative images results in excessive learning or conditions where learning cannot proceed correctly.

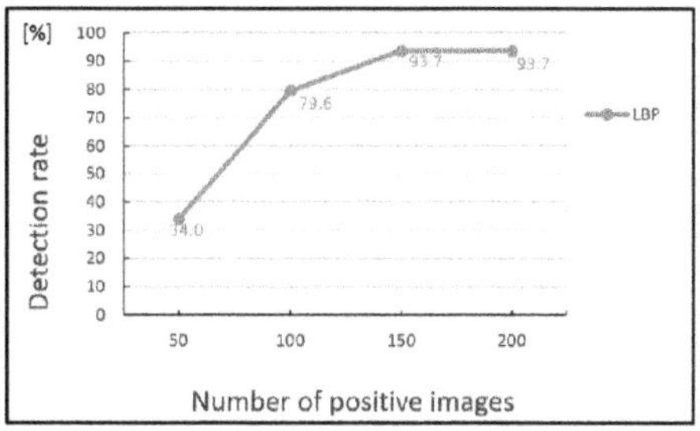

Figure 7. Change in detection rate when 1000 negative images were used and the positive images were increased from 50 to 200.

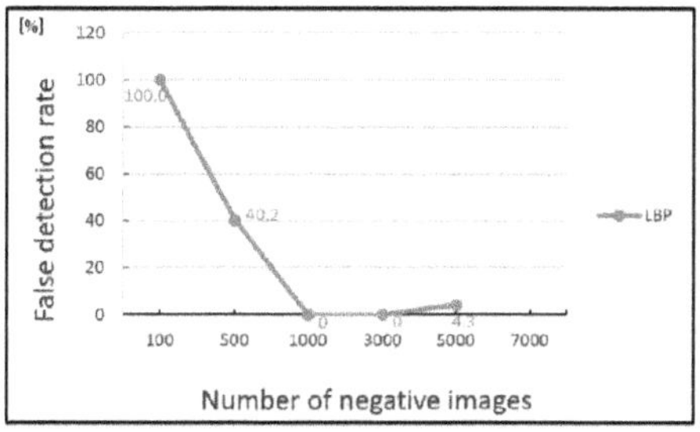

Figure 8. Changes in false detection rate when 200 positive images are used and the number of negative images is increased from 100 to 7000.

Figure 9 shows the detection results when an AdaBoost classifier that was created using 200 positive images and 1000 negative images was used with a testing set of 191 images that were different from the training set. The detection rate was 93.7(%) and the false detection rate was 0 (%). This means that out of the 191 images that this method was used with, there were 12 images in which the speed-limit sign was not detected, however, all 179 detected areas were actually speed-limit signs. Images a–g in Figure 9 show cases when the speed-limit sign was detected, and images (h) and (i) in Figure 9 show cases when the speed-limit sign was not detected.

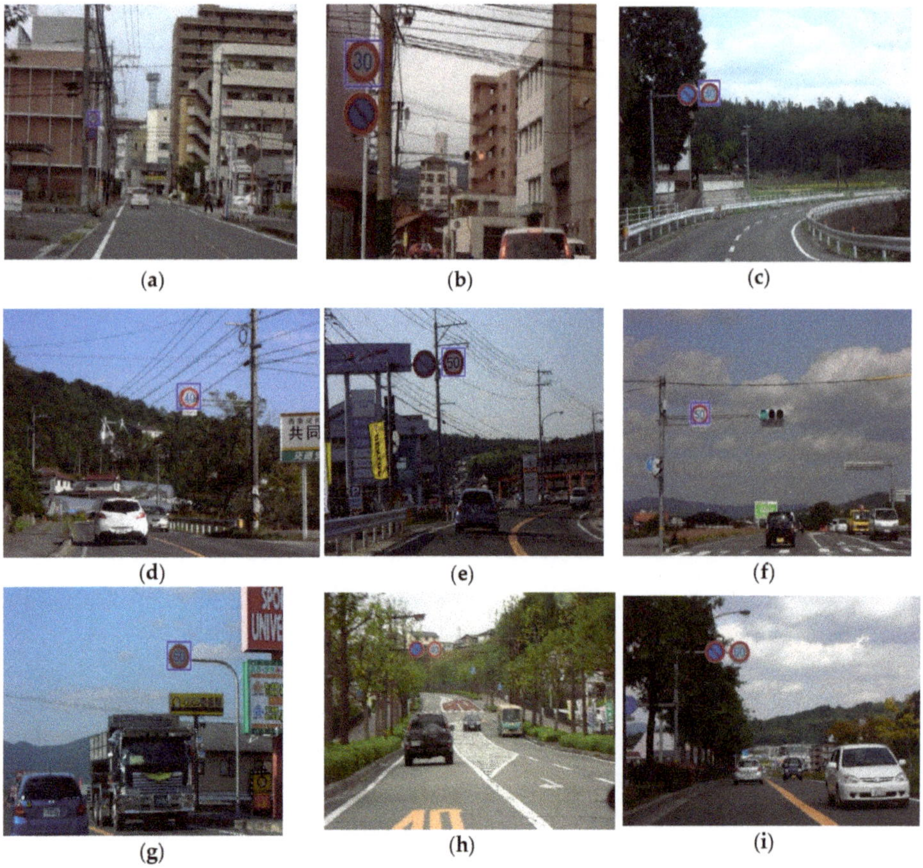

Figure 9. Examples of detected and undetected speed-limit signs. Images (**a**–**g**) show cases when the speed-limit sign was detected. Images (**h**) and (**i**) show cases when the speed-limit sign was not detected.

3. Recognition of Speed Limits on Speed-Limit Signs

The speeds on speed limit signs on ordinary roads in Japan are mostly speeds of 30 km, 40 km, and 50 km per hour. This section describes the method of recognizing the numbers on the speed-limit signs that were extracted using the method in Section 2.

The methods for identifying numbers and letters include a statistical approach and a parsing approach. The statistical approach is a method that evaluates the class to which the observed pattern most resembles based on the statistical quantities (average, variance, etc.) of feature quantities from a large quantity of data. The parsing approach considers the pattern of each class to be generated in accordance with rules, and identifies the class of the observed pattern by judging which of the class rules were used to generate it. Recently, the use of neural networks as a method of machine learning based on statistical processing makes it relatively simple to configure a classifier with good classifying performance based on learning data [38,39]. As a result, this method is frequently used for character recognition, voice recognition, and the other pattern recognition. It is said that a neural network makes it possible to acquire a non-linear identification boundary through learning, and delivers pattern recognition capabilities that are superior to previous technologies. Therefore, for the method in this study, a neural network was used to learn the probabilistic relationship between the number feature vectors and classes based on learning data, and to judge which class a target belonged to, being based on the unknown feature vectors acquired by measurement.

In this study, a 3-layer neural network was created, with the input layer, output layer, and intermediately layer each composed of four neurons.

The neural network learning data was prepared as follows. Individual numbers were extracted from images of speed-limit signs captured under a variety of conditions as shown in Figure 10, and were converted to grayscale images. The images were then resized to create four images: $5 \times 5, 10 \times 10$, 15×15, and 20×20. 35 images of the number 0, 20 images of the number 3, 35 images of the number 4, and 20 images of the number 5 were prepared.

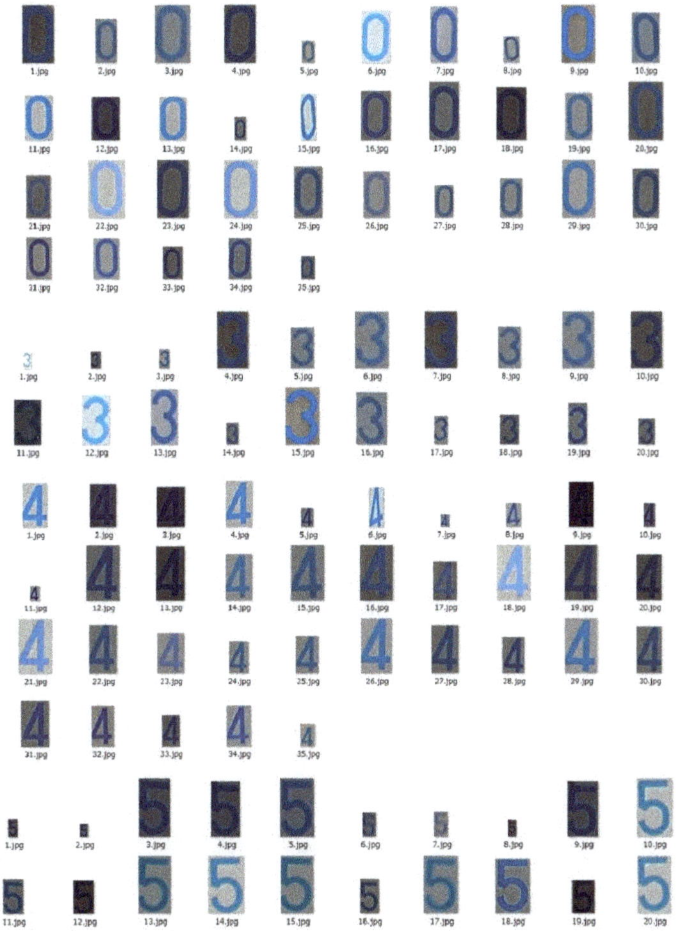

Figure 10. Training images for Number 0, 3, 4, 5 cut out from the images acquired from the camera.

For each image, the following three feature quantities were calculated to create the feature quantity file: brightness, horizontal and vertical projection. Examples of feature quantities for each number are shown in Figure 11.

Figure 11. Examples of three feature quantities for each number 0, 3, 4, and 5.

Using the fact that the numbers are printed on speed-limit signs in blue, the number areas are extracted based on color information. Because the Hue, Saturation, Value (HSV) color specification system is closer to human color perception than the red, green, blue (RGB) color specification system, the image is converted from a RGB image to a HSV image, and the Hue, Saturation, and Value parameters are adjusted to extract only the blue number. As shown in Figure 12, there are cases when the two-digit numbers cannot be separated, such as when the same parameters are used to process images acquired in different light environments, or when the image is too small. This study attempted to resolve this problem by using a variable average binarization process. If the image was acquired at a distance of 60–70 m from the sign, because the size of the image was 40 pixels or less, this problem was resolved by automatically doubling the image size before the number extraction process was performed. When the distance to the sign is approximately 70 m or more, the sign image itself is too small and it is not possible to extract the numbers.

Figure 12. When two numbers are combined and detected as a single area.

The following is a detailed description of the processing used to extract each of the numbers on speed-limit signs.

As shown in Figure 4a, a speed limit sign consists of a red circle on the periphery with the numbers printed in blue on a white background inside the circle. Rather than directly extracting the numbers from the blue area, the speed limit sign extracted from the color image is converted to a grayscale image, and that image is subjected to variable average binarization to produce the image shown in Figure 13a. The parts that are higher than a threshold for the Value inside and outside the red circle are converted to white, and all other parts are converted to black.

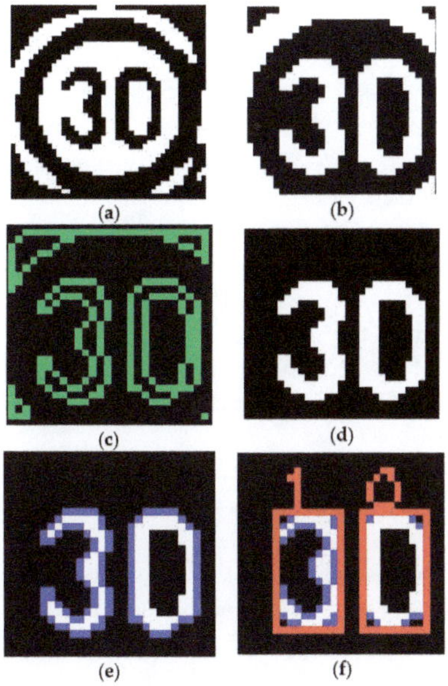

Figure 13. Detailed description of the processing used to extract the numbers on speed-limit signs. (**a**) Image following application of a variable average binarization processing to the grayscale image of the detected speed-limit sign; (**b**) Image with central area of image (**a**) extracted and black/white reversed; (**c**) Contours image of boundary between white areas and black areas in image (**b**); (**d**) Number candidate area obtained from mask processing of image (**b**) using a mask image obtained by using contours image (**c**); (**e**) Image following application of contour processing after dilation-erosion (closing) processing of the number candidate areas to eliminate small black areas inside the white objects; (**f**) Number image delivered to recognition processing when the contour area satisfies the aspect ratio of numbers.

Because the area of the speed-limit sign detected in the image is a square that contacts the outer edges of the red circle, the numbers are located approximately in the center of the square. In consideration of this, processing is focused only on the square area at the center. Since the surrounding areas contain large amounts of white area (noise), this means that the process for removing the noise can be omitted. Extracting the area near the center and performing reverse processing yields the image shown in Figure 13b.

Figure 13c shows the contours extracted from the borders between the white areas and black areas in Figure 13b. The areas inside the multiple extracted contours are calculated. For the number 0 part, there are both inner and outer contours. Therefore, with both the number 3 contour and the two number 0 contours, there are a total of three contours in the number candidate images, and they occupy a large area in the image. After acquiring the contours with the three largest areas and filling them in, this yields two filled-in areas. For these two areas, the centroid of each area is found and the vertical coordinate value (Y coordinate) of each is calculated. The number 3 and number 0 on the speed-limit sign are described using approximately the same vertical coordinate of position. Therefore, when the Y coordinates of the filled-in areas described above are approximately the same, the number 3 and number 0 areas are used as the mask image for the image in Figure 13b. This mask image is used to perform mask processing of Figure 13b to produce the number candidate areas shown in Figure 13d.

In addition, the dilation-erosion processing was performed to eliminate small black areas inside white objects in case that they exit in Figure 13d, and then a contour processing was performed to separate the areas again, producing Figure 13e. From this contour image (in this case, two contours), bounding rectangles with no inclination are found. Only the areas with aspect ratios close to those of the number 3 and number 0 are surrounded with the red bounding rectangles as shown in Figure 13f and are delivered to the recognition processing. The numbers 0 and 1 are assigned to indicate the delivery order to the recognition processing.

The area is divided using the rectangles contacting the peripheries of the number candidate areas as shown in Figure 13f. Each of these rectangular areas is separately identified as a number area. Therefore, the coordinate information for the number areas inside the speed limit sign is not used during the processing.

Following is an explanation of the application of this method to video. When image acquisition begins 150–180 m in front of the speed limit sign, detection begins in the distance range of 100–110 m. The distance at which recognition of the sign numbers begins is 70–90 m. Sufficient recognition was confirmed at 60–70 m. In this test, the processing time was on average 1.1 s/frame. The video camera image size was 1920 × 1080. The PC specs consisted of an IntelR CoreTM i7-3700 CPU @ 3.4 GHz and main memory of 8.00 GB.

At a vehicle speed of 50 km/h, a distance of 70 m is equivalent to the distance travelled in 5 s. Because the processing time for this method at the current stage is approximately 1 s, it is possible to inform the driver of the speed limit during that time.

In order to identify what number each input image corresponds to, the 3 feature quantities of brightness, vertical and horizontal projection were calculated for each separated and extracted number image. These were then input into a neural network that had completed learning. Number recognition was then performed at each of the image resolutions (5 × 5, 10 × 10, 15 × 15, 20 × 20). Images of the feature quantities at each resolution for the extracted numbers 30, 40, and 50 are shown in Figures 14–16. Identification results for 30 km, 40 km, and 50 km per hour speed limits are shown in Figures 17–19.

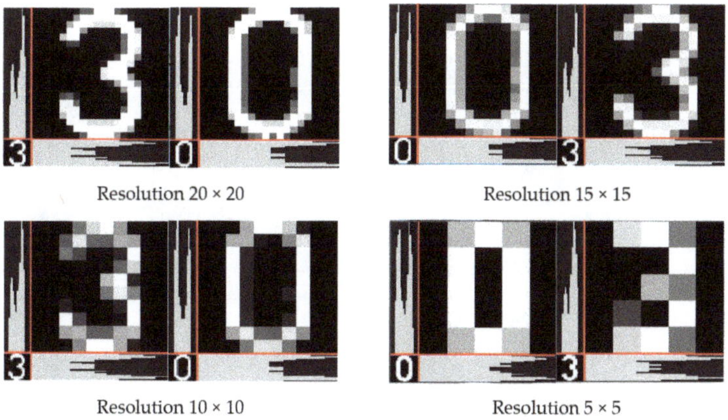

Figure 14. Three feature quantities at each resolution for the extracted number 30.

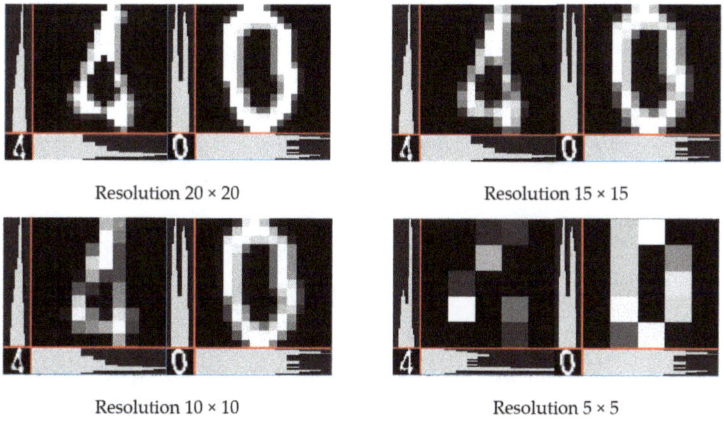

Figure 15. Three feature quantities at each resolution for the extracted number 40.

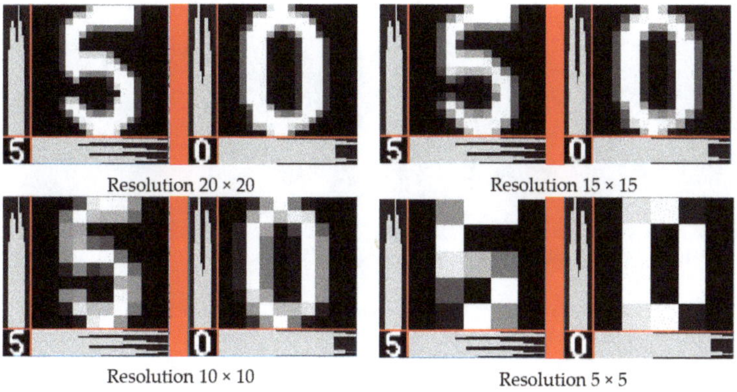

Figure 16. Three feature quantities at each resolution for the extracted number 50.

Resolution 20 × 20 (recognizing 00 for 30)

Resolution 15 × 15 (recognizing 30 for 30)

Resolution 10 × 10 (recognizing 30 for 30)

Resolution 5 × 5 (recognizing 30 for 30)

Figure 17. Example of recognition results for 30 km speed-limit sign.

Resolution 20 × 20 (recognizing 43 for 40)

Resolution 15 × 15 (recognizing 43 for 40)

Resolution 10 × 10 (recognizing 43 for 40)

Resolution 5 × 5 (recognizing 40 for 40)

Figure 18. Example of recognition results for 40 km speed-limit sign.

Resolution 20 × 20 (recognizing 55 for 50)

Resolution 15 × 15 (recognizing 53 for 50)

Resolution 10 × 10 (recognizing 53 for 50)

Resolution 5 × 5 (recognizing 50 for 50)

Figure 19. Example of recognition results for 50 km speed-limit sign.

Figure 17 shows the recognition results for the 30 km/h speed limit. When the resolution was 20 × 20, the 30 km/h speed limit was falsely recognized as 00 km/h. At all other resolutions, it was recognized correctly as 30 km/h.

Figure 18 shows the recognition results for the 40 km/h speed limit. It was recognized correctly as 40 km/h only at the 5 × 5 resolution. At other resolutions, it was falsely recognized as 43 km/h.

Figure 19 shows the recognition results for the 50 km/h speed limit. It was correctly recognized as 50 km/h at the 5 × 5 recognition. It was falsely recognized as 55 km/h at the 20 × 20 resolution, and 53 km/h at the 15 × 15 and 10 × 10 resolutions.

When the recognition conditions of 191 speed-limit signs were investigated, although there was almost no false recognition of 4 and 5 as other numbers, there were many cases of the numbers 0 and 3 each being falsely recognized as the other. Using the feature quantities in Figure 11, it was separately confirmed that when the Value of certain pixels in the number 3 is deliberately changed, it may be falsely recognized as 0. Conversely, when the same is done with the number 0 it may be falsely recognized as 3. Changing the Value of certain pixels means that depending on the threshold settings for Value, Saturation, and Hue when the sign numbers are extracted during actual processing, the grayscale Value of those pixels may vary. It is thought that this is the reason why the numbers 0 and 3 are falsely recognized as each other. When the lower limit values for Value and Saturation were set to 60 for the 38 images shown in Figure 20, then it was not possible to correctly extract the blue area numbers and false recognition occurred in the resolution 5 × 5 recognition results for 2.jpg, 10.jpg, 20.jpg, 26.jpg, and 36.jpg. For the 5 images where false recognition occurred, when the Saturation lower limit was set to 55 and the Value lower limit was set to 30, false recognition occurred in only 1 image (20.jpg). This shows that a slight difference in the threshold settings for Value, Saturation, and Hue will result in a small difference in the number feature quantities, and this has an effect on neural network recognition.

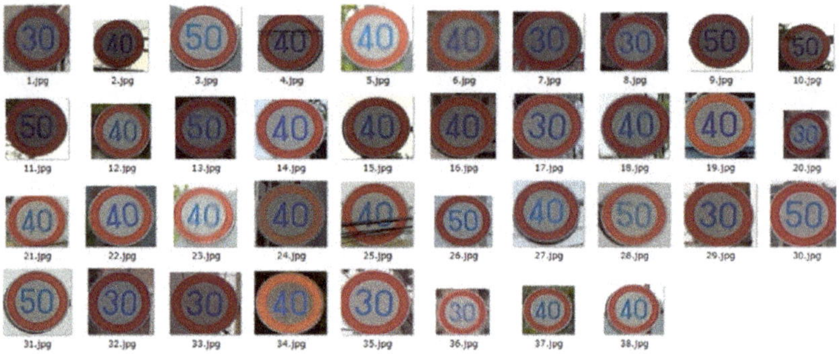

Figure 20. Effects of different lower limit values for Saturation and Value on the recognition results.

The speed recognition rate results for 191 extracted speed limit signs at each of the 4 resolutions (5×5, 10×10, 15×15, 20×20) are shown in Table 1.

Table 1. Recognition results for the four kinds of resolution.

Resolution (Pixels)	Recognition (%)
5×5	97.1
10×10	77.8
15×15	85.6
20×20	77.8

The methodology of this study involved identifying the four numbers 0, 3, 4, and 5 individually for speed limits 30 km/h, 40 km/h, and 50 km/h. Therefore, the recognition rate for a two-digit number was considered to be the recognition rate when both of the numbers were correctly identified, and this recognition rate was presented for each resolution. Because the numbers were recognized individually, there remained the problem of false recognition due to the effects of the extraction process for the number 0 and number 3. As a result, as shown in Figure 17, Figure 18, and Figure 19, speed limits which do not actually exist (00 km/h, 43 km/h, and 53 km/h) were indicated. When such speed limits that do not exist are present in the recognition results, they can be easily identified as false recognition. Finally, it will be necessary to improve the methodology by performing a repeated recognition process when such results occur.

It shows that the recognition rate was highest with the 5×5 resolution image, and decreased at higher resolutions. This is believed to be because when the resolution is higher, the discrimination ability increases and numbers can only be recognized when the feature quantities are a nearly perfect match. Conversely, when the resolution is lower, the allowable range for each feature quantity increases, increasing the invariability of the feature quantities. This means that identification at low resolutions is not largely affected by some degree of difference in feature quantities between the evaluation image and learning images.

4. Discussion

The recognition method previously proposed by the author and others had a number of problems that needed to be resolved, including the inability to recognize the speed printed on the signs, and the need for a process to eliminate all parts other than the detected speed sign because the method uses only color information. This study proposed a method which converts the color image to a grayscale image in order to convert color differences to differences in grayscale levels, and uses the resulting pattern information during the speed limit sign detection process. During the process for recognizing the speed printed on a speed limit sign, first the RGB color space is converted to a HSV color space, and a variable average binarization process is applied to that image, allowing the individual speed numbers printed in blue to be reliably extracted. A neural network is then used to recognize the speed numbers in the extracted number areas.

Detection of the speed limit signs was performed using an AdaBoost classifier based on image LBP feature quantities. Feature quantities that can be used include Haar-like feature quantities, HOG feature quantities, and others, however, LBP feature quantities are the most suitable for speed limit sign detection because the LPB feature quantity learning time is shortest and because the number of false detection target types is smaller than with other feature quantities.

For number recognition, four image resolutions (5×5, 10×10, 15×15, 20×20) were prepared, and of these the 5×5 resolution showed the highest recognition rate at 97.1%. Although the 5×5 resolution is lower than the 20×20 resolution, it is thought that false detections were fewer because the feature quantities for each number have a certain invariability at lower resolutions. The percentage in which the speed limit sign was detected and the number part of the sign was correctly identified was 85%. Based on this, it will be necessary to improve the speed limit sign detection accuracy in order to improve the overall recognition rate.

The remaining issues include cases when the speed limit sign is not detected using only the feature quantities used in this study, and cases when number extraction was not successful due to signs with faded colors, signs that were backlit, or signs that were partially hidden by trees or other objects. In the future, it will be necessary to improve detection and recognition for the issues above. Because the processing time at the current stage is 1.1 s, and is too long for appropriately informing the driver, steps to shorten the processing time will also be necessary. It is possible that the use of a convoluted neural network or similar means may result in further improvement, and it will be necessary to perform comparison studies to determine the relative superiority compared to the method in this study. It is thought that improvements to these issues will be of increasing utility in driver support when automated driving is used.

Acknowledgments: I would like to thank Kenji OKA who provides great support for this study and to everyone who discussed this study at the conference or the meeting.

Author Contributions: Shigehrau Miyata contributed to the design of the study, and to write the initial draft and the final version of the manuscript. I have contributed to data collection, data analysis and interpretation in the preparation of the manuscript. I am accountable for all aspects of the work in ensuring that questions related to the accuracy or integrity of any part of the work are appropriately investigated and resolved.

Conflicts of Interest: I or my institution at any time did not receive or services from a third party (government, commercial, private foundation, etc.) for any aspect of the submitted work (including but not limited to grants, data monitoring board, study design, manuscript preparation, statistical analysis, etc.).

References

1. Yamaguchi, H.; Kojima, A.; Miyamoto, T.; Takahashi, H.; Fujunaga, K. A Robust Road Sign Recognition by Superposing Extracted Regions from Successive Frame. *IEICE Trans. D-II* **2007**, *90*, 494–502. (In Japanese)
2. Uchida, S.; Iwamura, M.; Omach, S.; Kise, K. Category Data Embedding for Camera-Based Character Recognition. *IEICE Trans. D-II* **2006**, *89*, 344–352. (In Japanese)
3. Kimura, S.; Tano, H.; Sakamoto, T.; Shiroma, H.; Ohashi, Y. Traversal Region and Obstacle Detection Using Tilted-Plane-Based Stereo Vision. *Robot. Soc. Jpn.* **2005**, *23*, 59–66. (In Japanese)
4. Mo, G.; Aoki, Y. A Recognition Method for Traffic Sign in Color Image. *IEICE Trans. D-II* **2004**, *87*, 2124–2135. (In Japanese)
5. Doermann, D.; Liang, J.; Li, H. Progress in Camera-Based Document Image Analysis. In Proceedings of the Seventh International Conference on Document Analysis and Recognition, Edinburgh, UK, 6 August 2003; pp. 606–616.
6. Hsu, S.; Huang, C. Road Sign Detection and Recognition Using Matching Pursuit Method. *Image Vis. Comput.* **2001**, *19*, 119–129. [CrossRef]

7. Uchimura, K.; Wakiyama, S.; Fujino, M. Extraction of Circular Traffic Sign Using Limited Color Indication. *IEICE Trans. D-II* **2000**, *83*, 855–858. (In Japanese)
8. Gavrila, D.M. Multi-Feature Hierarchical Template Matching Using Distance Transforms. In Proceedings of the 19th International Conference on Pattern Recognition (ICPR 2008), Tampa, FL, USA, 8–11 Decmber 1998; pp. 439–444.
9. Piccioli, G.; de Michieli, E.; Parodi, P.; Campani, M. Robust Method for Road Sign Detection and Recognition. *Image Vis. Comput.* **1996**, *14*, 209–223. [CrossRef]
10. Frejlichwski, D. Application of the Polar-Fourier Greyscale Descriptor to the Automatic Traffic Sign Recognition. In Proceedings of the International Conference Image Analysis and Recognition (ICIAR), Niagara Falls, Canada, 22–24 July 2015; pp. 506–513.
11. Gao, J.; Fang, Y.; Li, X. Learning Local Histogram Representaion for Efficient Traffic Sign Recognition. In Proceedings of the 2015 8th International Congress on Image and Signal Processing (CISP 2015), Shenyang, China, 14–16 October 2015; pp. 631–635.
12. Sun, Z.-L.; Wang, H.; Lau, W.-S.; Seet, G.; Wang, D. Application of BW-ELM model on traffic sign recognition. *Neurocomputing* **2014**, *128*, 153–159. [CrossRef]
13. Liu, H.; Liu, Y.; Sun, F. Traffic sign recognition using group sparse coding. *Inf. Sci.* **2014**, *266*, 75–89. [CrossRef]
14. Zaklouta, F.; Stanciulescu, B. Real-time traffic sign recognition in three stages. *Rob. Auton. Syst.* **2014**, *62*, 16–24. [CrossRef]
15. Stallkamp, J.; Schlipsing, M.; Salmen, J.; Igel, C. Man vs. computer: Benchmarking machine learning algorithms for traffic sign recognition. *Neural Netw.* **2012**, *32*, 323–332. [CrossRef] [PubMed]
16. Berger, M.; Forechi, A.; De Souza, A.F.; Neto, J.D.O. Traffic Sign Recognition with VG-RAM Weightless Neural Networks. In Proceedings of the 2012 12th International Conference on Intelligent Systems Design and Applications (ISDA 2012), Kochi, India, 27–29 November 2012; pp. 315–319.
17. Miyata, S.; Ishikawa, T.; Nakamura, H.; Takehara, S. Traffic Sign Recognition utilizing an Eigen Space Method Based on the KL Transform. *Adv. Mater. Res.* **2012**, *452–453*, 876–882. [CrossRef]
18. Gu, Y.; Yendo, T.; Tehrani, M.P.; Fujii, T.; Tanimoto, M. Traffic sign detection in dual-focal active camera system. In Proceedings of the 2011 IEEE Intelligent Vehicles Symposium (IV 2011), Baden-Baden, Germany, 5–9 June 2011; pp. 1054–1059.
19. Zaklouta, F.; Stanciulescu, B. Real-time traffic sign recognition using spatially weighted HOG trees. In Proceedings of the 2011 15th International Conference on Advanced Robotics (ICAR 2011), Tallinn, Estonia, 20–23 June 2011; pp. 61–66.
20. Ruta, A.; Li, Y.; Liu, X. Real-time traffic sign recognition from video by class-specific discriminative features. *Pattern Recognit.* **2010**, *43*, 416–430. [CrossRef]
21. Miyata, S.; Yanou, A.; Nakamura, H.; Takehara, S. Road Sign Feature Extraction and Recognition sing Dynamic Image Processing. *Innov. Comput. Inf. Control* **2009**, *5*, 4105–4113.
22. Chen, C.C.; Kao, D.S. DCT-Based Zero Replacement Reversible Image Watermarking Approach. *Innov. Comput. Inf. Control* **2008**, *4*, 3027–3036.
23. Zhang, X. Improvement of a Feature-Based Image Mosaics Algorithm. *Innov. Comput. Inf. Control* **2008**, *4*, 2759–2764.
24. Zhang, Z.; Ahao, Y. Multiple Description Image Coding Based on Fractal. *Innov. Comput. Inf. Control* **2007**, *3*, 1615–1623.
25. Zin, T.T.; Hama, H. A Robust Road Sign Recognition Using Segmentation with Morphology and Relative Color. *Inst. Image Inf. Telev. Eng.* **2005**, *59*, 1333–1342. [CrossRef]
26. Kohashi, Y.; Ishikawa, N.; Nakajima, M. *Automatic Recognition of Road Signs in Bad Lighting Environment*; IEICE Techical Report; IEICE (Institute of Electronics, Information and Communication Engineers): Tokyo, Japan, 2003; Volume 103, pp. 57–62. (In Japanese)
27. Yamauchi, H.; Takahashi, H. A Road Sign Recognition Technique by Tracing Outline Vectors. *Inst. Image Inf. Telev. Eng.* **2003**, *57*, 847–853. [CrossRef]
28. Matsuura, D.; Yamauchi, H.; Takahashi, H. Extracting Circular Road Sign Using Specific Color Distinction and Region Limitation. *Syst. Comput. Jpn.* **2002**, *38*, 90–99. [CrossRef]

29. Yabuki, N.; Matsuda, Y.; Kataoka, D.; Sumi, Y.; Fukui, Y.; Miki, S. *A Study on an Automatic Stop of Computation in Active Net*; Technical Report; IEICE (Institute of Electronics, Information and Communication Engineers): Tokyo, Japan, 2000; Volume 99, pp. 69–76. (In Japanese)
30. Uchimura, K.; Kimura, H.; Wakiyama, S. Extraction and Recognition of Circular Road Signs Using Road Scene Color Image. *IEICE Trans. A* **1998**, *J81–A*, 546–553. (In Japanese)
31. Rehrmann, V.; Priese, L. Fast and Robust Segmentation of Natural Color Scenes. In Proceedings of the 3rd Asian Conference on Computer Vision, Hongkong, China, 8–10 January 1998; pp. 598–606.
32. Zhang, Y.J.; Yao, Y.R. Color Image Segmentation Based on HSI model. *High Technol. Lett.* **1998**, *4*, 28–31.
33. Miyahara, M.; Yoshida, Y. Mathematical Transform of (R,G,B) Color Space to Munsel (H,V,C) Color Data. *Proc. SPIE* **1988**. [CrossRef]
34. Schapire, R.E. Explaining AdaBoost. In *Empirical Inference*; Springer: Berlin/Heidelberg, Germany, 2013; pp. 37–52.
35. Takala, V.; Ahonen, T.; Pietikainen, M. Block-Based Methods for Image Retrieval Using Local Binary Patterns. In Proceedings of the SCIA 2005, Joensuu, Finland, 19–22 June 2005; pp. 882–891.
36. Ojala, T.; Pietikainen, M.; Maenpaa, T. Multiresolution Gray-Scale and Rotaion Invariant Texture Classification with Local Binary Pattern. *IEEE Trans. Pattern Anal. Mach. Intell.* **2002**, *24*, 971–987. [CrossRef]
37. Ojala, T.; Pietikainen, M.; Hardwood, D. A Comparative Study of Texture Measures with Classification Based on Feature Distribution. *Pattern Recognit.* **1996**, *29*, 51–59. [CrossRef]
38. Kecman, V. *Learning and Soft Computing*; The MIT Press: Cambridge, MA, USA, 2001.
39. Bishop, C.M. *Neural Networks for Pattern Recognition*; Oxford University Press: Oxford, UK, 1995.

Journal of
Imaging

MDPI

Article

Color Consistency and Local Contrast Enhancement for a Mobile Image-Based Change Detection System

Marco Tektonidis * and David Monnin

Advanced Visionics and Processing Group, French-German Research Institute of Saint-Louis,
5 rue du Général Cassagnou, 68300 Saint-Louis, France; David.MONNIN@isl.eu
* Correspondence: tektonidis@gmail.com; Tel.: +33-(0)389-69-5161

Received: 30 June 2017; Accepted: 8 August 2017; Published: 23 August 2017

Abstract: Mobile change detection systems allow for acquiring image sequences on a route of interest at different time points and display changes on a monitor. For the display of color images, a processing approach is required to enhance details, to reduce lightness/color inconsistencies along each image sequence as well as between corresponding image sequences due to the different illumination conditions, and to determine colors with natural appearance. We have developed a real-time local/global color processing approach for local contrast enhancement and lightness/color consistency, which processes images of the different sequences independently. Our approach combines the center/surround Retinex model and the Gray World hypothesis using a nonlinear color processing function. We propose an extended gain/offset scheme for Retinex to reduce the halo effect on shadow boundaries, and we employ stacked integral images (SII) for efficient Gaussian convolution. By applying the gain/offset function before the color processing function, we avoid color inversion issues, compared to the original scheme. Our combined Retinex/Gray World approach has been successfully applied to pairs of image sequences acquired on outdoor routes for change detection, and an experimental comparison with previous Retinex-based approaches has been carried out.

Keywords: color consistency; local contrast enhancement; Retinex; change detection

1. Introduction

Detecting changes in outdoor scenes is of high importance for video surveillance and security applications, allowing for identifying suspicious objects or threats such as IEDs (improvised explosive devices) in route clearance operations or detecting intrusions in secured areas. Image-based change detection systems on mobile platforms such as vehicles, unmanned aerial vehicles (UAVs), or unmanned ground vehicles (UGVs), can improve the detection of changes on a route of interest between a reference and the current time point. Based on the feature correspondences between the two time points, frames of a reference and a current image sequence that correspond to the same scene are aligned using a registration approach and can be alternately displayed on a monitor [1]. The alternate display of reference and current frames allows for visualizing changes, which can be evaluated by a human operator.

For the alternate display of color images in a change detection system, it is required to enhance *local contrast* for improving the visibility in image areas with low contrast (e.g., shadows), and to compensate *lightness and color inconsistencies* due to different illumination conditions (e.g., see Figure 1) in consecutive frames of an image sequence (i.e., achieving intra-sequence consistency), as well as in corresponding frames of a reference and a current image sequence that depicts the same scene (i.e., achieving inter-sequence consistency). In particular, achieving inter-sequence lightness/color consistency allows unchanged areas of corresponding frames of a reference and a current image sequence to appear constant during the alternate display, whereas changes are blinking and are thus better detectable by the observer. In addition, natural *color rendition* of the computed colors is required to facilitate the scene understanding. Color rendition describes the fidelity of colors w.r.t. the scene and the human color perception [2]. However, simultaneously achieving intra-sequence and inter-sequence color consistency as well as color rendition in real time is a challenging task. *Color constancy* approaches compensate the colors of a light source and can be used to achieve color consistency for images acquired under different illumination conditions. Most color constancy approaches (e.g., pre-calibrated, Gamut-based, statistical-based, or machine learning approaches) are either computationally expensive, unstable, or require sensor calibration [3]. On the other hand, certain *Retinex-based* approaches that have been previously used for lightness/color constancy, as well as for image enhancement, are automatic, robust, and suitable for real-time applications.

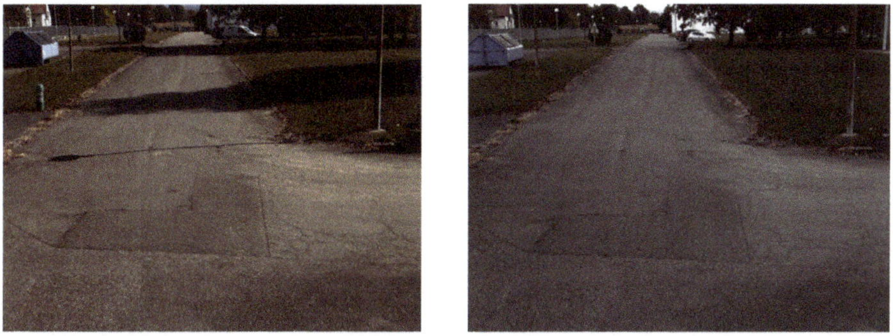

Figure 1. Two images of the same scene acquired at different time points under different illumination conditions.

Retinex was proposed by Land as a model of the human vision system in the context of lightness and color perception [4]. The Retinex model was initially based on a random walk concept to compute the global maximum estimates of each chromatic channel and to determine the relative lightness of surfaces of an image [5]. In [6], Land proposed determining the relative lightness using a local center/surround approach that computes the ratio of the intensity of a pixel and the average intensity of the surrounding. Jobson et al. [2] extended the center/surround Retinex with a surround Gaussian function and a log operation, achieving local contrast enhancement and lightness/color constancy simultaneously; however, the color rendition was often poor. To improve color rendition, Jobson et al. [7] proposed a nonlinear color processing function to partially restore the original colors. In [8], Barnard and Funt presented a luminance-based Retinex model to preserve the chromaticity of the original image. However, the use of the original color information in [7,8] affects color constancy. Recent extensions of the Retinex model have been mainly proposed for color enhancement (e.g., [9,10]). For a detailed discussion on Retinex models, we refer to, for example, [11,12]. Moreover, variational approaches related to the original Retinex formulation have been proposed for local contrast enhancement and color enhancement or color correction (e.g., [13,14]).

In this work, we have developed an automatic real-time local contrast enhancement and lightness/color consistency approach for the display of images in a mobile change detection system. Our approach combines the center/surround Retinex model [2] and the Gray World hypothesis [15] using a nonlinear color processing function previously used for color restoration [7]. Frames of the different image sequences are processed independently, and thus an image with reference colors is not required. Our approach takes advantage of the combination of *local* and *global* color processing: Retinex allows for locally achieving lightness/color consistency in unchanged areas of corresponding images regardless of changes in the depicted scenes, and the global Gray World allows for improving color rendition compared to Retinex. We have extended the gain/offset function used by the center/surround Retinex to reduce the halo effect on shadow boundaries. In our approach, we have applied the gain/offset function before the color processing function, avoiding color inversion issues compared to the original scheme in [7]. For Gaussian convolution, we employed stacked integral images (SII) [16] to reduce computation times and to allow the display of more than 20 frames per second in the change detection system. A previous version of our approach has been presented in [17]. We have performed an experimental evaluation of our approach based on image sequences of outdoor routes acquired by a vehicle-mounted color camera. In addition, a comparison with previous Retinex-based approaches in the context of color rendition, inter-sequence color consistency, and change detection has been conducted.

2. Methods

In this section, we first describe the employed center/surround Retinex model and introduce an extended gain/offset function. Then, we describe the Gray World hypothesis and present the nonlinear color processing function combining Retinex and Gray World. Finally, we describe the stacked integral images (SII) scheme for efficient Gaussian convolution.

2.1. Center/Surround Retinex Using an Extended Gain/Offset Function

Our image processing approach is based on the single-scale center/surround Retinex model [2], previously used for local image enhancement and lightness/color constancy and processes each channel of an RGB image independently. For each channel I_i ($i \in \{R, G, B\}$) and for each position $\mathbf{x} = (x, y)$ of an image, a Retinex value $R_i(\mathbf{x}, c)$ is computed:

$$R_i(\mathbf{x}, c) = \log(I_i(\mathbf{x})) - \log(F(\mathbf{x}, c) * I_i(\mathbf{x})), \tag{1}$$

where $I_i(\mathbf{x})$ denotes the intensity of the channel I_i at position $\mathbf{x}$, $*$ is the symbol for spatial convolution, and $F(\mathbf{x}, c)$ is the surround function defined by a Gaussian function:

$$F(\mathbf{x}, c) = Ke^{-(x^2+y^2)/c^2}, \tag{2}$$

where c denotes the scale of the Gaussian function and K is a value such that $\int F(\mathbf{x}, c)\, d\mathbf{x} = 1$.

The computed Retinex describes the logarithm of the ratio of the intensity of a pixel and the local mean intensity defined by the surround function, since Equation (1) can be rewritten to $R_i(\mathbf{x}, c) = \log\left(\frac{I_i(\mathbf{x})}{F(\mathbf{x},c)*I_i(\mathbf{x})}\right)$. The intensity ratios allows for locally enhancing details, in particular for image areas with uniform intensities and low contrast (e.g., shadows). Simultaneously, the intensity ratios describe the reflectance of surfaces, and can be used for compensating the illumination in a scene.

The scale c of the surround function in Equation (2) represents a trade-off between the level of local contrast enhancement and color rendition [2]. Decreasing c stronger enhances details in an image, and increasing c improves color rendition. The value of c cannot be determined automatically, and should be chosen based on the image data and the application. For larger image areas with uniform colors, Retinex computes colors with low saturation (i.e., colors close to gray) regardless of the colors in the original image.

To map the computed Retinex values from the logarithmic domain to output intensity values $I_i^{Ret}(\mathbf{x})$, we employ a gain/offset function [2]:

$$I_i^{Ret}(\mathbf{x}) = A + CR_i(\mathbf{x}, c), \tag{3}$$

where A denotes the offset and C is the gain that regulates the contrast of the output image. For example, increasing C increases the contrast; however, the halo effect is increased as well. The function in Equation (3) in conjunction with Equation (1) describes a transfer function that maps the local intensity ratios $\frac{I_i(\mathbf{x})}{F(\mathbf{x},c)*I_i(\mathbf{x})}$ to intensities of the output image (see Figure 2). Note that negative values in Equation (3) are clipped to zero, and values larger than 255 are clipped to 255 (for 8-bit images).

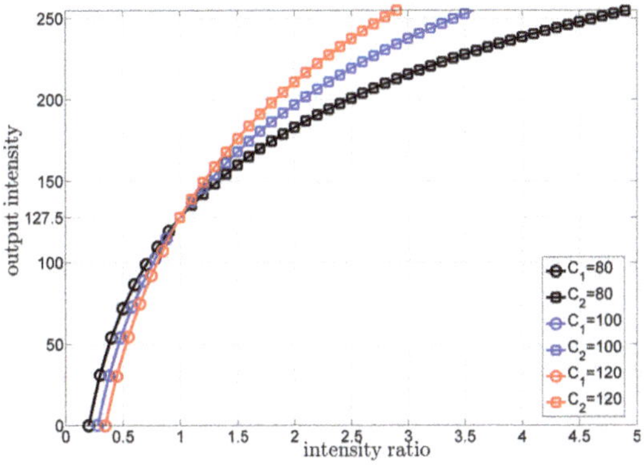

Figure 2. Intensity transfer function for the Retinex model using a gain/offset function with $A = 127.5$ and two contrast gains (C_1 and C_2). For $C_1 = C_2$, the transfer function corresponds to the gain/offset function with a single gain value.

(**a**) (**b**)

Figure 3. Processed images using (**a**) the original Retinex [2] with a single gain and (**b**) Retinex [2] extended with the proposed two gains. Enlarged sections for the marked regions are shown on the bottom. (**a**) $C = 120$; (**b**) $C_1 = 80, C_2 = 120$.

In our approach, we use two different gain values C_1 and C_2. For each pixel, we choose among the two gain values based on the value of the intensity ratio $\frac{I_i(\mathbf{x})}{F(\mathbf{x},c)*I_i(\mathbf{x})}$:

$$C = \begin{cases} C_1, & \text{if } \frac{I_i(\mathbf{x})}{F(\mathbf{x},c)*I_i(\mathbf{x})} \leq 1, \\ C_2, & \text{otherwise.} \end{cases} \tag{4}$$

The use of two different gain values allows for regulating the steepness of the upper and the lower part of the intensity transfer function independently. A lower value for C_1 compared to C_2 reduces the halo effect on shadow boundaries (cf. Figure 3a,b), while retaining the number of bright pixels and thus avoiding the decrease of contrast. Note that the transfer function in Equation (3) is continuous at $\frac{I_i(\mathbf{x})}{F(\mathbf{x},c)*I_i(\mathbf{x})} = 1$, since $I_i^{Ret}(\mathbf{x}) = A$ for any value of C.

2.2. Gray World

In our approach, we also employ the Gray World hypothesis [15]. This global color constancy method is based on the assumption that the average reflectance in a scene is achromatic, thus the average intensity values for the three RGB channels should be equal (gray) under a neutral illuminant (light source). Based on this assumption, any deviation of the average colors of an image from gray is due to the colors of the illuminant. This deviation can be used to compute the colors $I_i^{GW}(\mathbf{x})$ of a scene under a neutral illuminant:

$$I_i^{GW}(\mathbf{x}) = \frac{I_{gray}}{\overline{I}_i} I_i(\mathbf{x}), \tag{5}$$

where I_{gray} denotes the defined gray value and $\overline{I}_i$ is the average intensity of each channel.

2.3. Color Processing Function Combining Retinex and Gray World

To combine the Retinex model with the Gray World hypothesis, we employed a nonlinear color processing function previously used for color restoration [7]. This function maps the colors computed using Gray World (Equation (5)) to the Retinex result. For each channel and for each image position, we compute a modified Retinex value $R_i^{GW}(\mathbf{x})$:

$$R_i^{GW}(\mathbf{x}) = C_i^{GW}(\mathbf{x})\, I_i^{Ret}(\mathbf{x}), \tag{6}$$

where $I_i^{Ret}(\mathbf{x})$ denotes the computed Retinex value after applying the gain/offset function in Equation (3) and $C_i^{GW}(\mathbf{x})$ is a nonlinear color processing function defined by:

$$C_i^{GW}(\mathbf{x}) = \beta(\mathbf{x}) \left(\log\left(\alpha I_i^{GW}(\mathbf{x})\right) - \log\left(\sum_{j\in R,G,B} I_j^{GW}(\mathbf{x}) \right) \right), \tag{7}$$

where α ($\alpha > 1$) denotes the nonlinearity strength that controls the influence of Gray World, and $\beta(\mathbf{x})$ controls the brightness levels. Note that Equation (6) is not used for pixels with zero brightness, i.e., $\sum_i I_i^{Ret}(\mathbf{x}) = 0$ for $i \in R, G, B$. We determine the value for $\beta(\mathbf{x})$ automatically for each pixel, such that using the color processing function in Equation (6) preserves the brightness of the Retinex result in Equation (3), i.e., $\sum_i R_i^{GW}(\mathbf{x}) = \sum_i I_i^{Ret}(\mathbf{x})$:

$$\beta(\mathbf{x}) = \frac{\sum_i I_i^{Ret}(\mathbf{x})}{\sum_i \left(\left(\log\left(\alpha I_i^{GW}(\mathbf{x})\right) - \log\left(\sum_{j\in R,G,B} I_j^{GW}(\mathbf{x}) \right) \right) I_i^{Ret}(\mathbf{x}) \right)}. \tag{8}$$

The differences of using the color processing function in Equation (6) compared to the original scheme in [7] are three. First, we use the colors computed by Gray World, instead of the original colors. Note that also different global constancy approaches can be used in Equation (7), for example, Shades of Gray [18] or Gray Edge [19]. Second, we compute the value of $\beta(\mathbf{x})$ in Equation (7) automatically for each pixel, compared to using a fixed value in [7]. This guarantees that the color processing function in Equation (6) is brightness-preserving and only modifies the chromaticities, compared to the result of the original Retinex [2] in Equation (3). Third, we apply the gain/offset function (Eqation (3)) before the color processing function, in contrast to [7] where the order of the two operations is reversed (i.e., we use $I_i^{Ret}(\mathbf{x})$ in Equation (6), instead of $R_i(\mathbf{x}, c)$). We found that this prevents from color conversion issues for images that contain saturated colors, compared to the original color processing scheme [7,20]. In Figure 4, we show an example of an image that contains pixels with saturated colors, for example, pixels of the blue balls with $I_R = 0$ (Figure 4a). For such pixels, the Retinex result in Equation (1) as well as the result of the color processing function in Equation (7) are negative. Following the original scheme and applying the color processing function directly on the Retinex result yields a positive value, which is mapped by the gain/offset function to a large value for $R_R^{GW}(\mathbf{x})$ (Figure 4c). In contrast, the gain/offset function in our approach maps the low negative Retinex values to negative image values in Equation (3) which are clipped to zero, and the subsequent application of the color processing function in Equation (6) yields a zero value $R_R^{GW}(\mathbf{x}) = 0$ (Figure 4b).

(a) (b) (c)

Figure 4. Example of preventing from color inversion issues for (**a**) an image containing saturated colors (**b**) using our approach, compared to (**c**) using the original scheme [7] with the gain/offset function applied after the color processing function. (image credit: J.L. Lisani CC BY)

For image regions with homogeneous colors, the weighting in Equation (6) increases the color saturation compared to Retinex. The reason is that, for such pixels, the variation of the values of $C_i^{GW}(\mathbf{x})$ is larger for the different $i \in R, G, B$, compared to the variation of the respective values of $R_i(\mathbf{x}, c)$. Thus, the combination with Gray World and the exploitation of global image statistics allows for overcoming a main drawback of the center/surround Retinex that uses local image information only.

2.4. Stacked Integral Images (SII)

To determine the local mean intensity $F(\mathbf{x}, c) * I_i(\mathbf{x})$ in Equation (1), we employ stacked integral images (SII) [16]. This computationally efficient scheme for Gaussian convolution is based on an *integral image* defined for each image position $\mathbf{x} = (x, y)$ as the cumulative sum of intensity values over all rows and all columns from position $(0, 0)$ to position (x, y) of a RGB channel:

$$II_i(x, y) = \sum_{x' \leq x, y' \leq y} I_i(x', y'), \tag{9}$$

where $II_i(x, y)$ denotes the computed integral image value for channel I_i. To construct the integral image, two additions per pixel are required. For a rectangle area box defined by the positions $\mathbf{x}_1$, $\mathbf{x}_2$, $\mathbf{x}_3$, and $\mathbf{x}_4$, the sum $S_i(box)$ of intensities in I_i can be computed using the integral image:

$$S_i(box) = II_i(\mathbf{x}_4) - II_i(\mathbf{x}_3) - II_i(\mathbf{x}_2) + II_i(\mathbf{x}_1), \tag{10}$$

requiring four pixel accesses and three additions/subtractions, regardless of the size of box. This significantly reduces computation times compared to iterating over all pixels in box (N_{box} pixel accesses, $N_{box} - 1$ additions, where N_{box} denotes the number of pixels in box). The local mean intensity can be obtained by dividing $S_i(box)$ with N_{box}. Gaussian convolution using SII is based on the weighted sum of K stacked boxes of different size [16]:

$$\bar{I}_i(\mathbf{x}, \sigma, K) = \sum_{k=1}^{K} w_k S_i(box_k), \tag{11}$$

where $\overline{I}_i(\mathbf{x}, \sigma, K)$ denotes the local mean intensity value of I_i defined by a Gaussian kernel centered at $\mathbf{x}$ with standard deviation σ. Note that the weights w_k and the size of each box_k depend on σ and on K. The local mean intensity $\overline{I}_i(\mathbf{x}, \sigma, K)$ in Equation (11) corresponds to the result of the spatial convolution $F(\mathbf{x}, c) * I_i(\mathbf{x})$ in Equation (1). SII yielded the lowest computation times among different investigated schemes for Gaussian convolution in [21].

3. Results

We have applied our combined Retinex/Gray World approach for color consistency and local contrast enhancement to color image sequences acquired on five outdoor itineraries using a vehicle-mounted camera system (Figure 5). For each itinerary, we have recorded image sequences at two different time points under different illumination conditions. The five pairs of reference and current image sequences consist of 786 to 2400 corresponding frames. We have aligned frames of the reference image sequences w.r.t. the corresponding frames of the current sequences using an affine homography-based registration approach [22]. For our combined Retinex/Gray World approach, the same parameter setting was used for all image data (see Table 1). We have chosen the values for C_1, C_2, and σ taking into account the contrast in the computed images, the visibility within shadows, and the halo effect on shadow boundaries. Taking into account the computation times, we used $K = 3$, since increasing K did not significantly improve the results. We used $\alpha = 125$ as suggested in [7], and the typical values for A and I_{gray}. The computation time for processing images with a resolution of 1280×960 pixel on a workstation under Linux Ubuntu 14.04 with a Xeon E5-1650v3 CPU (3.5 GHz) is 35 ms using an OpenGL-based implementation, allowing the display of more than 20 frames per second in the change detection system. We investigated the performance of our approach w.r.t. color rendition, inter-sequence color consistency, and change detection. We have also performed an experimental comparison with previous Retinex-based approaches: the original center/surround Retinex model [2], a Gray World [15] extension based on an intensity-based Retinex [23], and a hue-preserving intensity-based Retinex approach [24] (without shadow detection). We have used the same values for the parameters of the previous Retinex-based approaches that are common with our approach, except for $C = 120$ in Equation (3). The Gray World extension and the hue-preserving extension use an affine function [25] to map the colors computed by Gray World and the original colors, respectively, to the intensities computed by the intensity-based Retinex [23].

Figure 5. Vehicle-mounted color camera system.

Table 1. Parameter setting for our combined Retinex/Gray World approach.

A in [3]	C_1, C_2 in [4]	I_{gray} in [5]	α in [7]	σ in [11]	K in [11]
127.5	80, 120	127.5	125	$30/\pi$	3

3.1. Color Rendition

First, we have evaluated the performance of our approach in the context of color rendition and the visibility in the processed images. In Figure 6, we show an example of processing an outdoor color image using our combined Retinex/Gray World approach as well as the three previous Retinex-based approaches. It can be seen that the original image (Figure 6a) contains large shadows with low local contrast, and thus visibility is poor. It can be also seen that the colors of the non-shadowed areas are strongly influenced by the warm colors of the light source. It turned out that the Retinex-based approaches enhanced local contrast (Figure 6b–e, improving the visibility within shadows. Compared to the traditional gain/offset function ($C = 120$) used by the previous Retinex-based approaches, the extended gain/offset function ($C_1 = 80$, $C_2 = 120$) in our approach reduces the halo effect on the shadow boundaries. In Figure 6b, it can be seen that the original center/surround Retinex approach strongly compensated the color of the illuminant. However, color rendition is poor since the processed image appears grayish (low saturation), particularly in larger areas with homogeneous colors (e.g., grass area) due to the local color processing. In contrast, the saturation is higher for smaller objects with different colors compared to the background, and thus visibility of these objects is good (Figure 6b, right). The saturation of the processed image applying the Gray World extension is significantly increased (Figure 6c). However, colors in certain areas of the image that have been more strongly enhanced w.r.t. the original image (e.g., shadows) appear unnatural. The reason is that for images with spatially inhomogeneous illumination (e.g., images containing both shadowed and directly illuminated areas) or for images with unbalanced color distribution (i.e., when the scene has a dominant color) the assumptions of the Gray World hypothesis do not hold. The hue-preserving Retinex approach avoids unnatural colors by preserving the original color information, however, the color of the illuminant is partially preserved as well (Figure 6d). In addition, due to the global decrease of saturation, the visibility of objects of different color is relatively low (Figure 6d, right). The result of applying our combined Retinex/Gray World approach is shown in Figure 6e. It can be seen that the colors of the illuminant have been strongly compensated and that the color appearance represents a combination of the results of Retinex and Gray World. Compared to Retinex, saturation is increased and the computed colors appear more natural (cf. the grass area in the images). We have determined the mean saturation (based on the HSI color model) averaged over 4800 corresponding frames of two image sequences and found that our approach yields 0.0997, which is an increase of 28% compared to Retinex (0.0776). Note that the differences in contrast between our approach and Retinex (cf. Figure 6b,e) are due to the proposed scheme with two gains in Equation (3), and the differences in chromaticities are due to the color processing function (Equation (6)). Compared to the Gray World extension, our approach avoids strong unnatural colors (cf. the shadowed areas and the non-shadowed road area in Figure 6c,e). Compared to the hue-preserving Retinex, saturation is increased, the colors of the illuminant are strongly compensated, and objects of locally distinctive colors are more visible due to the use of local color processing (cf. Figure 6d,e, right).

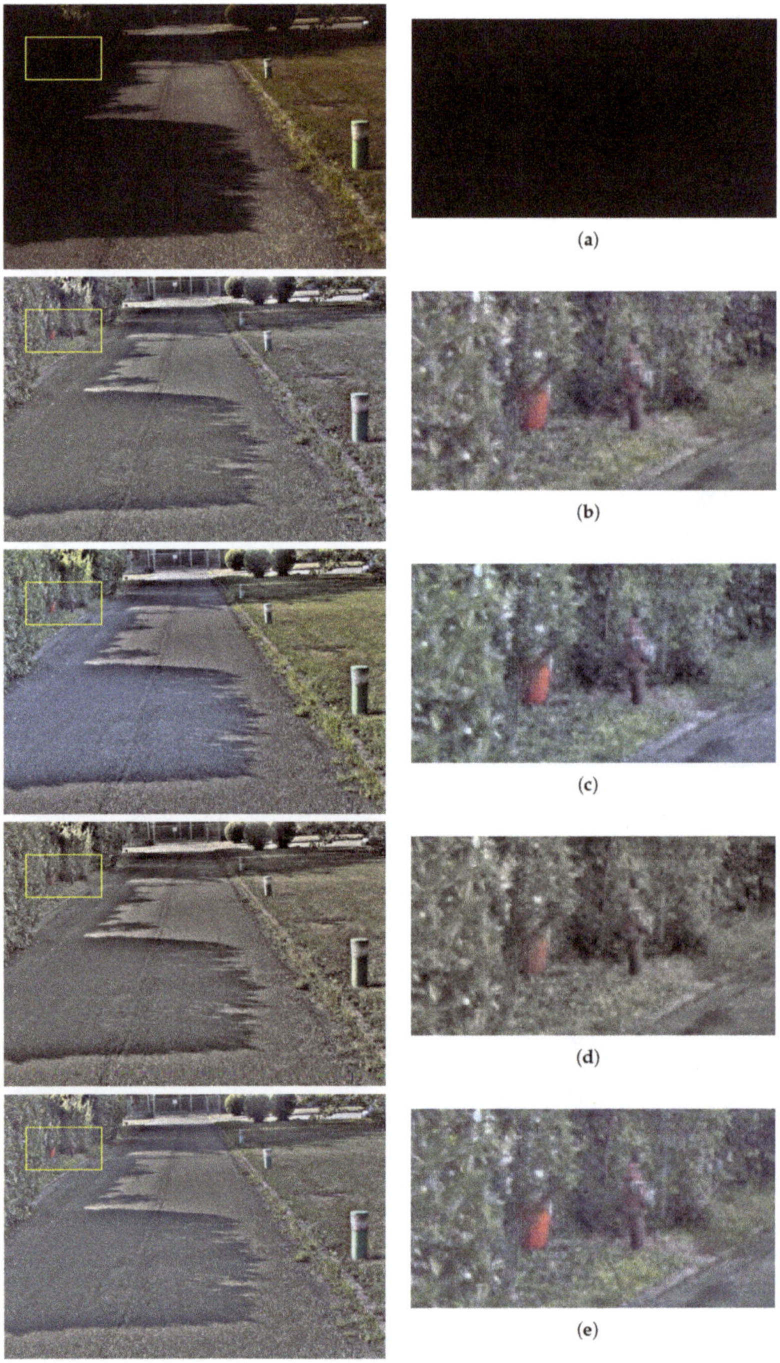

Figure 6. (**a**) original and processed images applying (**b–d**) previous Retinex-based approaches and (**e**) our approach. Enlarged sections for the marked regions are shown on the right. (**a**) unprocessed image; (**b**) Retinex [2]; (**c**) Gray World [15] extension; (**d**) hue-preserving Retinex [24]; (**e**) new combined Retinex/Gray World.

3.2. Inter-Sequence Color Consistency

We have also evaluated our combined Retinex/Gray World approach in the context of inter-sequence color consistency, that is, the color consistency between corresponding frames of a reference and a current image sequence recorded on the same itinerary at different time points. In Figure 7, we show an example of applying our approach to two corresponding frames of the same scene acquired at different time points with different illumination conditions (Figure 7a). It can be seen that our approach strongly compensates lightness and color inconsistencies between the two images (Figure 7c). In contrast, the Gray World extension yields images with visible color inconsistencies (Figure 7b).

Figure 7. (**a**) original and processed images applying (**b**) the Gray World extension and (**c**) our approach for a scene acquired at two different time points. The second image for each example has been registered w.r.t. the first image (top) and the two images have been combined using a checkerboard (bottom). (**a**) unprocessed images; (**b**) Gray World extension; (**c**) new combined Retinex/Gray World.

To quantitatively assess inter-sequence color consistency, we have computed the mean *RGB angular error* and the mean *rg chromaticity endpoint error* between corresponding frames of pairs of reference and current image sequences. These error metrics are typically used in combination with ground truth images for evaluation of color constancy approaches [26]. Note that, in addition to the different illumination conditions, changes in the scene and registration errors cause lightness and color differences as well, and thus influence the error values. However, we assume that the influence of these factors is similar for each investigated approach and can be ignored. In Figure 8, we show an example of the mean RGB angular error over time for corresponding frames of two image sequences of an outdoor route. It can be seen that for all time points our approach yields a lower error compared to the unprocessed images. The mean error averaged over all time points computes to 2.43°, which is an improvement of 46% compared to the respective error (4.51°) for the unprocessed images. It can be also seen that our approach outperforms the Retinex approach (3.21°) as well as the Gray World extension (3.36°). The worse performance of Retinex is due to inconsistencies caused by the stronger lightness halo effect on shadow boundaries and the color halo effect on edges of color change. The reasons for the worse performance of Gray World are two. First, the unnatural colors in certain areas of images with inhomogeneous illumination (see also Section 3.1 above) introduce color inconsistencies. Second, changes in the scene influence the global statistics of an image and affect the computed colors for unchanged areas, introducing color inconsistencies as well. Our approach yields a lower mean error also compared to the hue-preserving Retinex approach (2.64°); however, for certain time points the error is higher. The reason is that for these time points the hue-preserving Retinex determines colors with low saturation due to the global saturation decrease, resulting in small color differences and low error values, as expected. The mean RGB angular errors and the mean rg endpoint errors averaged over all time points and over the five pairs of corresponding image sequences are shown in Table 2. It turns out that our approach yields significantly lower errors compared to the unprocessed images, as well as compared to Retinex and Gray World. The hue-preserving Retinex yields the lowest errors, however, due to the lower saturation of the computed images.

We have also used the Shades of Gray [18] and the Gray Edge [19] global color constancy approaches in Eqution (6), as alternatives for Gray World. For both variants, we used $p = 5$ for the Minkowski norm. We applied these two variants to the two corresponding image sequences used for the example in Figure 8, and we computed the mean RGB angular error over time. The errors for the combined Retinex/Shades of Gray and the combined Retinex/Gray Edge approaches compute to 2.6° and 2.85°, respectively, which are an increase of 6.9% and 17.3%, compared to the proposed Retinex/Gray World approach. The reason for the better performance of Gray World for this image data is that the image mean used by Gray World is more stable w.r.t. changes in the scenes, compared to the Minkowski norm and the image derivatives used by Shades of Gray and Gray Edge.

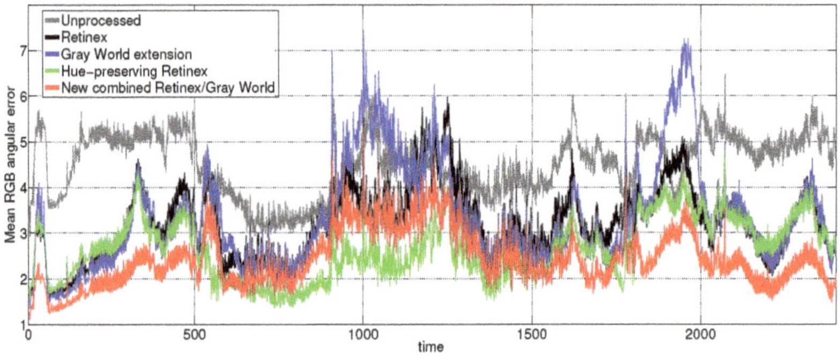

Figure 8. Mean RGB angular error over time for corresponding images of two image sequences for our approach and three previous Retinex-based approaches.

Table 2. Mean errors averaged over corresponding time points of five pairs of image sequences for our approach and three previous Retinex-based approaches. Percentages indicate the change compared to the unprocessed image data.

Approach	Mean RGB Angular Error	Mean rg Endpoint Error
Unprocessed	4.06°	0.029
Retinex [2]	$3.32^{\circ}_{-18\%}$	$0.027_{-8\%}$
Gray World [15] extension	$3.93^{\circ}_{-3\%}$	$0.029_{-1\%}$
Hue-preserving Retinex [24]	$2.39^{\circ}_{-41\%}$	$0.016_{-47\%}$
New combined Retinex/Gray World	$2.56^{\circ}_{-37\%}$	$0.020_{-31\%}$

3.3. Change Detection

Finally, we have evaluated our combined Retinex/Gray World approach in terms of change detection. In Figure 9, we show an example of a pair of images corresponding to a changed scene at two different time points. However, it is difficult to quantitatively evaluate the performance of image processing approaches for change detection systems that are based on a human operator. The reason is that there are different factors influencing change detection that are related to the perception of the human operator as well as on the properties of the displayed images. Factors related to the images are lightness and color rendition, local contrast, and lightness and color consistency. In general, images with relatively natural lightness and color rendition facilitate the operator to more quickly understand a scene and focus on areas of interest, compared to, for example, overenhanced images containing too many details, or, images containing unnatural colors. Changes are better visible when the differences between the two images for the changed area are larger compared to the differences for the background (unchanged area), within a spatial neighborhood. Note that the latter criterion is used by image differencing approaches for automatic change detection [27]. Based on this, we computed the *CIELAB color differences* between corresponding images. Note that the CIELAB color differences better describe the sensitivity of the human eye to color differences, compared to RGB color differences. Then, we determined the ratio of the mean differences for the changed area to the mean differences for the background, within a spatial neighborhood around the change. The changed area and the background have been obtained based on manual segmentation.

An example of the CIELAB color differences for a neighborhood around a change is shown in Figure 9a (right). It can be seen that the magnitude of the differences for the area of the change are similar with the magnitudes of the differences for the background, and the ratio computes to 1.3. Applying our approach, the differences for the area of the change increased in contrast to the differences for the background that decreased (Figure 9b, right), compared to the respective differences for the unprocessed images. The ratio of the mean differences computes to 11.2. This is a significant increase compared to the unprocessed images, and, therefore, the change is potentially better detectable by a human operator. In Table 3 we show the ratio of CIELAB color differences for six examples of changed objects in different scenes, applying our approach as well as the three previous Retinex-based approaches. It can be seen that our approach yields the largest ratio for most changes, as well as the largest mean ratio.

(a)

(b)

Figure 9. (**a**) original and (**b**) processed images applying our approach for a scene with a change. The second image has been registered w.r.t. the first image. On the right, the magnitudes of the CIELAB color differences between the two images are visualized with the contour of the segmented change. (**a**) original images; (**b**) new combined Retinex/Gray World.

Table 3. Ratios of the mean CIELAB color differences for changed areas to unchanged areas within a spatial neighborhood around the change. Our combined Retinex/Gray World approach and three previous Retinex-based approaches have been applied to six pairs of corresponding images of changed scenes.

	Changed Objects						
Approach	**Traffic Pole**	**Stone**	**Stone2**	**Barrel**	**Fake Stone**	**Car**	**Mean**
Unprocessed	1.3	3.6	3.6	1.3	0.8	0.6	1.9
Retinex [2]	9.5	11.7	4.5	2.6	21.7	8.8	9.8
Gray World [15] extension	8.0	11.3	4.7	2.1	20.3	5.5	8.6
Hue-preserving Retinex [24]	11.1	13.1	4.5	4.0	17.9	6.6	9.6
New combined Retinex/Gray World	11.2	28.4	5.1	2.9	22.1	8.4	13

4. Discussion

From the results, it turns out that our combined Retinex/Gray World approach enhances local contrast and reduces inter-sequence color inconsistencies between corresponding images acquired under different illumination conditions, taking advantage of the combination of local and global color processing. The advantages of the local center/surround Retinex are that the computation of the colors of a pixel is neither influenced by information of distant image regions with different illumination nor by changes in other regions of the image. The advantage of the global Gray World is the generally higher color saturation. Simultaneously, our approach overcomes typical issues of both local and global color processing.

The experimental comparison showed that our approach yields better results than previous Retinex-based approaches. Compared to the center/surround Retinex that often computes grayish colors due to the use of local image information only, our approach increases the saturation and improves color rendition. Our approach outperforms Retinex also in the context of color consistency, since the color halo effect on the edges of color change is reduced. Compared to Gray World, which yields strong unnatural colors for areas of images with inhomogeneous illumination, our approach determines more natural colors. This improves color rendition and, in addition, improves color consistency as well, since unnatural colors introduce color differences between corresponding images. The second reason for the worse performance of Gray World that influences color consistency is that changes in a scene affect the computed colors of the whole image. Compared to a hue-preserving Retinex approach, objects of locally distinctive color are better visible due to the local color processing. In addition, our approach yields images with higher saturation, compared to the hue-preserving Retinex, which is based on the global saturation decrease.

Furthermore, the proposed gain/offset function allows for reducing halo effect on shadow boundaries, without decreasing the contrast in other areas of an image. In future work, we plan to further reduce the halo effect on shadow boundaries, using, for example, an edge-preserving filter or a guided filter (e.g., [9,10]). This is, however, a challenging task, due to the high computational complexity of these processing schemes. Another alternative could be to extend a previous shadow-based Retinex approach [24] by improving the consistency of the employed shadow detection method.

5. Conclusions

We have presented an automatic approach for local contrast enhancement and color consistency for the display of images in a mobile change detection system. Our approach is based on the center/surround Retinex model for local contrast enhancement and lightness/color constancy, and on the Gray World method for global color constancy. We employed a nonlinear color processing function to combine Retinex and Gray World and to take advantage of both local and global color processing. We proposed a gain/offset scheme that uses two gains, in order to reduce lightness halo effect on shadow boundaries and to improve visibility within shadows. The use of stacked integral images (SII) for

Gaussian convolution allows for achieving low computation times and using the proposed approach in real-time applications. We have successfully applied our approach to color image sequences acquired from outdoor itineraries using a vehicle-mounted camera, and an experimental comparison demonstrated that our approach outperforms previous Retinex-based approaches and Gray World w.r.t. color rendition, inter-sequence color consistency, and change detection.

Author Contributions: Marco Tektonidis developed the approach, performed the experiments, analyzed the data, and wrote the paper. David Monnin contributed to the design of the approach and the experiments, and proofread the paper.

Conflicts of Interest: The authors declare no conflict of interest.

References

1. Monnin, D.; Schneider, A.L.; Bieber, E. Detecting suspicious objects along frequently used itineraries. In Proceedings of the SPIE, Security and Defence: Electro-Optical and Infrared Systems: Technology and Applications VII, Toulouse, France, 20 September 2010; Volume 7834, pp. 1–6.
2. Jobson, D.; Rahman, Z.; Woodell, G.A. Properties and performance of a center/surround Retinex. *IEEE Trans. Image Process.* **1997**, *6*, 451–462.
3. Agarwal, V.; Abidi, B.R.; Koschan, A.; Abidi, M.A. An overview of color constancy algorithms. *J. Pattern Recognit. Res.* **2006**, *1*, 42–54.
4. Land, E.H. The Retinex Theory of Color Vision. *Sci. Am.* **1977**, *237*, 108–129.
5. Land, E.H.; McCann, J.J. Lightness and retinex theory. *Opt. Soc. Am.* **1971**, *61*, 1–11.
6. Land, E.H. An alternative technique for the computation of the designator in the Retinex theory of color vision. *Proc. Natl. Acad. Sci. USA* **1986**, *83*, 3078–3080.
7. Jobson, D.; Rahman, Z.; Woodell, G.A. A multiscale Retinex for bridging the gap between color images and the human observation of scenes. *IEEE Trans. Image Process.* **1997**, *6*, 965–976.
8. Barnard, K.; Funt, B. Investigations into multi-scale Retinex. In *Colour Imaging: Vision and Technology*; John Wiley & Sons: Hoboken, NJ, USA, 1999; pp. 9–17.
9. Pei, S.C.; Shen, C.T. High-dynamic-range parallel multi-scale retinex enhancement with spatially-adaptive prior. In Proceedings of the IEEE International Symposium on Circuits and Systems, Melbourne, Australia, 1–5 June 2014; pp. 2720–2723.
10. Liu, H.; Sun, X.; Han, H.; Cao, W. Low-light video image enhancement based on multiscale Retinex-like algorithm. In Proceedings of the IEEE Chinese Control and Decision Conference, Yinchuan, China, 28–30 May 2016; pp. 3712–3715.
11. McCann, J.J. Retinex at 50: Color theory and spatial algorithms, a review. *J. Electron. Imaging* **2017**, *26*, 1–14.
12. Provenzi, E. Similarities and differences in the mathematical formalizations of the Retinex model and its variants. In Proceedings of the International Workshop on Computational Color Imaging, Milan, Italy, 29–31 March 2017; Springer: Berlin, Germany, 2017; pp. 55–67.
13. Bertalmío, M.; Caselles, V.; Provenzi, E. Issues about retinex theory and contrast enhancement. *Int. J. Comput. Vis.* **2009**, *83*, 101–119.
14. Provenzi, E.; Caselles, V. A wavelet perspective on variational perceptually-inspired color enhancement. *Int. J. Comput. Vis.* **2014**, *106*, 153–171.
15. Buchsbaum, G. A spatial processor model for object colour perception. *J. Frankl. Inst.* **1980**, *310*, 1–26.
16. Bhatia, A.; Snyder, W.E.; Bilbro, G. Stacked integral image. In Proceedings of the 2010 IEEE International Conference on Robotics and Automation, Anchorage, AK, USA, 3–7 May 2010; pp. 1530–1535.
17. Tektonidis, M.; Monnin, D. Image enhancement and color constancy for a vehicle-mounted change detection system. In Proceedings of the SPIE, Security and Defense: Electro-Optical Remote Sensing, Edinburgh, UK, 26 September 2016; Volume 9988, pp. 1–8.
18. Finlayson, G.D.; Trezzi, E. Shades of gray and colour constancy. In Proceedings of the Color and Imaging Conference, Scottsdale, AZ, USA, 9 November 2004; Society for Imaging Science and Technology: Springfield, VA, USA, 2004; pp. 37–41.
19. Van De Weijer, J.; Gevers, T.; Gijsenij, A. Edge-based color constancy. *IEEE Trans. Image Process.* **2007**, *16*, 2207–2214.

20. Petro, A.B.; Sbert, C.; Morel, J.M. Multiscale Retinex. *Image Process. On Line* **2014**, *4*, 71–88.
21. Getreuer, P. A survey of Gaussian convolution algorithms. *Image Process. On Line* **2013**, *3*, 286–310.
22. Gond, L.; Monnin, D.; Schneider, A. Optimized feature-detection for on-board vision-based surveillance. In Proceedings of the SPIE, Defence and Security: Detection and Sensing of Mines, Explosive Objects, and Obscured Targets XVII, Baltimore, MA, USA, 23 April 2012; Volume 8357, pp. 1–12.
23. Funt, B.; Barnard, K.; Brockington, M.; Cardei, V. Luminance-based multi-scale Retinex. In Proceedings of the AIC Color, Kyoto, Japan, 25–30 May 1997; Volume 97, pp. 25–30.
24. Tektonidis, M.; Monnin, D.; Christnacher, F. Hue-preserving local contrast enhancement and illumination compensation for outdoor color images. In Proceedings of the SPIE, Security and Defense: Electro-Optical Remote Sensing, Photonic Technologies, and Applications IX, Toulouse, France, 21 September 2015; Volume 9649, pp. 1–13.
25. Yang, C.C.; Rodríguez, J.J. Efficient luminance and saturation processing techniques for color images. *J. Vis. Commun. Image Represent.* **1997**, *8*, 263–277.
26. Barnard, K.; Cardei, V.; Funt, B. A comparison of computational color constancy algorithms—Part I: Methodology and experiments with synthesized data. *IEEE Trans. Image Process.* **2002**, *11*, 972–984.
27. Radke, R.J.; Andra, S.; Al-Kofahi, O.; Roysam, B. Image change detection algorithms: A systematic survey. *IEEE Trans. Image Process.* **2005**, *14*, 294–307.

Journal of
Imaging

MDPI

Article

Improved Color Mapping Methods for Multiband Nighttime Image Fusion

Maarten A. Hogervorst * and Alexander Toet

TNO, Perceptual and Cognitive Systems, Kampweg 5, 3769DE Soesterberg, The Netherlands; lex.toet@tno.nl
* Correspondence: maarten.hogervorstt@tno.nl; Tel.: +31-6-2246-9545

Received: 30 June 2017; Accepted: 24 August 2017; Published: 28 August 2017

Abstract: Previously, we presented two color mapping methods for the application of daytime colors to fused nighttime (e.g., intensified and longwave infrared or thermal (LWIR)) imagery. These mappings not only impart a natural daylight color appearance to multiband nighttime images but also enhance their contrast and the visibility of otherwise obscured details. As a result, it has been shown that these colorizing methods lead to an increased ease of interpretation, better discrimination and identification of materials, faster reaction times and ultimately improved situational awareness. A crucial step in the proposed coloring process is the choice of a suitable color mapping scheme. When both daytime color images and multiband sensor images of the same scene are available, the color mapping can be derived from matching image samples (i.e., by relating color values to sensor output signal intensities in a sample-based approach). When no exact matching reference images are available, the color transformation can be derived from the first-order statistical properties of the reference image and the multiband sensor image. In the current study, we investigated new color fusion schemes that combine the advantages of both methods (i.e., the efficiency and color constancy of the sample-based method with the ability of the statistical method to use the image of a different but somewhat similar scene as a reference image), using the correspondence between multiband sensor values and daytime colors (sample-based method) in a smooth transformation (statistical method). We designed and evaluated three new fusion schemes that focus on (i) a closer match with the daytime luminances; (ii) an improved saliency of hot targets; and (iii) an improved discriminability of materials. We performed both qualitative and quantitative analyses to assess the weak and strong points of all methods.

Keywords: sensor fusion; visualization; night vision; image intensifier; thermal sensor; color mapping

1. Introduction

The increasing availability and use of co-registered imagery from sensors with different spectral sensitivities have spurred the development of image fusion techniques [1]. Effective combinations of complementary and partially redundant multispectral imagery can visualize information that is not directly evident from the individual sensor images. For instance, in nighttime (low-light) outdoor surveillance applications, intensified visual (II) or near-infrared (NIR) imagery often provides a detailed representation of the spatial layout of a scene, while targets of interest like persons or cars may be hard to distinguish because of their low luminance contrast. While thermal infrared (IR) imagery typically represents these targets with high contrast, their background (context) is often washed out due to low thermal contrast. In this case, a fused image that clearly represents both the targets and their background can significantly enhance the situational awareness of the user by showing the location of targets relative to landmarks in their surroundings (i.e., by providing more information than either of the input images alone). Additional benefits of image fusion are a wider spatial and temporal coverage, decreased uncertainty, improved reliability, and increased system robustness.

Fused imagery that is intended for human inspection should not only combine the information from two or more sensors into a single composite image but should also present the fused imagery in an intuitive format that maximizes recognition speed while minimizing cognitive workload. Depending on the task of the observer, fused images should preferably use familiar representations (e.g., natural colors) to facilitate scene or target recognition or should highlight details of interest to speed up the search (e.g., by using color to make targets stand out from the clutter in a scene). This consideration has led to the development of numerous fusion schemes that use color to achieve these goals [2–5].

In principle, color imagery has several benefits over monochrome imagery for human inspection. While the human eye can only distinguish about 100 shades of gray at any instant, it can discriminate several thousand colors. By improving feature contrast and reducing visual clutter, color may help the visual system to parse (complex) images both faster and more efficiently, achieving superior segmentation into separate, identifiable objects, thereby aiding the semantic 'tagging' of visual objects [6]. Color imagery may, therefore, yield a more complete and accurate mental representation of the perceived scene, resulting in better situational awareness. Scene understanding and recognition, reaction time, and object identification are indeed faster and more accurate with realistic and diagnostically (and also—though to a lesser extent—non-diagnostically [7]) colored imagery than with monochrome imagery [8–10].

Color also contributes to ultra-rapid scene categorization or gist perception [11–14] and drives overt visual attention [15]. It appears that color facilitates the processing of color diagnostic objects at the (higher) semantic level of visual processing [10], while it facilitates the processing of non-color diagnostic objects at the (lower) level of structural description [16,17]. Moreover, observers can selectively attend task-relevant color targets and to ignore non-targets with a task-irrelevant color [18–20]. Hence, simply mapping multiple spectral bands into a three-dimensional (false) color space may already serve to increase the dynamic range of a sensor system [21]. Thus, it may provide immediate benefits such as improved detection probability, reduced false alarm rates, reduced search times, and increased capability to detect camouflaged targets and to discriminate targets from decoys [22,23].

In general, the color mapping should be adapted to the task at hand [24]. Although general design rules can be applied to assure that the information available in the sensor image is optimally conveyed to the observer [25], it is not trivial to derive a mapping from the various sensor bands to the three independent color channels. In practice, many tasks may benefit from a representation that renders fused imagery in realistic colors. Realistic colors facilitate object recognition by allowing access to stored color knowledge [26]. Experimental evidence indicates that object recognition depends on stored knowledge of the object's chromatic characteristics [26]. In natural scene recognition, optimal reaction times and accuracy are typically obtained for realistic (or diagnostically) colored images, followed by their grayscale version, and lastly by their (nondiagnostically) false colored version [12–14].

When sensors operate outside the visible waveband, artificial color mappings inherently yield false color images whose chromatic characteristics do not correspond in any intuitive or obvious way to those of a scene viewed under realistic photopic illumination [27]. As a result, this type of false-color imagery may disrupt the recognition process by denying access to stored knowledge. In that case, observers need to rely on color contrast to segment a scene and recognize the objects therein. This may lead to a performance that is even worse compared to single band imagery alone [28,29]. Experiments have indeed demonstrated that a false color rendering of fused nighttime imagery which resembles realistic color imagery significantly improves observer performance and reaction times in tasks that involve scene segmentation and classification [30–33], and the simulation of color depth cues by varying saturation can restore depth perception [34], whereas color mappings that produce counter-intuitive (unrealistically looking) results are detrimental to human performance [30,35,36]. One of the reasons often cited for inconsistent color mapping is a lack of physical color constancy [35]. Thus, the challenge is to give night vision imagery an intuitively meaningful ('realistic' or 'natural') color appearance, which is also stable for camera motion and changes in scene composition and lighting conditions. A realistic and stable color representation serves to improve the viewer's scene comprehension and enhance object recognition and discrimination [37]. Several different techniques have been proposed to render night-time imagery in color [38–43]. Simply mapping the signals from different nighttime sensors (sensitive in different spectral wavebands) to the individual channels of a standard RGB color display or to the individual components of a perceptually decorrelated color space (sometimes preceded by a principal component transform or followed by a linear transformation of the color pixels to enhance color contrast) usually results in imagery with an unrealistic color appearance [36,43–46]. More intuitive color schemes may be obtained by opponent processing through feedforward center-surround shunting neural networks similar to those found in vertebrate color vision [47–55]. Although this approach produces fused nighttime images with appreciable color contrast, the resulting color schemes remain rather arbitrary and are usually not strictly related to the actual daytime color scheme of the scene that is registered.

We, therefore, introduced a method to give fused multiband nighttime imagery a realistic color appearance by transferring the first order color statistics of color daylight images to the nighttime imager [41]. This approach has recently received considerable attention [3,5,56–66], and has successfully been applied to colorize fused intensified visual and thermal imagery [5,57,58,60,61,64], FLIR imagery [67], SAR and FLIR imagery [38], remote sensing imagery [68], and polarization imagery [63]. However, color transfer methods based on global or semi-local (regional) image statistics typically do not achieve color constancy and are computationally expensive.

To alleviate these drawbacks, we recently introduced a look-up-table transform-based color mapping to give fused multiband nighttime imagery a realistic color appearance [69–71]. The transform can either be defined by applying a statistical transform to the color table of an indexed false color night vision image, or by establishing a color mapping between a set of corresponding samples taken from a daytime color reference image and a multi-band nighttime image. Once the mapping has been defined, it can be implemented as a color look-up-table transform. As a result, the color transform is extremely simple and fast and can easily be applied in real-time using standard hardware. Moreover, it yields fused images with a realistic color appearance and provides object color constancy, since the relation between sensor output and colors is fixed. The sample-based mapping is highly specific for different types of materials in the scene and can therefore easily be adapted to the task at hand, such as optimizing the visibility of camouflaged targets. In a recent study [72], we observed that multiband nighttime imagery that has been recolored using this look-up-table based color transform conveys the gist of a scene better (i.e., to a larger extent and more accurately) than each of the individual infrared and intensified image channels. Moreover, we found that this recolored imagery conveys the gist of a scene just as well as regular daylight color photographs. In addition, targets of interest such as persons or vehicles were fixated faster [72].

In the current paper, we present and investigate various alterations on the existing color fusion schemes. We will compare various methods and come up with fusion schemes that are suited for different tasks: (i) target detection; (ii) discrimination of different materials; and (iii) easy, intuitive interpretation (using natural daytime colors).

2. Overview of Color Fusion Methods

Broadly speaking, one can distinguish two types of color fusion:

1. Statistical methods, resulting in an image in which the statistical properties (e.g., average color, width of the distribution) match that of a reference image;
2. Sample-based methods, in which the color transformation is derived from a training set of samples for which the input and output (the reference values) are known.

Both of these types of methods have their advantages and disadvantages. One advantage of the statistical methods is that they require no exact match of a multiband sample image to derive the color transformation, and an image showing a scene with similar content suffices to derive the color transformation. The outcome is a smooth transformation that uses a large part of the color space, which is advantageous for the discrimination of different materials, as it leads to a smooth transformation this method may also generalize better to untrained scenes. The downside is that, since no correspondence between individual samples is used (only statistical properties of the color distribution are used), it results in somewhat less naturalistic colors.

On the other hand, the sample-based method derives the color transformation from the direct correspondence between input sensor values and output daytime colors and therefore leads to colors that match the daytime colors well. Hence, this method requires a multiband image and a perfectly matching daytime image of the same scene. It can handle a highly nonlinear relationship between input (sensor values) and output (daytime colors). This also means that the transformation is not as smooth as that of the statistical method. Also, a limited part of the color range is used (available in the training set). Therefore, the discrimination of materials is more difficult than with the statistical method. We have seen [69] that it generalizes well to untrained scenes of a similar environment and sensor settings. However, it remains to be seen how well it generalizes to different scenes and sensor settings.

In this study we investigated new methods that combine the advantages of both types of methods. We are looking for methods that lead to improvement on military relevant tasks: intuitive, natural colors (for good situational awareness, easy and fast interpretation), good discriminability of materials, good detectability of (hot) targets and a fusion scheme that generalizes well to untrained scenes. Note that these properties may not necessarily be combined in a single fusion scheme. Depending on the task at hand different fusion schemes can be optimal (and selected). Therefore, we have designed three new methods based on the existing method, that focus on (i) naturalistic coloring; (ii) detection of hot targets; (iii) discriminability of materials.

In this study we use the imagery obtained by the TRI-band color low-light observation (TRICLOBS) prototype imaging system for a comparative evaluation of the different fusion algorithms [73,74]. The TRICLOBS system provides co-axially registered visual, NIR (near infrared), and LWIR (longwave infrared or thermal) imagery (for an example, see Figure 1). The visual and NIR supply information about the context, while the LWIR is particularly suited for depicting (hot) targets, and allows for looking through smoke. Images have been recorded with this system in various environments. This makes it possible to investigate how well a fusion scheme derived from one image (set) and reference (set) transfers to untrained images recorded in the same environment (and with the same sensor settings) to an untrained, new scene recorded in the same environment or in a different environment. The main training set consists of six images recorded in the MOUT (Military Operations in Urban Terrain) village of Marnehuizen in the Netherlands [74].

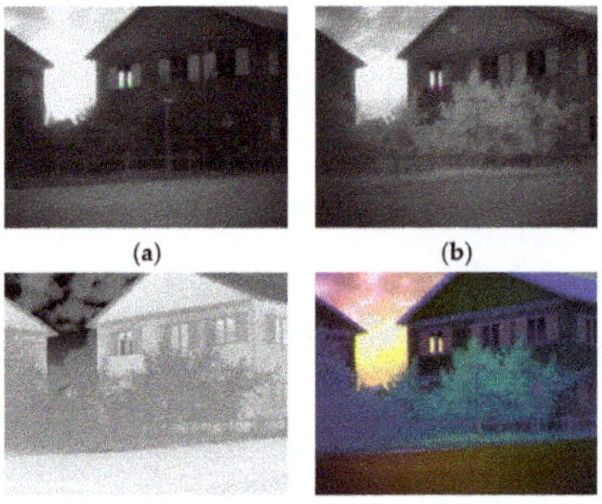

Figure 1. Example from the total training set of six images. (**a**) Visual sensor band; (**b**) near infra-red (NIR) band; (**c**) longwave infrared or thermal (LWIR) (thermal) band; and (**d**) RGB-representation of the multiband sensor image (in which the 'hot = dark' mode is used).

3. Existing and New Color Fusion Methods

In this section, we give a short description of existing color fusion methods, and present our proposals for improved color fusion mappings.

3.1. Existing Color Fusion Methods

3.1.1. Statistics Based Method

Toet [41] presented a statistical color fusion method (or SCF) in which the first order statistics of the color distribution of the transformed multiband sensor image (Figure 1d) is matched to that of a daytime reference color image (Figure 2a). In the proposed scheme the false color (source) multiband image representation and the daytime color reference (target) image are first transformed to the perceptually decorrelated quasi-uniform CIELAB ($L^*a^*b^*$) color space. Next, the mean (–) and standard deviation (σ) of each of the color channels (L^*, a^*, b^*) of the multiband source image are set to the corresponding values of the daytime color reference image as follows:

$$\begin{aligned} L_s^{*\prime} &= \frac{\sigma_t^{L^*}}{\sigma_s^{L^*}} \left(L_s^* - \overline{L_s^*} \right) + \overline{L_t^*} \\ a_s^{*\prime} &= \frac{\sigma_t^{a^*}}{\sigma_s^{a^*}} \left(a_s^* - \overline{a_s^*} \right) + \overline{a_t^*} \\ b_s^{*\prime} &= \frac{\sigma_t^{b^*}}{\sigma_s^{b^*}} \left(b_s^* - \overline{b_s^*} \right) + \overline{b_t^*} \end{aligned} \quad (1)$$

Finally, the colors are transformed back to RGB for display (e.g., Figure 3a).

Another example of a statistical method is described by Pitié et al. [75]. Their method allows one to match the complete 3D distribution by performing histogram equalization in three dimensions. However, we found that various artifacts result when this algorithm is applied to convert the input sensor values (RGB) into the output RGB values of the daytime reference images (e.g., Figure 3b). It has to be noted that Pitié et al. [75] designed their algorithm to account for (small) color changes in daytime photographs, and therefore it may not apply to an application in which the initial image is formed by sensor values outside the visible range.

3.1.2. Sample Based Method

Hogervorst and Toet [70] have shown that a color mapping similar to Toet's statistical method [41] can also be implemented as a color-lookup table transformation (see also [4]). This makes the color transform computationally cheap and fast and thereby suitable for real-time implementation. In addition, by using a fixed lookup-table based mapping, object colors remain stable even when the image content (and thereby the distributions of colors) changes (e.g., when processing video sequences or when the multiband sensor suite pans over a scene).

In the default (so-called *color-the-night* or CTN) sample-based color fusion scheme [70] the color mapping is derived from the combination of a multiband sensor image and a corresponding daytime reference image. Each pair of corresponding pixels in both images is used as a training sample. Therefore, the multiband sensor image and its daytime color reference image need to be perfectly matched (i.e., they need to represent the same scene and need to have the same pixel dimensions). An optimized color transformation between the input (multiband sensor values) and the output (the corresponding daytime color) can then be derived in a training phase that consists of the following steps (see Figure 4):

1. The individual bands of the multiband sensor images and the daytime color reference image are spatially aligned.
2. The different sensor bands are fed into the R, G, B, channels (e.g., Figure 1d) to create an initial false-color fused representation of the multiband sensor image. In principle, it is not important

which band feeds into which channel. This is merely a first presentation of the image and has no influence on the final outcome (the color fused multiband sensor image). To create an initial representation that is closest to the natural daytime image we adopted the 'black-is-hot' setting of the LWIR sensor.

3. The false-color fused image is transformed to an indexed image with a corresponding $CLUT_1$ (color lookup table) that has a limited number of entries N. This comes down to a cluster analysis in 3-D sensor space, with a predefined number of clusters (e.g., the standard k-means clustering techniques may be used for implementation, thus generalizing to N-band multiband sensor imagery).
4. A new $CLUT_2$ is computed as follows. For a given index d in $CLUT_1$ all pixels in the false-color fused image with index d are identified. Then, the median RGB color value is computed over the corresponding set of pixels in the daytime color reference image, and is assigned to index d. Repeating this step for each index in $CLUT_1$ results in a new $CLUT_2$ in which each entry represents the daytime color equivalent of the corresponding false color entry in $CLUT_1$. Thus, when $I = \{1,\ldots,N\}$ represents the set of indices used in the indexed image representation, and $d \in I$ represents a given index in I, then the support Ω_d of d in the source (false-colored) image S is given by

$$\Omega_d = \{\{i,j\} \mid \mathrm{Index}\,(S_{i,j}) = d\} \tag{2}$$

and the new RGB color value $S'_{i,j}$ for index d is computed as the median color value over the same support Ω_d in the daytime color reference image R as follows:

$$S'_{i,j} = \mathrm{Median}\{\{R_{i,j}\} \mid \{i,j\} \in \Omega_d\} \tag{3}$$

5. The color fused image is created by swapping the $CLUT_1$ of the indexed sensor image to the new daytime reference $CLUT_2$. The result from this step may suffer from 'solarizing effects' when small changes in the input (i.e., the sensor values) lead to a large jump in the output luminance. This is undesirable, unnatural and leads to clutter (see Figure 2b).
6. To eliminate these undesirable effects, a final step was included in which the luminance channel is adapted such that it varies monotonously with increasing input values. The luminance of the entry is thereto made proportional to the Euclidean distance in RGB space of the initial representation (the sensor values; see Figure 2c).

(a) (b) (c)

Figure 2. (**a**) Daytime reference image; (**b**) Intermediate (result at step 5); and (**c**) final result (after step 6) of the color-the-night (CTN) fusion method, in which the luminance is determined by the input sensor values (rather than by the corresponding daytime reference).

(a)

(b)

Figure 3. Examples of (**a**) Toet [41]; and (**b**) Pitié et al. [76].

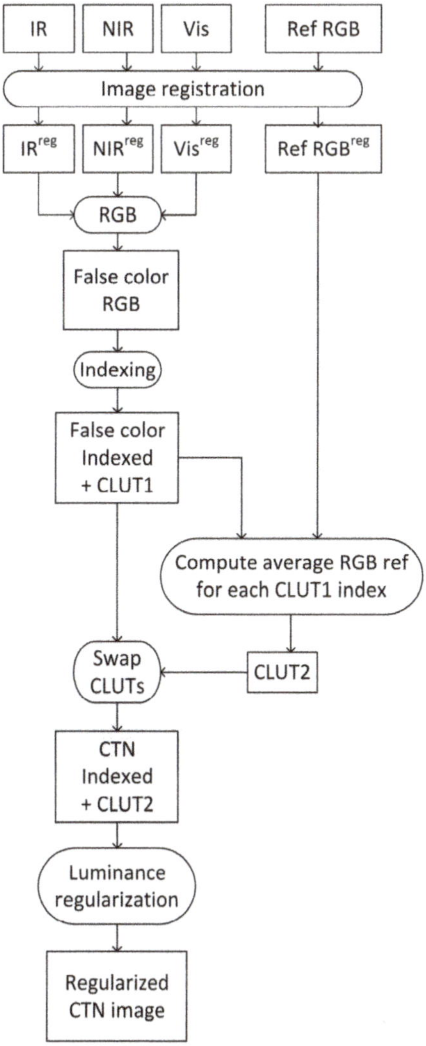

Figure 4. Processing scheme of the CTN sample-based color fusion method.

3.2. *New Color Fusion Methods*

3.2.1. Luminance-From-Fit

In the original CTN scheme, the luminance of the fused image was regularized using only on the input sensor values (see step 6, Section 3.1.2, and Figure 4). This regularization step was introduced to remove unwanted solarizing effects and to assure that the output luminance is a smooth function of the input sensor values. To make the appearance of the fused result more similar to the daytime reference image we derived a smooth luminance-from-fit (LFF) transformation between the input colors and the output luminance of the training samples. This step was implemented in the original CTN scheme as a transformation between two color lookup tables (the general processing scheme is shown Figure 5). Therefore, we converted the RGB data of the reference (daytime) image to HSV (hue, saturation, value) and derived a smooth transformation between input RGB colors and output value using a pseudo-inverse transform:

$$V = V_0 + M \cdot x \tag{4}$$

where $x' = (r, g, b)$. We tried fitting higher polynomial functions as well as a simple linear relationship and found that the latter gave the best results (as judged by eye). The results of the LFF color fusion scheme derived from the standard training set and applied to this same set are depicted in Figure 6c. This figure shows that the LFF results show more resemblance with the daytime reference image (Figure 3a) than the result of the CTN method (Figure 6b). This is most apparent for the vegetation, which is darker in the LFF result than in the CTN fusion result, in line with the daytime luminance.

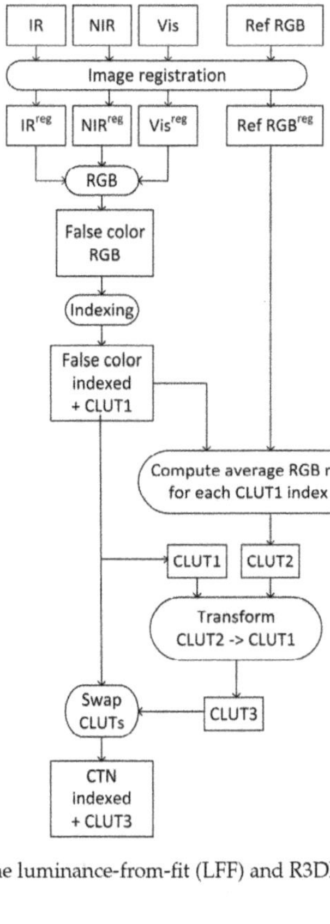

Figure 5. Processing scheme of the luminance-from-fit (LFF) and R3DF sample based color fusion methods.

(a)

(b)

(c)

(d)

Figure 6. (**a**) Standard training set of daytime reference images; (**b**) Result of the CTN algorithm using the images in (**a**) for reference; (**c**) Result from the LFF method; and from (**d**) the SHT method (the training and test sets were the same in these cases).

3.2.2. Salient-Hot-Targets

For situations in which it is especially important to detect hot targets, we derived a color scheme intended to make hot elements more salient while showing the environment in natural colors. This result was obtained by mixing the result from the CTN method (see Figure 7a) with the result from a two-band color transform (Figure 7b) using a weighted sum of the resulting CLUTs in which the weights depend on the temperature to the power 6 (see Figure 8 for the processing scheme of this transformation). This salient-hot-target (SHT) mapping results in colors that are the same as the CTN scheme except for hot elements, which are depicted in the color of the two-band system. We chose a mix with a color scheme that depends on the visible and NIR sensor values using the colors depicted in the inset of Figure 7b, with visible sensor values increasing from left to right, and NIR sensor values increasing from top to bottom. An alternative would be to depict hot elements in a color that does not depend on the sensor values of the visible and NIR bands. However, the proposed scheme also allows for discrimination between hot elements that differ in the values of the two other sensor bands.

Figure 7. Results from (**a**) the CTN scheme (trained on the standard reference image set from Figure 6a); (**b**) a two-band color transformation in which the colors depend on the visible and NIR sensor values (using the color table depicted in the inset); and (**c**) the salient-hot-target (SHT) method in which hot elements are assigned their corresponding color from (**b**).

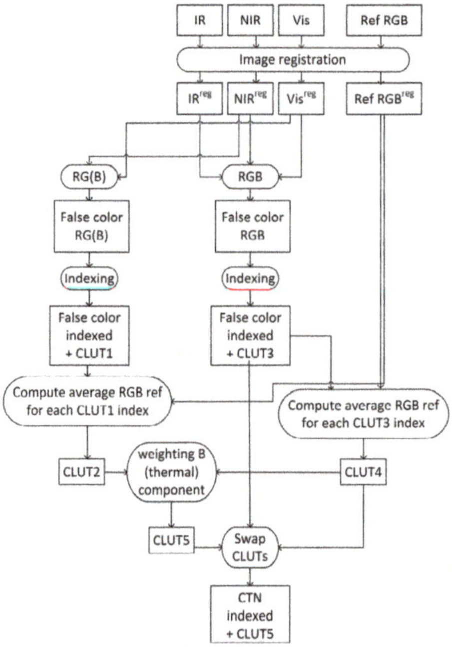

Figure 8. Processing scheme of the SHT sample based color fusion method.

3.2.3. Rigid 3D-Fit

In our quest for a smooth color transformation we first tried to fit an affine (linear) transformation to convert the input RGB triples x into output RGB triples y:

$$y = M \cdot x + t \tag{5}$$

where M is a linear transformation and t is a translation vector. However, although this resulted in a smooth transformation, it also gave images a rather grayish appearance (see Figure 9a). By introducing higher-order terms, the result was smoother (see Figure 9b) and approached the CTN result, but the range of colors that was used remained limited. This problem may be due to the fact that the range of colors in the training set (the reference daytime images) is also limited. As a result, only a limited part of the color space is used, which hinders the discrimination of different materials (and is undesirable). This problem may be solved by using a larger variety of training sets.

To prevent the transformation from leading to a collapse of the color space, we propose to use a 3D rigid transformation. We have fitted a rigid 3D transformation describing the mapping from the input values corresponding to the entries of the initial CLUT_1 to the values held in the output CLUT_2 (see step 4 in Section 3.1.2), by finding the rigid transformation (with rotation R and translation t) that best describes the relationship (with ζ a deviation term that is minimized) using the method described by Arun et al. [76]:

$$\mathrm{CLUT}_2 = t + R \cdot \mathrm{CLUT}_1 + \zeta \tag{6}$$

Next, the fitted values of the new CLUT_3 were obtained by applying the rigid transformation to the input CLUT_1:

$$\mathrm{CLUT}_3 = t + R \cdot \mathrm{CLUT}_1 \tag{7}$$

As in the LFF method, this step was implemented in the original CTN scheme as a transformation between two color lookup tables (the general processing scheme is shown in Figure 5). Figure 9c shows an example in which the rigid-3D-fit (R3DF) transformation has been applied. Figure 10b shows the results obtained by applying the fitted 3-D transformation derived from the standard training set to the test set, which are the same in this case. The result shows some resemblance with the result of the statistical method (Figure 10a). However, R3DF results in a broader range of colors, and therefore a better discriminability of different materials. The colors look somewhat less natural than those resulting from the CTN method (Figure 6b). An advantage of the R3DF method over the SCF statistical method is that the color transformation is derived from a direct correspondence between the multiband sensor values and the output colors of individual samples (pixels), and therefore does not depend on the distribution of colors depicted in the training scene.

(a) (b) (c)

Figure 9. Results of (**a**) an affine fit-transform; (**b**) a 2nd order polynomial fit; and (**c**) a R3DF transformation fit.

(a)

(b)

Figure 10. Results from (**a**) the SCF method and (**b**) rigid-3D-fit (R3DF) method. The training and test sets were the same in these cases.

4. Qualitative Comparison of Color Fusion Methods

First, we performed a qualitative comparison between the different color fusion schemes. In our evaluation we included the CTN method (Figure 6b), the SCF method (Figure 10a), and the three newly proposed schemes: (1) the LFF method (see Figure 6c); (2) the SHT method (Figure 6d); and (3) the R3DF method (Figure 10b). We also included the result of the CTN method using only two input bands: the visible and the NIR band (CTN2: for the processing scheme see Figure 11). This last condition was added to investigate whether the colors derived from this two-band system transfer better to untrained scenes than when the LWIR band is used as well. The idea behind this is that only bands close to the visible range can be expected to show some correlation with the visible daytime colors. However, the LWIR values are probably relatively independent of the daytime color and therefore may not help in inferring the daytime color (and may even lead to unnatural colors). Next, we present some examples of the various images that were evaluated.

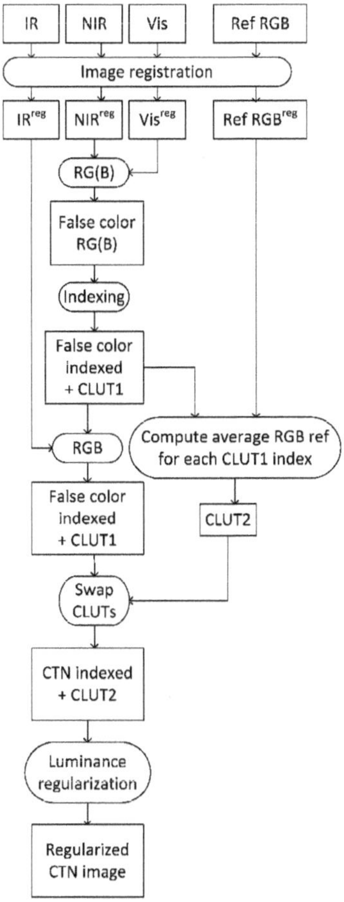

Figure 11. Processing scheme of the CTN2 sample-based color fusion method.

Figures 6 and 7 show the results of the various color methods that were derived from the standard training set (of six images) and were applied to multiband images of the same scenes (except for the two-band system). In line with our expectations, the LFF method (Figure 6c) leads to results that are more similar to the daytime reference image (Figure 6a), with for instance vegetation shown in dark green instead of in light green in the CTN scheme (Figure 6b). As intended, the SHT method (Figure 6d) leads to hot elements (the engine of the vehicle) depicted in bright blue, which makes them more salient and thus easier to detect. The elements are shown in blue because the sensor values in the visible and NIR bands are close to zero in this case. The result of the R3DF method are shown in Figure 10b. As mentioned before, in this case, the results show some resemblance with the statistical method. However, the colors are more outspoken, due to the fact that the range of colors is not reduced in the transformation. Therefore, the discriminability of materials is quite good. The downside is that the colors are somewhat less natural than the CTN result (Figure 6b), although they are still quite intuitive. Figure 12 shows an example in which the color transformations derived from the standard training set were applied to a new (untrained) scene taken in the same environment. As expected, the CTN scheme transfers well to the untrained scene. Again, the LFF result matches the daytime reference slightly better than the CTN scheme, and the R3DF result shows somewhat less natural colors, but still yields good discriminability of the different materials in the scene. Surprisingly, the two-band system (Figure 12e) does not lead to more naturalistic colors than the three-band (CTN) method (Figure 12a).

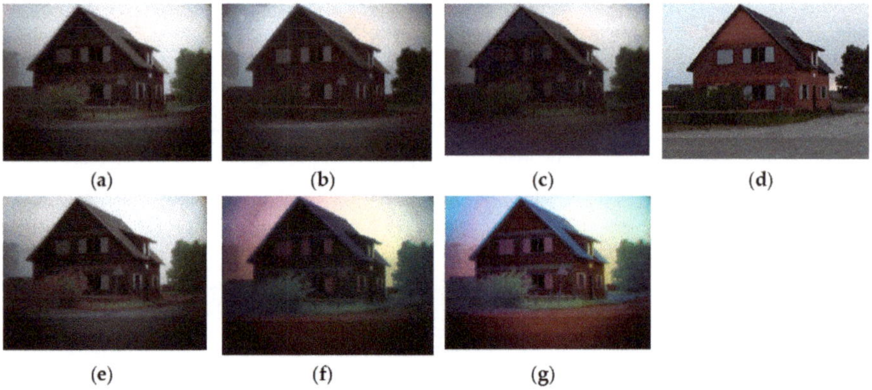

Figure 12. Results from color transformations derived from the standard training set (see Figures 6 and 7) and applied to a different scene in the same environment: (**a**) CTN; (**b**) LFF method; (**c**) SHT method; (**d**) daytime reference (not used for training); (**e**) CTN2; (**f**) statistical color fusion (SCF) method; (**g**) R3DF.

Figure 13 shows yet another example of applying the methods to an image that was taken in the same environment but not used in the training set. In this case, the light level was lower than the levels that occur in the training set, which also led to differences in the sensor settings. No daytime reference is available in this case. Most of the color fusion methods lead to colors that are less outspoken. Again, the colors in the R3DF result are the most vibrant and lead to the best discriminability of materials. Figure 14 shows an example in which the SHT method leads to a yellow hot target, due to the fact that the sensor values in the visible and NIR bands are both high (see the inset Figure 7b for the color scheme that was used). In this case this leads to *lower* target saliency, since the local background is white. This indicates that this method is not yet optimal for all situations.

Figure 15 shows an example in which the color transformations were applied to an image recorded in a totally different environment (and with different sensor settings). Again, the CTN method transfers quite well to this new environment, while the two-band method performs less well. Finally, Figure 16 shows the results of applying the different color mapping schemes to a multiband image recorded in the standard environment, after they were trained on scenes representing a different environment (see Figure 16d). In this case, the resulting color appearance is not so natural as when the mapping schemes were trained in the same environment (see e.g., Figure 7a,c). Also here, the R3DF method (Figure 16g) appears to transfer well to this untrained situation (environment and sensor settings).

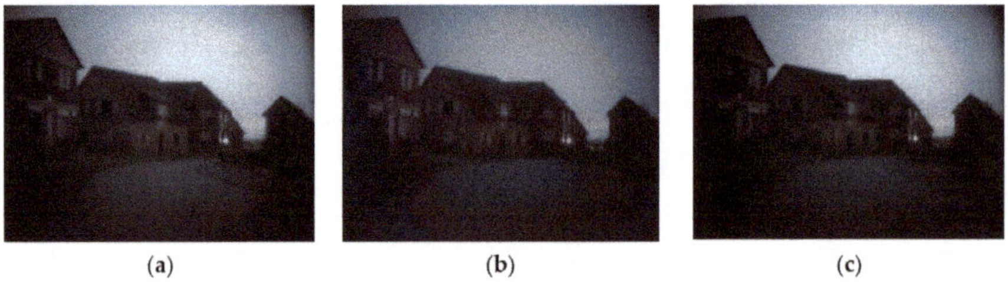

Figure 13. *Cont.*

(d) (e) (f)

Figure 13. Results from color transformations derived from the standard training set applied to a different scene with different sensor settings, registered in the same environment: (**a**) CTN method; (**b**) LFF method; (**c**) SHT method; (**d**) CTN2 method; (**e**) SCF method; (**f**) R3DF method.

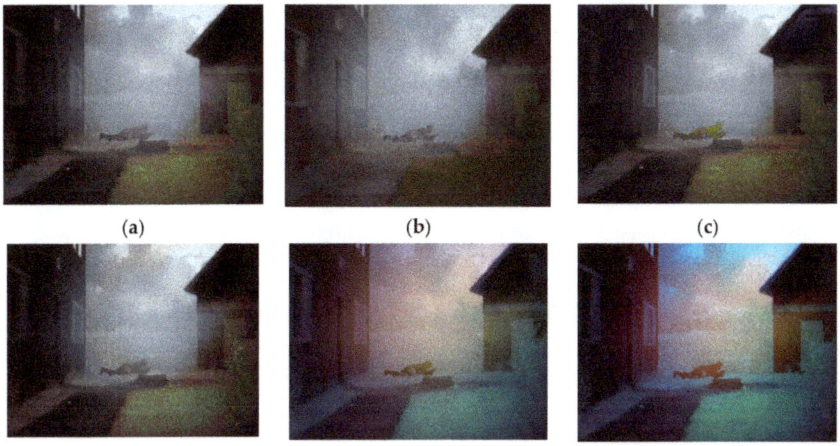

(a) (b) (c)

(d) (e) (f)

Figure 14. Results from color transformations derived from the standard training set applied to a different scene with different sensor settings in the same environment: (**a**) CTN method; (**b**) LFF method; (**c**) SHT method; (**d**) CTN2 method; (**e**) SCF method; (**f**) R3DF method.

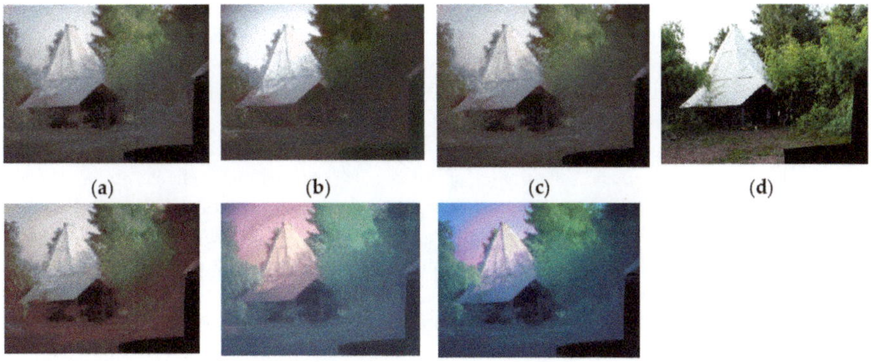

(a) (b) (c) (d)

(e) (f) (g)

Figure 15. Results from color transformations derived from the standard training set applied to a different environment with different sensor settings: (**a**) CTN method; (**b**) LFF method; (**c**) SHT method; (**d**) daytime reference image; (**e**) CTN2 method; (**f**) SCF method; (**g**) R3DF method.

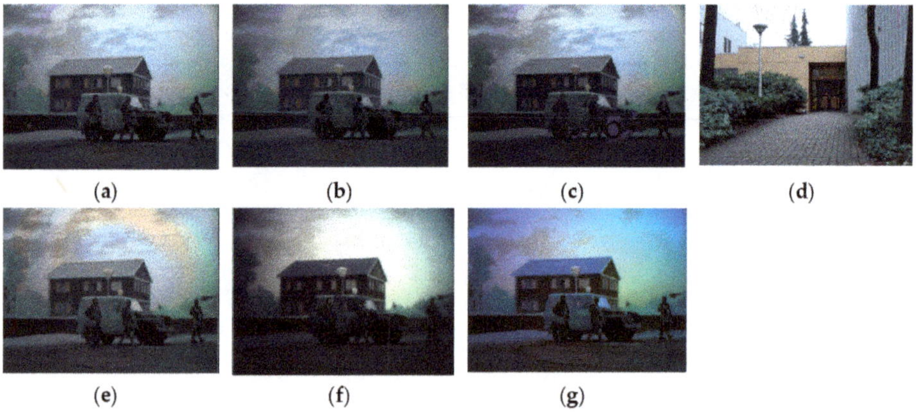

Figure 16. Results from color transformations derived from the scene shown on the right (**d**) with different sensor settings: (**a**) CTN method; (**b**) LFF method; (**c**) SHT method; (**d**) scene used for training the color transformations; (**e**) CTN2 method; (**f**) SCF method; (**g**) R3DF method.

5. Quantitative Evaluation of Color Fusion Methods

To quantitatively compare the performance of the different color fusion schemes discussed in this study we performed both a subjective ranking experiment and a computational image quality evaluation study. Both evaluation experiments were performed with the same set of 138 color fused multiband images. These images were obtained by fusing 23 three-band (visual, NIR, LWIR) TRICLOBS images (each representing a different scene, see [74]) with each of the six different color mappings investigated in this study (CTN, CTN-2 band, statistical, luminance-from-fit, salient-hot-targets, and rigid-3D-fit).

5.1. Subjective Ranking Experiment

5.1.1. Methods

Four observers (two males and two females, aged between 31 and 61) participated in a subjective evaluation experiment. The observers had (corrected to) normal vision and no known color deficiencies. They were comfortably seated at a distance of 50 cm in front of a Philips 231P4QU monitor that was placed in a dark room. The images were 620 × 450 pixels in size, and were presented on a black background with a size of 1920 × 1080 pixels in a screen area of 50.8 × 28.8 cm^2. For each scene, the observers ranked its six different fused color representations (resulting from the six different color fusion methods investigated in this study) in terms of three criteria: image naturalness (color realism, how natural the image appears), discriminability (the amount of different materials that can be distinguished in the image), and the saliency of hot targets (persons, cars, wheels, etc.) in the scene. The resulting rank order was converted to a set of scores, ranging from 1 (corresponding to the worst performing method) to 6 (denoting the best performing method). The entire experiment consisted of three blocks. In each block, the same ranking criterium was used (either naturalness, discriminability or saliency) and each scene was used only once. The presentation order of the 23 different scenes was randomized between participants and between blocks. On each trial, a different scene was shown and the participant was asked to rank order the six different color representations of that given scene from "best performing" (leftmost image) to "worst performing" (rightmost image). The images were displayed in pairs. The participant was instructed to imagine that the display represented a window showing two out of six images that were arranged on a horizontal row. By selecting the right (left) arrow on the keyboard the participant could slide this virtual window from left to right (vice versa) over the row of images, corresponding to higher (lower) ratings. By pressing the up-arrow the left-right order of the two images on the screen could be reversed. By repeatedly comparing successive image pairs and switching their left-right order the participant could rank order the entire row of six images. When the participant was satisfied with the result, he/she pressed the Q-key to proceed to the next trial.

5.1.2. Results

Figure 17 shows the mean observer ranking scores for naturalness, discriminability, and hot target saliency for each of the six color fusion methods tested (CTN, SCF, CTN2, LFF, SHT, and R3DF). This figure shows the scores separately both for images that were included and excluded from the training sets.

To measure the inter-rater agreement (also called inter-rater reliability or IRR) between our observers we computed Krippendorff's alpha, using the R package *'irr'* [77]. The IRR analysis showed that the observers had a substantial agreement in their ratings on naturalness ($\alpha = 0.51$), and a very high agreement in their ratings of the Saliency of hot targets in the scenes ($\alpha = 0.95$). However, they did not agree in their ratings of the discriminability of different materials in the scene ($\alpha = 0.03$).

Figure 17a shows that the CTN and LFF methods score relatively high on naturalness, while the results of the R3DF method were rated as least natural by the observers. This figure also shows that the LFF method yields more natural looking results especially for images that were included in the training set. For scenes that were not included in the training set the naturalness decreases, meaning that the relation between daytime reference colors and nighttime sensor values does not extrapolate very well to different scenes.

Figure 17b shows that the ratings on discriminability vary largely between observers, resulting in a low inter rater agreement. This may be a result of the fact that different observers used different details in the scene to make their judgments. In a debriefing, some observers remarked that they had paid more attention to the distinctness of vegetation, while others stated that they had fixated more on details of buildings. On average, the highest discriminability scores are given to the R3DF, SHT and SCF color fusion methods (in descending order).

Figure 17c shows that the SHT method, which was specifically designed to enhance the saliency of hot targets in a scene, appears to perform well in this respect. The R3DF method also appears to represent the hot targets at high contrast.

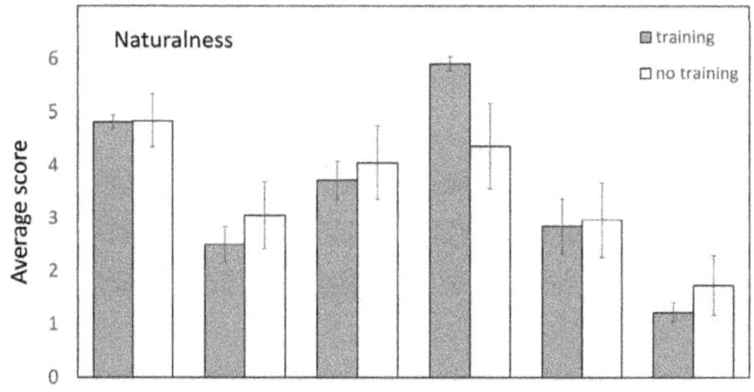

(a)

Figure 17. *Cont.*

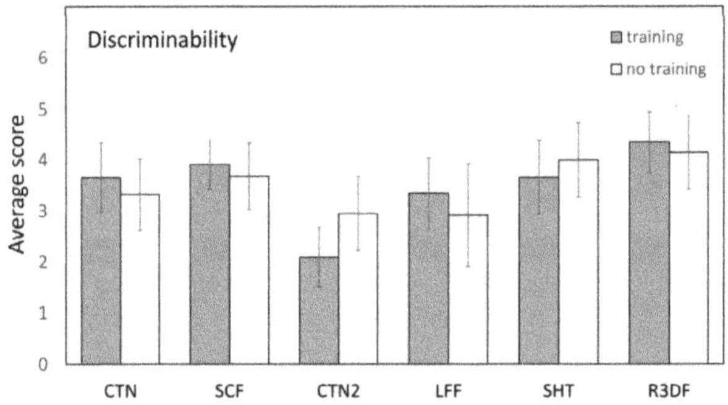

(b)

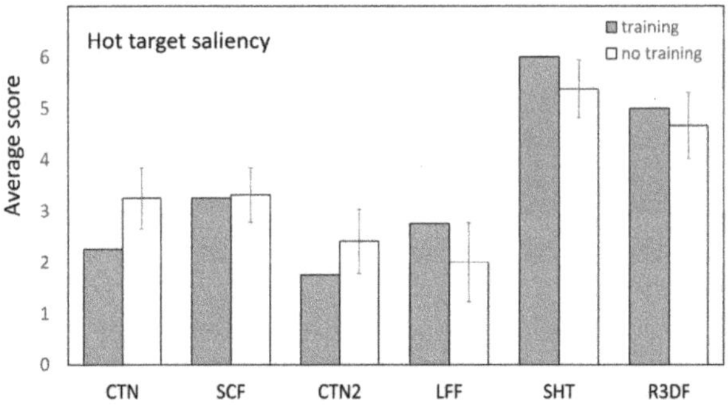

(c)

Figure 17. Mean ranking scores for (**a**) Naturalness; (**b**) Discriminability; and (**c**) Saliency of hot targets, for each of the six color fusion methods (CTN, SCF, CTN2, LFF, SHT, R3DF). Filled (empty) bars represent the mean ranking scores for methods applied to images that were (not) in their training set. Filled (open) bars represent the scores when the methods were applied to images that were (not) included their training set. Error bars represent the standard error of the mean.

5.2. *Objective Quality Metrics*

5.2.1. Methods

We used three no-reference and three full-reference computational image quality metrics to objectively assess and compare the performance of the six different color fusion schemes discussed in this study.

The first no-reference metric is the global color image contrast metric (ICM) that measures the global image contrast [78]. The ICM computes a weighted estimate of the dynamic ranges of both the graylevel and color luminance (L* in CIELAB *L*a*b** color space) histograms. The range of ICM is [0,1]. Larger ICM values correspond to higher perceived image contrast.

The second no-reference metric is the color colorfulness metric (CCM) that measures the color vividness of an image as a weighted combination of color saturation and color variety [78]. Larger CCM values correspond to more colorful images.

The third no-reference metric is the number of characteristic colors in an image (NC). We obtained this number by converting the RGB images to indexed images using minimum variance quantization [79] with an upper bound of 65,536 possible colors. NC is then the number of colors that are actually used for the indexed image representation.

The first full-reference metric is the color image feature similarity metric (FSIMc) that combines measures of local image structure (computed from phase congruency) and local image contrast (computed as the gradient magnitude) in YIQ color space to measure the degree of correspondence between a color fused image and a daylight reference color image [80]. The range of FSIMc is [0,1]. The larger the FSIMc value of a colorized image is, the more similar it is to the reference image. Extensive evaluation studies on several color image quality databases have shown that FSIMc predicts human visual quality scores for color images [80].

The second full-reference metric is the color natural metric (CNM: [78,81,82]). The CNM measures the similarity of the color distributions of a color fused image and a daylight reference color image in Lab color space using Ma's [83] gray relational coefficients. The range of CNM is [0,1]. The larger the CNM value of a colorized image is, the more similar its color distribution is to that of the reference image.

The third full-reference metric is the objective evaluation index (OEI: [81,82]). The OEI measures the degree of correspondence between a color fused image and a daylight reference color image by effectively integrating four established image quality metrics in CIELAB $L^*a^*b^*$ color space: phase congruency (representing local image structure; [84]), gradient magnitude (measuring local image contrast or sharpness), image contrast (ICM), and color naturalness (CNM). The range of OEI is [0,1]. The larger the OEI value of a colorized image is, the more similar it is to the reference image.

5.2.2. Results

Table 1 shows the mean values (with their standard error) of the computational image metrics for each of the six color fusion methods investigated in this study. The full-reference FSIMc, CNM and OEI metrics all assign the largest values to the LFF method. This result agrees with our subjective observation that LFF produces color fused imagery with the most natural appearance (Section 5.1.2). The original CTN method also appears to perform well overall, with the highest mean image contrast (ICM), image colorfulness (CCM) and color naturalness (CNM) values. In addition, the sample-based CTN method outperforms the statistical SCF method (which is computationally more expensive and yields a less stable color representation) on all quality metrics. The low CNM value for the R3DF method confirms our subjective observation that the imagery produced by this method yields looks less natural. CTN and CTN2 both have the same CNM values, and do not differ much in their CCM values, supporting our qualitative and somewhat surprising observation that CTN2 does not lead to more naturalistic colors than the three band CTN color mapping.

To assess the overall agreement between the observer judgements and the computational metrics we computed Spearman's rank correlation coefficient between all six computational image quality metrics and the observer scores for naturalness, discriminability and saliency of hot targets (Table 2). Most computational metrics show a significant correlation with the human observer ratings for naturalness. It appears that the OEI metric most strongly predicts the human observer ratings on all three criteria. This agrees with a previous finding in the literature that the OEI metric ranked different color fused images similar as human observers [81]. The correlation between the OEI and perceived naturalness is specially high (0.95).

Table 1. Results of the computational image quality metrics (with their standard error) for each of the color fusion methods investigated in this study. Overall highest values are printed in bold.

Method	No-Reference Metric			Full-Reference Metric		
	ICM	CCM	NC	FSIMc	CNM	OEI
CTN	**0.382** (0.008)	**4.6** (1.2)	815 (43)	0.78 (0.02)	**0.82** (0.02)	0.73 (0.02)
SCF	0.370 (0.006)	3.6 (1.0)	1589 (117)	0.77 (0.02)	0.76 (0.02)	0.72 (0.02)
CTN2	0.380 (0.008)	4.4 (1.2)	938 (102)	0.78 (0.02)	**0.82** (0.02)	0.73 (0.02)
LFF	0.358 (0.009)	4.5 (1.2)	860 (64)	**0.80** (0.01)	**0.82** (0.02)	**0.74** (0.02)
SHT	0.349 (0.007)	4.4 (1.2)	1247 (241)	0.78 (0.01)	**0.82** (0.02)	0.72 (0.02)
R3DF	0.341 (0.011)	4.1 (1.1)	2630 (170)	0.77 (0.01)	0.74 (0.02)	0.70 (0.03)

ICM = image contrast metric, CCM = color colorfulness metric, NC = number of colors, FSIMc = feature similarity metric, CNM = color natural metric, OEI = objective evaluation index.

Table 2. Pearson's correlation coefficient between the computational image quality metrics and the observer ratings for naturalness, discriminability and the saliency of hot targets. Overall highest values are printed in bold.

Method	No-Reference Metric			Full-Reference Metric		
	ICM	CCM	NC	FSIMc	CNM	OEI
Naturalness	0.66	0.64	0.92	0.81	0.81	**0.95**
Discriminability	0.68	0.45	0.77	0.67	0.66	**0.84**
Saliency hot targets	0.68	0.16	0.58	0.65	0.32	**0.77**

6. Discussion and Conclusions

We have proposed three new methods that focus on improving performance in different military tasks. The *luminance-fit* (LFF) method was intended to give a result in which the luminance more closely matches the daytime situation (compared to the result of the CTN method). Both our qualitative and quantitative (observer experiments and computational image quality metrics) evaluations indicate that this is indeed the case. This method is especially suited for situations in which natural daytime colors are required (leading to good situational awareness, and fast and easy interpretation) and for systems that need to be operated by untrained users. The disadvantage of this method over the CTN-scheme is that it leads to a somewhat lower discriminability of different materials. Again, the choice between the two fusion schemes has to be based on the application (i.e., adapted to the task and situation).

Secondly, we proposed a salient-hot-targets (SHT) fusion scheme, intended to render hot targets as more salient by painting the hot elements in more vibrant colors. The results of the quantitative evaluation tests show that this method does indeed represent hot elements as more salient in the fused image in most situations. However, in some cases a decrease in saliency may result. This suggests that this fusion scheme may be improved, e.g., by adapting the luminance of the hot elements to that of their local background (i.e., by enhancing local luminance contrast), or by using a different mixing scheme (e.g., by replacing the scheme depicted in the inset of Figure 10b by a different one).

Our third proposal was to create a color fusion method (rigid-3D-fit or R3DF method) that combines the advantages of the sample-based method (the fact that the direct correspondence between sensor values and output colors is used to create a result that closely matches the daytime image) along with the advantages of the statistical method (the fact that this method leads to a smooth transformation in which a fuller range of colors is used, leading to better discriminability of materials). A rigid-3D-fit was used to transform the input (sensor values) and output (daytime colors) (by mapping their CLUTs) to assure that the color space did not collapse under the transformation. The results of this fusion scheme look somewhat similar to that of the statistical method, although the colors are somewhat less naturalistic (but still intuitive). However, this method results in better discriminability of materials and has good generalization properties (i.e., it transfers well to untrained scenes). This is probably due to the fact that the transformation is constrained by the direct correspondence between input and output (and not only by the widths of the distributions). This fusion method is especially suited for applications in which the discriminability of different materials is important while the exact color is somewhat less important. Still, the colors that are generated are quite intuitive. Another advantage of this method is that the transformation may be derived from a very limited number of image samples (e.g., $N = 4$), and does not rely on a large training set spanning the complete set of possible multiband sensor values. The transformation can be made to yield predefined colors for elements of interest in the scene (e.g., vegetation, certain targets).

The results from both our qualitative and quantitative evaluation studies show that the original CTN method works quite well and that it shows good transfer to untrained imagery taken in the same environment and with similar sensor settings. Even in cases in which the environment or sensor settings are different, it still applies reasonably well.

Surprisingly, the CTN2 two-band mapping does not performs as well as expected, even when applied to untrained scenes. This suggests that there may be a relationship between the daytime colors and the LWIR sensor values which the fusion method utilizes and that this also applies to the untrained situations. It may be the case that there are different types of environments in which this relationship differs and that we happen to have recorded (and evaluated) environments in which this relationship was quite similar. Given the limited dataset we used (and which is freely available for research purposes: [74]) this may not be surprising. It suggests that the system should be trained with scenes representing different types of environments in which LWIR can be used to infer the daytime color.

One of the reasons why the learned color transformation does not always transfer well to untrained situations is probably that, in a new situation, the sensor settings can differ considerably. When, for instance, the light level changes, the (auto)gain settings may change, and one may end up in a very different location in 3D sensor/color space, which may ultimately result in very different output colors (for the same object). This can be only be prevented by using the sensor settings to recalculate (recalibrate) the values to those that would have been obtained if sensor settings had been used that corresponded to the training situation.

The set of images that is available for testing is still rather limited. Therefore, we intend to extend our dataset to include more variation in backgrounds, environmental conditions (weather, light conditions, etc.), which can serve as a benchmark set for improving and testing new color fusion schemes in the future.

Acknowledgments: Effort sponsored by the Air Force Office of Scientific Research, Air Force Material Command, USAF, under grant number FA9550-17-1-0079. The U.S. Government is authorized to reproduce and distribute reprints for Governmental purpose notwithstanding any copyright notation thereon. The authors thank Yufeng Zheng and Erik Blasch for providing the Matlab code of the OEI metric.

Author Contributions: The two authors contributed equally to the paper.

Conflicts of Interest: The authors declare no conflict of interest. The founding sponsors had no role in the design of the study; in the collection, analyses, or interpretation of data; in the writing of the manuscript, and in the decision to publish the results.

References

1. Li, S.; Kang, X.; Fang, L.; Hu, J.; Yin, H. Pixel-level image fusion: A survey of the state of the art. *Inf. Fusion* **2017**, *33*, 100–112. [CrossRef]
2. Mahmood, S.; Khan, Y.D.; Khalid Mahmood, M. A treatise to vision enhancement and color fusion techniques in night vision devices. *Multimed. Tools Appl.* **2017**, *76*, 1–49. [CrossRef]
3. Zheng, Y. An Overview of Night Vision Colorization Techniques Using Multispectral Images: From Color Fusion to Color Mapping. In Proceedings of the IEEE International Conference on Audio, Language and Image Processing (ICALIP), Shanghai, China, 16–18 July 2012; pp. 134–143.
4. Toet, A.; Hogervorst, M.A. Progress in color night vision. *Opt. Eng.* **2012**, *51*, 010901. [CrossRef]
5. Zheng, Y. An exploration of color fusion with multispectral images for night vision enhancement. In *Image Fusion and Its Applications*; Zheng, Y., Ed.; InTech Open: Rijeka, Croatia, 2011; pp. 35–54.
6. Wichmann, F.A.; Sharpe, L.T.; Gegenfurtner, K.R. The contributions of color to recognition memory for natural scenes. *J. Exp. Psychol. Learn. Mem. Cognit.* **2002**, *28*, 509–520. [CrossRef]
7. Bramão, I.; Reis, A.; Petersson, K.M.; Faísca, L. The role of color information on object recognition: A review and meta-analysis. *Acta Psychol.* **2011**, *138*, 244–253. [CrossRef] [PubMed]
8. Sampson, M.T. *An Assessment of the Impact of Fused Monochrome and Fused Color Night Vision Displays on Reaction Time and Accuracy in Target Detection*; Report AD-A321226; Naval Postgraduate School: Monterey, CA, USA, 1996.
9. Gegenfurtner, K.R.; Rieger, J. Sensory and cognitive contributions of color to the recognition of natural scenes. *Curr. Biol.* **2000**, *10*, 805–808. [CrossRef]
10. Tanaka, J.W.; Presnell, L.M. Color diagnosticity in object recognition. *Percept. Psychophys.* **1999**, *61*, 1140–1153. [CrossRef] [PubMed]

11. Castelhano, M.S.; Henderson, J.M. The influence of color on the perception of scene gist. *J. Exp. Psychol. Hum. Percept. Perform.* **2008**, *34*, 660–675. [CrossRef] [PubMed]
12. Rousselet, G.A.; Joubert, O.R.; Fabre-Thorpe, M. How long to get the "gist" of real-world natural scenes? *Vis. Cognit.* **2005**, *12*, 852–877. [CrossRef]
13. Goffaux, V.; Jacques, C.; Mouraux, A.; Oliva, A.; Schyns, P.; Rossion, B. Diagnostic colours contribute to the early stages of scene categorization: Behavioural and neurophysiological evidence. *Vis. Cognit.* **2005**, *12*, 878–892. [CrossRef]
14. Oliva, A.; Schyns, P.G. Diagnostic colors mediate scene recognition. *Cognit. Psychol.* **2000**, *41*, 176–210. [CrossRef] [PubMed]
15. Frey, H.-P.; Honey, C.; König, P. What's color got to do with it? The influence of color on visual attention in different categories. *J. Vis.* **2008**, *8*, 6. [CrossRef] [PubMed]
16. Bramão, I.; Inácio, F.; Faísca, L.; Reis, A.; Petersson, K.M. The influence of color information on the recognition of color diagnostic and noncolor diagnostic objects. *J. Gen. Psychol.* **2011**, *138*, 49–65. [CrossRef] [PubMed]
17. Spence, I.; Wong, P.; Rusan, M.; Rastegar, N. How color enhances visual memory for natural scenes. *Psychol. Sci.* **2006**, *17*, 1–6. [CrossRef] [PubMed]
18. Ansorge, U.; Horstmann, G.; Carbone, E. Top-down contingent capture by color: Evidence from RT distribution analyses in a manual choice reaction task. *Acta Psychol.* **2005**, *120*, 243–266. [CrossRef] [PubMed]
19. Green, B.F.; Anderson, L.K. Colour coding in a visual search task. *J. Exp. Psychol.* **1956**, *51*, 19–24. [CrossRef] [PubMed]
20. Folk, C.L.; Remington, R. Selectivity in distraction by irrelevant featural singletons: Evidence for two forms of attentional capture. *J. Exp. Psychol. Hum. Percept. Perform.* **1998**, *24*, 847–858. [CrossRef] [PubMed]
21. Driggers, R.G.; Krapels, K.A.; Vollmerhausen, R.H.; Warren, P.R.; Scribner, D.A.; Howard, J.G.; Tsou, B.H.; Krebs, W.K. Target detection threshold in noisy color imagery. In *Infrared Imaging Systems: Design, Analysis, Modeling, and Testing XII*; Holst, G.C., Ed.; The International Society for Optical Engineering: Bellingham, WA, USA, 2001; Volume 4372, pp. 162–169.
22. Horn, S.; Campbell, J.; O'Neill, J.; Driggers, R.G.; Reago, D.; Waterman, J.; Scribner, D.; Warren, P.; Omaggio, J. Monolithic multispectral FPA. In *International Military Sensing Symposium*; NATO RTO: Paris, France, 2002; pp. 1–18.
23. Lanir, J.; Maltz, M.; Rotman, S.R. Comparing multispectral image fusion methods for a target detection task. *Opt. Eng.* **2007**, *46*, 1–8. [CrossRef]
24. Martinsen, G.L.; Hosket, J.S.; Pinkus, A.R. Correlating military operators' visual demands with multi-spectral image fusion. In *Signal Processing, Sensor Fusion, and Target Recognition XVII*; Kadar, I., Ed.; The International Society for Optical Engineering: Bellingham, WA, USA, 2008; Volume 6968, pp. 1–7.
25. Jacobson, N.P.; Gupta, M.R. Design goals and solutions for display of hyperspectral images. *IEEE Trans. Geosci. Remote Sens.* **2005**, *43*, 2684–2692. [CrossRef]
26. Joseph, J.E.; Proffitt, D.R. Semantic versus perceptual influences of color in object recognition. *J. Exp. Psychol. Learn. Mem. Cognit.* **1996**, *22*, 407–429. [CrossRef]
27. Fredembach, C.; Süsstrunk, S. Colouring the near-infrared. In *IS&T/SID 16th Color Imaging Conference*; The Society for Imaging Science and Technology: Springfield, VA, USA, 2008; pp. 176–182.
28. Krebs, W.K.; Sinai, M.J. Psychophysical assessments of image-sensor fused imagery. *Hum. Factors* **2002**, *44*, 257–271. [CrossRef] [PubMed]
29. McCarley, J.S.; Krebs, W.K. Visibility of road hazards in thermal, visible, and sensor-fused night-time imagery. *Appl. Ergon.* **2000**, *31*, 523–530. [CrossRef]
30. Toet, A.; IJspeert, J.K. Perceptual evaluation of different image fusion schemes. In *Signal Processing, Sensor Fusion, and Target Recognition X*; Kadar, I., Ed.; The International Society for Optical Engineering: Bellingham, WA, USA, 2001; Volume 4380, pp. 436–441.
31. Toet, A.; IJspeert, J.K.; Waxman, A.M.; Aguilar, M. Fusion of visible and thermal imagery improves situational awareness. In *Enhanced and Synthetic Vision 1997*; Verly, J.G., Ed.; International Society for Optical Engineering: Bellingham, WA, USA, 1997; Volume 3088, pp. 177–188.
32. Essock, E.A.; Sinai, M.J.; McCarley, J.S.; Krebs, W.K.; DeFord, J.K. Perceptual ability with real-world nighttime scenes: Image-intensified, infrared, and fused-color imagery. *Hum. Factors* **1999**, *41*, 438–452. [CrossRef] [PubMed]

33. Essock, E.A.; Sinai, M.J.; DeFord, J.K.; Hansen, B.C.; Srinivasan, N. Human perceptual performance with nonliteral imagery: Region recognition and texture-based segmentation. *J. Exp. Psychol. Appl.* **2004**, *10*, 97–110. [CrossRef] [PubMed]
34. Gu, X.; Sun, S.; Fang, J. Coloring night vision imagery for depth perception. *Chin. Opt. Lett.* **2009**, *7*, 396–399.
35. Vargo, J.T. *Evaluation of Operator Performance Using True Color and Artificial Color in Natural Scene Perception*; Report AD-A363036; Naval Postgraduate School: Monterey, CA, USA, 1999.
36. Krebs, W.K.; Scribner, D.A.; Miller, G.M.; Ogawa, J.S.; Schuler, J. Beyond third generation: A sensor-fusion targeting FLIR pod for the F/A-18. In *Sensor Fusion: Architectures, Algorithms, and Applications II*; Dasarathy, B.V., Ed.; International Society for Optical Engineering: Bellingham, WA, USA, 1998; Volume 3376, pp. 129–140.
37. Scribner, D.; Warren, P.; Schuler, J. Extending Color Vision Methods to Bands Beyond the Visible. In Proceedings of the IEEE Workshop on Computer Vision Beyond the Visible Spectrum: Methods and Applications, Fort Collins, CO, USA, 22 June 1999; pp. 33–40.
38. Sun, S.; Jing, Z.; Li, Z.; Liu, G. Color fusion of SAR and FLIR images using a natural color transfer technique. *Chin. Opt. Lett.* **2005**, *3*, 202–204.
39. Tsagaris, V.; Anastassopoulos, V. Fusion of visible and infrared imagery for night color vision. *Displays* **2005**, *26*, 191–196. [CrossRef]
40. Zheng, Y.; Hansen, B.C.; Haun, A.M.; Essock, E.A. Coloring night-vision imagery with statistical properties of natural colors by using image segmentation and histogram matching. In *Color Imaging X: Processing, Hardcopy and Applications*; Eschbach, R., Marcu, G.G., Eds.; The International Society for Optical Engineering: Bellingham, WA, USA, 2005; Volume 5667, pp. 107–117.
41. Toet, A. Natural colour mapping for multiband nightvision imagery. *Inf. Fusion* **2003**, *4*, 155–166. [CrossRef]
42. Wang, L.; Jin, W.; Gao, Z.; Liu, G. Color fusion schemes for low-light CCD and infrared images of different properties. In *Electronic Imaging and Multimedia Technology III*; Zhou, L., Li, C.-S., Suzuki, Y., Eds.; The International Society for Optical Engineering: Bellingham, WA, USA, 2002; Volume 4925, pp. 459–466.
43. Li, J.; Pan, Q.; Yang, T.; Cheng, Y.-M. Color Based Grayscale-Fused Image Enhancement Algorithm for Video Surveillance. In Proceedings of the IEEE Third International Conference on Image and Graphics (ICIG'04), Hong Kong, China, 18–20 December 2004; pp. 47–50.
44. Howard, J.G.; Warren, P.; Klien, R.; Schuler, J.; Satyshur, M.; Scribner, D.; Kruer, M.R. Real-time color fusion of E/O sensors with PC-based COTS hardware. In *Targets and Backgrounds VI: Characterization, Visualization, and the Detection Process*; Watkins, W.R., Clement, D., Reynolds, W.R., Eds.; The International Society for Optical Engineering: Bellingham, WA, USA, 2000; Volume 4029, pp. 41–48.
45. Scribner, D.; Schuler, J.M.; Warren, P.; Klein, R.; Howard, J.G. *Sensor and Image Fusion*; Driggers, R.G., Ed.; Marcel Dekker Inc.: New York, NY, USA; pp. 2577–2582.
46. Schuler, J.; Howard, J.G.; Warren, P.; Scribner, D.A.; Klien, R.; Satyshur, M.; Kruer, M.R. Multiband E/O color fusion with consideration of noise and registration. In *Targets and Backgrounds VI: Characterization, Visualization, and the Detection Process*; Watkins, W.R., Clement, D., Reynolds, W.R., Eds.; The International Society for Optical Engineering: Bellingham, WA, USA, 2000; Volume 4029, pp. 32–40.
47. Waxman, A.M.; Gove, A.N.; Fay, D.A.; Racamoto, J.P.; Carrick, J.E.; Seibert, M.C.; Savoye, E.D. Color night vision: Opponent processing in the fusion of visible and IR imagery. *Neural Netw.* **1997**, *10*, 1–6. [CrossRef]
48. Waxman, A.M.; Fay, D.A.; Gove, A.N.; Seibert, M.C.; Racamato, J.P.; Carrick, J.E.; Savoye, E.D. Color night vision: Fusion of intensified visible and thermal IR imagery. In *Synthetic Vision for Vehicle Guidance and Control*; Verly, J.G., Ed.; The International Society for Optical Engineering: Bellingham, WA, USA, 1995; Volume 2463, pp. 58–68.
49. Warren, P.; Howard, J.G.; Waterman, J.; Scribner, D.A.; Schuler, J. *Real-Time, PC-Based Color Fusion Displays*; Report A073093; Naval Research Lab: Washington, DC, USA, 1999.
50. Fay, D.A.; Waxman, A.M.; Aguilar, M.; Ireland, D.B.; Racamato, J.P.; Ross, W.D.; Streilein, W.; Braun, M.I. Fusion of multi-sensor imagery for night vision: Color visualization, target learning and search. In *Third International Conference on Information Fusion, Vol. I-TuD3*; IEEE Press: Piscataway, NJ, USA, 2000; pp. 3–10.
51. Aguilar, M.; Fay, D.A.; Ross, W.D.; Waxman, A.M.; Ireland, D.B.; Racamoto, J.P. Real-time fusion of low-light CCD and uncooled IR imagery for color night vision. In *Enhanced and Synthetic Vision 1998*; Verly, J.G., Ed.; The International Society for Optical Engineering: Bellinggam, WA, USA, 1998; Volume 3364, pp. 124–135.

52. Waxman, A.M.; Aguilar, M.; Baxter, R.A.; Fay, D.A.; Ireland, D.B.; Racamoto, J.P.; Ross, W.D. Opponent-Color Fusion of Multi-Sensor Imagery: Visible, IR and SAR. Available online: http://www.dtic.mil/docs/citations/ADA400557 (accessed on 28 August 2017).
53. Aguilar, M.; Fay, D.A.; Ireland, D.B.; Racamoto, J.P.; Ross, W.D.; Waxman, A.M. Field evaluations of dual-band fusion for color night vision. In *Enhanced and Synthetic Vision 1999*; Verly, J.G., Ed.; The International Society for Optical Engineering: Bellingham, WA, USA, 1999; Volume 3691, pp. 168–175.
54. Fay, D.A.; Waxman, A.M.; Aguilar, M.; Ireland, D.B.; Racamato, J.P.; Ross, W.D.; Streilein, W.; Braun, M.I. Fusion of 2-/3-/4-sensor imagery for visualization, target learning, and search. In *Enhanced and Synthetic Vision 2000*; Verly, J.G., Ed.; SPIE—The International Society for Optical Engineering: Bellingham, WA, USA, 2000; Volume 4023, pp. 106–115.
55. Huang, G.; Ni, G.; Zhang, B. Visual and infrared dual-band false color image fusion method motivated by Land's experiment. *Opt. Eng.* **2007**, *46*, 1–10. [CrossRef]
56. Li, G. Image fusion based on color transfer technique. In *Image Fusion and Its Applications*; Zheng, Y., Ed.; InTech Open: Rijeka, Croatia, 2011; pp. 55–72.
57. Zaveri, T.; Zaveri, M.; Makwana, I.; Mehta, H. An Optimized Region-Based Color Transfer Method for Night Vision Application. In Proceedings of the 3rd IEEE International Conference on Signal and Image Processing (ICSIP 2010), Chennai, India, 15–17 December 2010; pp. 96–101.
58. Zhang, J.; Han, Y.; Chang, B.; Yuan, Y. Region-based fusion for infrared and LLL images. In *Image Fusion*; Ukimura, O., Ed.; INTECH: Rijeka, Croatia, 2011; pp. 285–302.
59. Qian, X.; Han, L.; Wang, Y.; Wang, B. Color contrast enhancement for color night vision based on color mapping. *Infrared Phys. Technol.* **2013**, *57*, 36–41. [CrossRef]
60. Li, G.; Xu, S.; Zhao, X. *Fast Color-Transfer-Based Image Fusion Method for Merging Infrared and Visible Images*; Braun, J.J., Ed.; The International Society for Optical Engineering: Bellingham, WA, USA, 2010; Volume 77100S, pp. 1–12.
61. Li, G.; Xu, S.; Zhao, X. An efficient color transfer algorithm for recoloring multiband night vision imagery. In *Enhanced and Synthetic Vision 2010*; Güell, J.J., Bernier, K.L., Eds.; The International Society for Optical Engineering: Bellingham, WA, USA, 2010; Volume 7689, pp. 1–12.
62. Li, G.; Wang, K. Applying daytime colors to nighttime imagery with an efficient color transfer method. In *Enhanced and Synthetic Vision 2007*; Verly, J.G., Guell, J.J., Eds.; The International Society for Optical Engineering: Bellingham, WA, USA, 2007; Volume 6559, pp. 1–12.
63. Shen, H.; Zhou, P. Near natural color polarization imagery fusion approach. In *Third International Congress on Image and Signal Processing (CISP 2010)*; IEEE Press: Piscataway, NJ, USA, 2010; Volume 6, pp. 2802–2805.
64. Yin, S.; Cao, L.; Ling, Y.; Jin, G. One color contrast enhanced infrared and visible image fusion method. *Infrared Phys. Technol.* **2010**, *53*, 146–150. [CrossRef]
65. Ali, E.A.; Qadir, H.; Kozaitis, S.P. Color night vision system for ground vehicle navigation. In *Infrared Technology and Applications XL*; Andresen, B.F., Fulop, G.F., Hanson, C.M., Norton, P.R., Eds.; SPIE: Bellingham, WA, USA, 2014; Volume 9070, pp. 1–5.
66. Jiang, M.; Jin, W.; Zhou, L.; Liu, G. Multiple reference images based on lookup-table color image fusion algorithm. In *International Symposium on Computers & Informatics (ISCI 2015)*; Atlantis Press: Amsterdam, The Netherlands, 2015; pp. 1031–1038.
67. Sun, S.; Zhao, H. Natural color mapping for FLIR images. In *1st International Congress on Image and Signal Processing CISP 2008*; IEEE Press: Piscataway, NJ, USA, 2008; pp. 44–48.
68. Li, Z.; Jing, Z.; Yang, X. Color transfer based remote sensing image fusion using non-separable wavelet frame transform. *Pattern Recognit. Lett.* **2005**, *26*, 2006–2014. [CrossRef]
69. Hogervorst, M.A.; Toet, A. Fast natural color mapping for night-time imagery. *Inf. Fusion* **2010**, *11*, 69–77. [CrossRef]
70. Hogervorst, M.A.; Toet, A. Presenting Nighttime Imagery in Daytime Colours. In Proceedings of the IEEE 11th International Conference on Information Fusion, Cologne, Germany, 30 June–3 July 2008; pp. 706–713.
71. Hogervorst, M.A.; Toet, A. Method for applying daytime colors to nighttime imagery in realtime. In *Multisensor, Multisource Information Fusion: Architectures, Algorithms, and Applications 2008*; Dasarathy, B.V., Ed.; The International Society for Optical Engineering: Bellingham, WA, USA, 2008; pp. 1–9.
72. Toet, A.; de Jong, M.J.; Hogervorst, M.A.; Hooge, I.T.C. Perceptual evaluation of color transformed multispectral imagery. *Opt. Eng.* **2014**, *53*, 043101. [CrossRef]

73. Toet, A.; Hogervorst, M.A. TRICLOBS portable triband lowlight color observation system. In *Multisensor, Multisource Information Fusion: Architectures, Algorithms, and Applications 2009*; Dasarathy, B.V., Ed.; The International Society for Optical Engineering: Bellingham, WA, USA, 2009; pp. 1–11.
74. Toet, A.; Hogervorst, M.A.; Pinkus, A.R. The TRICLOBS Dynamic Multi-Band Image Data Set for the development and evaluation of image fusion methods. *PLoS ONE* **2016**, *11*, e0165016. [CrossRef] [PubMed]
75. Pitié, F.; Kokaram, A.C.; Dahyot, R. Automated colour grading using colour distribution transfer. *Comput. Vis. Image Underst.* **2007**, *107*, 123–137. [CrossRef]
76. Arun, K.S.; Huang, T.S.; Blostein, S.D. Least-squares fitting of two 3-D point sets. *IEEE Trans. Pattern Anal. Mach. Intell.* **1987**, *5*, 698–700. [CrossRef]
77. Gamer, M.; Lemon, J.; Fellows, I.; Sing, P. Package 'irr': Various Coefficients of Interrater Reliability and Agreement (Version 0.84). 2015. Available online: http://CRAN.R-project.org/package=irr (accessed on 25 August 2017).
78. Yuan, Y.; Zhang, J.; Chang, B.; Han, Y. Objective quality evaluation of visible and infrared color fusion image. *Opt. Eng.* **2011**, *50*, 1–11. [CrossRef]
79. Heckbert, P. Color image quantization for frame buffer display. *Comput. Gr.* **1982**, *16*, 297–307. [CrossRef]
80. Zhang, L.; Zhang, L.; Mou, X.; Zhang, D. FSIM: A feature similarity index for image quality assessment. *IEEE Trans. Image Process.* **2011**, *20*, 2378–2386. [CrossRef] [PubMed]
81. Zheng, Y.; Dong, W.; Chen, G.; Blasch, E.P. *The Objective Evaluation Index (OEI) for Evaluation of Night Vision Colorization Techniques*; Miao, Q., Ed.; New Advances in Image Fusion; InTech Open: Rijeka, Croatia, 2013; pp. 79–102.
82. Zheng, Y.; Dong, W.; Blasch, E.P. Qualitative and quantitative comparisons of multispectral night vision colorization techniques. *Opt. Eng.* **2012**, *51*, 087004. [CrossRef]
83. Ma, M.; Tian, H.; Hao, C. New method to quality evaluation for image fusion using gray relational analysis. *Opt. Eng.* **2005**, *44*, 1–5.
84. Kovesi, P. Image features from phase congruency. *Videre J. Comput. Vis. Res.* **1999**, *1*, 2–26.

Journal of
Imaging

Article

Illusion and Illusoriness of Color and Coloration

Baingio Pinna [1,*], Daniele Porcheddu [2] and Katia Deiana [1]

1 Department of Biomedical Sciences, University of Sassari, 07100 Sassari, Italy; katiadeiana@yahoo.it
2 Department of Economics and Business, University of Sassari, 07100 Sassari, Italy; daniele@uniss.it
* Correspondence: baingio@uniss.it; Tel.: +39-3926315770

Received: 24 November 2017; Accepted: 22 January 2018; Published: 30 January 2018

Abstract: In this work, through a phenomenological analysis, we studied the perception of the chromatic illusion and illusoriness. The necessary condition for an illusion to occur is the discovery of a mismatch/disagreement between the geometrical/physical domain and the phenomenal one. The illusoriness is instead a phenomenal attribute related to a sense of strangeness, deception, singularity, mendacity, and oddity. The main purpose of this work is to study the phenomenology of chromatic illusion vs. illusoriness, which is useful for shedding new light on the no-man's land between "sensory" and "cognitive" processes that have not been fully explored. Some basic psychological and biological implications for living organisms are deduced.

Keywords: color vision; visual illusions; watercolor illusion; neon color spreading; illusion and illusoriness

1. Introduction: From Real to Illusory Colors

The perception of colors is one of the most interesting psychological experiences. A world without colors, as partially experienced in black-and-white photos, appears to be missing important qualities. For example, the play of colors during a spring walk in a public park reveals the light reflected in the still water, filling in the basin of a fountain and showing leaves, plants, and trees reflected on the water underneath a morning blue sky. It can be easily imagined that the same experience with achromatic colors would contribute to missing some of the very essential perceptual qualities. For example, the spring colors of plants would disappear, the morning blue sky would be absent, the vibrations and reflexes of the light on the water and on the surrounding plants and objects would be barely perceived, the water itself would be almost invisible, and more, the different phenomenal depth planes revealed by the colors, e.g., among the trees and the sky, and above the still water where the leaves and plants are reflected, would not be clearly defined and segregated. On the contrary, the play of chromatic colors shows the beauty and expressiveness of the components and of the reflected lights that are totally absent in the achromatic picture. More importantly, colors not only add important information to define what a specific portion of the visual world is, but also they create unique and emerging expressive qualities that cannot be otherwise induced.

Impressionist artists, like Claude Monet, knew how to create "impressions" by using colors. In fact, he depicted the transient effects of water and sky and the reflected colors of its ripples by employing many colors and varied textures. His purpose was to reproduce the manifold and animated effects of sunlight and shadow, as well as recreate those other ones perceived under direct and reflected light. This was done by reproducing immediate visual impressions and by using complementary colors in shadows, in place of using whites and blacks. Furthermore, he also made large use of discrete flecks of pure harmonizing or contrasting colors, in order to evoke the broken-hued brilliance and the hue variations produced by sunlight and its reflections. As a consequence, objects in his paintings appear dematerialized, shimmering, and vibrating within their multiplicity of chromatic colors. Indeed, his atmospheres and reflexes would be completely invisible and inaccessible if painted achromatically.

Despite the extraordinary experience of color perception, all colors are mere illusions, in the sense that, although naive people normally think that objects appear colored because they are colored, this belief is mistaken. Neither objects nor lights are colored, but colors are the result of neural processes. In fact, the physical properties of colors are different from the way colors are perceived. Newton first knew it when in 1704 wrote: "for the rays to speak properly are not colored". The perception of colors starts with the spectral content of the light that triggers a long chain of neural reactions, starting from three classes of cone photoreceptors in the retina and, finally, involving specialized areas in the brain. Only at an unknown end or at some extremes of the physiological chain in the brain, the experience of color emerges. The study of color perception allows the understanding of the chain of neural processes and lastly of the brain that creates this unique and extraordinary result that is not necessarily restricted to the mere primary and sensorial processes.

Stating that colors are mere illusions means that they are related to a mismatch/disagreement between the geometrical/physical domain and the phenomenal one [1–3]. More generally, the basic condition for the perception of an illusion is the discovery of this mismatch. Despite the presence of the mismatch, it can be considered as necessary (I, if M; I whenever M and I when M, where I is illusion and M mismatch) to perceive an illusion, although it is not sufficient (if I then M or I implies M); in fact, not all the kinds of mismatches reveal the attribute of being illusory with the same strength.

These preliminary observations point to a first set of problems that is useful for revealing the complexity of the notion of visual illusion. More specifically, the illusion and the attribute of being illusory (the sense of strangeness, deception, singularity, mendacity, and oddity), i.e., the "illusoriness", can be considered as belonging to different perceptual issues. As a matter of fact, when the illusion is ascertained, the phenomenal attribute of being illusory does not necessary emerge. Even by perceiving the mismatch between different domains, the illusoriness can be weak or totally absent. On the other hand, a strong illusoriness is not necessarily related to the presence of an illusion and, thus, to a mismatch between different domains.

A second set of problems suggests that the perception of an illusion is a process that mostly occurs over time. Accordingly, the perception of the property of being illusory requires free observations, comparisons, afterthoughts, many ways of seeing, etc.

Although all colors are illusions, they usually do not appear illusory. Nevertheless, under certain sensory conditions colors and their attributes can also appear illusory. The appearance of being illusory is a perceptual attribute, whose values are arranged in a long phenomenal gradient going from the condition where colors are not perceived as illusory (e.g., the color of the objects in front of us) to the condition where colors are immediately perceived as a mere result of the visual process. As shown in the next sections, in between these two extremes there is a complex set of possible color appearances that deserve to be studied to understand the phenomenology of color perception and the way color is coded and processed by the sensory and cognitive systems.

Even though there is a very large literature related to the huge and increasing number of visual illusions and to their possible classifications [4], actually there are no studies that focus on the inner complexity of visual illusions as previously described. More particularly, the term "illusoriness" and its phenomenal notion have been totally unnoticed up to now.

The purpose of this work is to show effective examples of illusory colors, to demonstrate the distinction between chromatic illusion and illusoriness, and to direct the scientific attention to the complexity of "sensory" and "cognitive" processes concerning the perception of colors in an extended acceptation similar to that previously described during a spring walk in a public park.

This distinction will be shown to have important psychological and biological implications for living organisms. In short, each organism uses visual strategies to create illusions and illusoriness, for example, to deceive prays or predators, to attract conspecifics, or to reject, dissuade, or repulse them. They play with illusions to change and manipulate their appearance and influence other organisms. To be effective, the biological illusory strategies should reduce the sense of illusoriness that could signal the deceiving structure or coloration. It is within a complex equilibrium between illusion and illusoriness that the adaptive fitness is played.

2. General Methods

2.1. Subjects

In the experiments described in the next sections, different groups of 12 naive subjects, each ranging from 20 to 27 years of age, were involved for each stimulus, if not otherwise reported. Subjects were about 50% males and 50% females and all had normal or corrected to normal vision. They also had normal color vision tested on the Ishihara 24 test plates and on the Farnsworth-Munsell 100 hue Color Vision Test. Subjects were recruited under previous informed consent signed by all the participants in compliance with the Helsinki declaration.

2.2. Stimuli

The stimuli were the same as the figures shown in the next sections. The luminance of the white background was ~122.3 cd/m^2. Black contours had a luminance value of ~2.6 cd/m^2). The stimuli were displayed on a 33 cm color CRT monitor (Sony GDM-F520; Sony: Tokyo, Japan; 1600 × 1200 R, G, and B subpixels, refresh rate 100 Hz), driven by a MacBook computer (Apple: California, USA) with an NVIDIA GeForce 8600M GT (NVIDIA: California, USA), in an ambient illuminated by a Osram Daylight fluorescent light (250 lux, 5600 K). They were viewed binocularly and in a front-parallel plane at a distance of 50 cm from the monitor.

2.3. Procedure

Phenomenological task—the subjects had to report spontaneously what they perceived for each stimulus. They were expected to give as much as possible, an exhaustive description and, if necessary, to answer the questions asked by the experimenter. Subjects were also instructed to scale the relative strength and salience in percent (100 is the maximal salience, while 0 is the minimal) of the perceived results, if there were any. The relative confidence and appropriateness of their responses were also taken into account.

Scaling task—subjects were asked to report if they perceived a chromatic/achromatic illusion and to scale the relative strength or salience of the perceived illusion in percent. The reference values of the scale were as follows: 0 represents the absence of the chromatic/achromatic illusion, in other words, not any illusion is perceived at all; 100 represents the maximal strength of the illusory appearance, indeed, here, the color is perceived illusory, and therefore not present in the pattern of stimuli but only as a product of vision.

Relevant descriptions are included in the following sections, within the main text, in order to help the reader in the stream of argumentations. The edited descriptions reported were judged as the best example to provide a fair representation among those provided by the observers. When not specified, the descriptions reported in the next sections were those spontaneously communicated by ten out of twelve subjects, considered highly appropriate (more than 90%) and receiving a scaling value higher than 90%.

During the experiments, subjects could make free comparisons, confrontations, and afterthoughts; additionally, they were allowed to see in different ways, and to make variations and comparisons in the observation distance, etc. Subjects could also receive suggestions/questions from the experimenter.

Subjects were tested one by one. No time limit was set on the descriptions and their scaling, which occurred spontaneously and fast. The stimuli were shown continuously during the description task. All the details and variations concerning the experiments and related to the subjects, the stimuli, and the procedure will be reported more in details in the next sections. Besides, together with the results of each phenomenological experiment, a theoretical discussion will also be provided.

3. Results

3.1. Illusory Colors and Colorations

As stated in the Introduction section, visual illusions are usually associated with a mismatch/disagreement between the geometrical/physical and the phenomenal domains [1–4]. The necessary condition for the occurrence and the perception of an illusion is the "discovery" of this mismatch, which is mostly a cognitive process. This general and well-accepted definition of "visual illusion" is postulated either explicitly or implicitly when a new phenomenon is discovered and first published in a scientific journal.

This simple definition of illusion, apparently clear, reveals several inner ambiguities that we start here to investigate phenomenologically. First of all, the ambiguities involve both domains. What is physical and what is phenomenal? In this work, the complexity of the meaning of the physical domain is skipped to prevent a regression to a too long and already explored epistemological issue [5] and, at the same time, to keep our rationale as simple and direct as possible. Here, we will especially focus on the phenomenology of a colored object starting from the most simple questions that could be asked to a subject during a visual experiment: "What is it?" and "What color is it?". Through these questions, apparently trivial, we will explore the phenomenology of illusory colors, when colors are or are not perceived as illusory.

The first question, "What is it?", was asked by showing Figure 1a. The most frequent answer was "an undulated circular shape" (9 subjects out of 12).

Within the phenomenological approach here adopted, the scientific interest not only is related to what emerges and is immediately described by the subjects, but is also related to what is implicit, taken for granted, and not described. The distinction between explicitness and implicitness, together with the gradient of phenomenal appearance, is absolutely crucial to understand the gradient of visibility and also the hierarchical organization of the object attributes [6]. In fact, the gradient of visibility not only shows what is perceived at first sight and most saliently, but also puts in evidence what is seen as secondary, what stands in the background, or what is completely invisible. However, what is implicit or not statistically relevant and meaningful can be phenomenally even more important for the understanding of the object organization than what is immediately perceived.

Under the conditions of Figure 1a, only shape attributes were spontaneously mentioned. The color was not acknowledged at all. This does not mean that it is invisible, but that color attributes fade into a lower level of visibility and are kept implicit in relation to the primacy of the shape. They become invisible on the foreground visual/linguistic level. Even more implicit/invisible was the color (white) of the surrounding background, which was never reported by the subjects and, thus, placed at the lowest level of visibility.

The second question "What color is it?" was rephrased as follows "What about the color?". This question was asked to minimize any suggestions and to leave subjects free to describe which, among the three colors (the white of the inside object, the white of the outside object, and the black of the boundary contours) is the one that spontaneously and explicitly emerges, and which instead remains implicit. The most immediate and common outcome (9 out of 12) was "the undulated object is white". As a consequence, only the inner color of the perceived shape emerged explicitly. Both the colors of the boundary contours and of the background were kept implicit/invisible. It is essential to highlight the reaction of 2 out of 12 subjects soon after they answered the question: they manifested surprise becoming aware that there might be other options. Despite these further options, mostly considered as related to the color of the boundary contours, they selected the following one: "the undulated object is white", and thus not black.

The explicitness and implicitness of visual attributes are more likely related to the biological need for an organism to organize the visual world into a gradient of visibility, eliciting the emergence of the most significant information in terms of adaptive fitness to the detriment of the less important visual information. This entails that the background is less important than the segregated object, and that the color of the inner edges of the object is more important than the color of its boundaries. Moreover, the shape of an object is more important than its color.

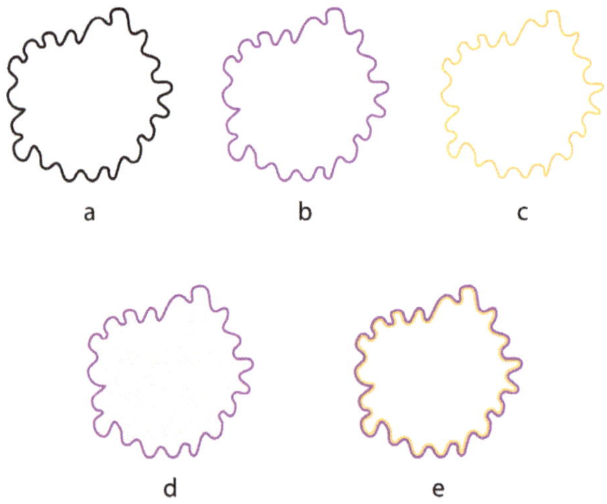

Figure 1. Undulated circular objects. (**a**) Black; (**b**) purple; (**c**) orange; (**d**) purple filled with orange; (**e**) illusory watercolored.

It is worthwhile mentioning that 3 out of 12 subjects answered the previous question as follows: "the undulated object is transparent (1 out of 12); it is empty or made only by its boundary contours like a wire (2 out of 12)". At a first sight, they appear as the most correct answers to the given conditions. However, the transparency and the absence of an inner surface of the object implies that the background can be perceived through it, but this also means that the color seen behind it is white. In other words, though these descriptions appear very different and more plausible than the most common ones, they are conceptually and structurally the same. As a matter of fact, not any subject mentioned the color of the boundary contours. Moreover, by saying "transparent" or "empty", they are describing an inner property of the object, not the properties of its contours, and this is exactly what the majority of subjects reported by stating that the undulated object is white. In short, since the inner property is in one case "white", in the other two, transparent or empty, the phenomenal structure is the same (*quod erat demonstrandum*).

The results of this first set of data can be summarized according to the next general statements: (i) Color can be implicit, unnoticed, or invisible; shape cannot (principle of primacy of shape against color); (ii) the color of the boundary contours is phenomenally "invisible".

By replacing the black contours of Figure 1a with purple ones (Figure 1b), the outcomes reported by a different group of subjects were similar to the previous ones, although 4 out of 12 subjects mentioned the purple color of the boundary contours. Nevertheless, the answer to the question "What about the color?" was again "white" not purple. These results suggest that chromatic attributes of the boundary contours are more easily noticed than the black ones, although they do not influence the outcome. This is *a fortiori* true when the boundary contours are orange as shown in Figure 1c. Now, 8 out of 12 subjects mentioned the orange contours but defined as white the color of the object. The larger number of subjects under this condition is related to the lower contrast of the orange with the white background. This point will be cleared up when the watercolor illusion will be discussed.

For the three previous stimuli, a third question was asked: "Do you perceive any color illusion?". The answer was definitively "no" for each stimulus. Actually, Figure 1a–c manifests two chromatic illusions of invisibility: the one concerning the contours and the other one related to the white background. The implicitness of the color of the boundaries and of the white background might be considered as a clear mismatch/disagreement between the geometrical/physical and the phenomenal domains. By making subjects aware of these mismatches, i.e., by pushing the color of the boundaries and of the background above threshold, they manifested a sense of illusoriness due to the unaware omission and oversight induced by the stimulus.

Under our work, the illusoriness can be defined as an independent perceptual attribute emerging from a multiplicity of visual outcomes within the same stimulus pattern, not necessarily related to the presence of a mismatch, and perceived by itself like a sense of strangeness, deception, singularity, mendacity, and oddity.

In Figure 1a–c, no illusoriness was perceived spontaneously; therefore, no illusion was reported in spite of the discrepancies previously described and mostly located within the linguistic domain of the description task due to unaware oversights. As previously stated, a sense of illusoriness emerged only when subjects were made aware of the invisibility of the two other colors. Though subjects considered this effect mainly as a lack within the description, a surprise was manifested by all of them. This suggests that the visual attribute of illusoriness of these figures can be perceived, even if weakly, and it is related to the mismatch related to the illusions of invisibility. This entails weak illusoriness with weak illusion.

Similar results were collected with a different group of subjects through the spontaneous descriptions of Figure 1d. Again, the answer to the first question ("What is it?") was "an undulated circular shape" (6 out of 12) and, only after the second question ("What about the color?"), the inner color of the shape emerged ("the undulated circular shape is orange"), but neither the color of the contours nor the one of the background was mentioned. As before, subjects did not perceive any color illusion nor any illusoriness, unless suggested by the experimenter. The illusoriness of Figure 1d is weaker than the one of the previous stimuli, since the inner color is not white but clearly different from the background.

The inferred general statements are now the followings: the color of the boundary contour plays a different role from the one of the inner surface, and their roles are, respectively, shape and color. Since the boundary contour defines the shape of the object, its color belongs to the shape domain and it assumes this role, i.e., it is only shape not color. On the contrary, the color of the inner surface is not shape, but it appears as the color or coloration of the object defined in its shape by its boundary contours. Therefore, the inner color is just color and not shape. From this statement, a corollary can be deduced: the color of the boundary contours is not the color of the object; therefore, as color, it is "invisible" or implicit. It is visible or "explicit" only as boundary or shape of the object.

These statements are based on a more general assumption of a specialization of roles, according to which a specific region can assume only one role. Therefore, if a role is assumed by one region, the other assumes a different and complementary role (principle of unicity and separation of roles). This role specialization fits with the more general tendency of the brain to create the simplest, most non-ambiguous, non-ambivalent, and economical phenomenal results in the basis of given stimulus conditions [7].

The principle of unicity and separation of roles can explain both the absence of illusoriness within the spontaneous outcomes and the emergence of the illusoriness when other possible options were suggested to the subjects. As a matter of fact, this principle works by switching one set of univocal roles to an alternative set of roles, both of which are mutually exclusive.

Finally, a further general statement related to all the previous results is: the color of the background is unnoticed, implicit and, thus, invisible.

Apparently similar but substantially different from the previous conditions is the next stimulus in which the purple and orange contours of Figure 1b,c are now combined and placed adjacent. By presenting Figure 1e to an independent group of subjects, the results were the same as those obtained in all the previous conditions. As a matter of fact, also under those circumstances, the answer to the first question was "an undulated circular shape" (8 out of 12) and, only after the second question, the inside color of the object was mentioned, but neither the one of the contours nor the one of the background emerged. As it happened in the previous case, subjects did not notice any color illusion nor any illusoriness. The inner color was perceived as it is physically depicted within the shape. These results are corroborated by the answers to the further question "What is the color of the inner region of the undulated circle?". The answer reported for this question was "an orange color slightly lighter than the one along the boundaries".

After these preliminary tasks and reports, subjects were informed that Figure 1e is a visual illusion. The new task was to disclose the hidden illusion. After a first momentary astonishment and confusion, subjects guessed about the inner coloration of the shapes, although all of them needed to check closely the figure in order to be sure and realize that indeed the illusion occurred. In spite of this feedback, they were not totally sure that the inner coloration was the illusion suggested by the experimenter. Four out of 12 subjects questioned and challenged the request of the experimenter. The inner coloration appeared to be real, as it was actually depicted on the computer screen. A few subjects (5 out of 12) proposed to consider as illusion the circular shape, which appears as such although it is very irregular and different from a circle. According to these results, the watercolor illusion, illustrated in Figure 1e, can be considered as an illusion without illusoriness.

The watercolor illusion [8] can be briefly described as a long-range coloration effect sending out from a thin colored line running parallel and contiguous to a darker chromatic contour and imparting a strong figural effect, similar to a bulging volumetric effect, across large areas [8–11]. Although the watercolor illusion manifests two main effects only, the coloration was noticed, and the second hidden "illusion" (the figural effect) was never detected, although the experimenter suggested the presence of a second illusion to be discovered. In between the two main effects, the long range homogeneous coloration is more prominent as an illusion. The primacy of the shape against the color [12,13] makes it more difficult to acknowledge its possibility to be an illusion and its illusoriness.

The figural effect can be made explicit by comparing the bulging volumetric effect of the undulated circular objects of Figure 1b,e or through the inverted figure-ground organizations illustrated in Figure 2, where the same pattern of stimuli, with purple and orange contours reversed, shows complementary figure-ground segregation, i.e., crosses vs. stars.

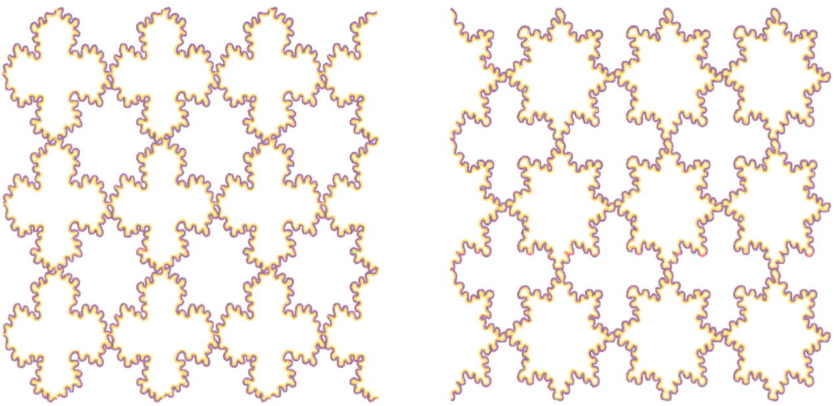

Figure 2. *The watercolor illusion*: purple and orange contours reversed show complementary figure-ground segregation, i.e., crosses vs. stars.

In Figure 3, transparent watercolor objects and holes placed on a gray rectangle are shown. The discovery of the illusoriness of this more complex condition required more time for our subjects, and it was related to the antinomic proximity of objects and holes, not to the figural and coloration effects of the watercolor illusion.

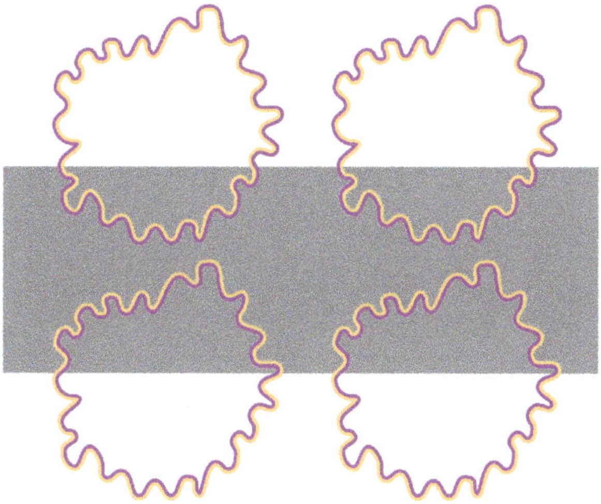

Figure 3. Transparent watercolor objects and holes.

The figural and coloration effects of the watercolor illusion was more salient in Supplementary Materials, in which the complementary peninsulas and the coloration are switched by reversing the contrast of adjacent contours and by changing their color at the same time. The dynamical switch is accompanied by a phenomenal enlarging and shrinking, a "breath" of the figures while emerging and fading. However, the true illusions are the figural and coloration effects; it was the breathing outcome that was considered by the subjects as the main source of the perceived illusoriness. Only partially (4 out of 12) the illusoriness was assigned to the "true" illusions.

Actually, the breathing is not a true illusion, since the contour with the highest contrast, perceived as the boundary contour of the figure, changes its spatial location from the position of one contour to the position of the other. As a consequence, the boundary contour location appears to move laterally, as it really does, by inducing the enlarging and shrinking effect along the vertical direction of the stimulus. Under these dynamic conditions, the illusoriness is stronger without the need to be revealed to the subjects. However, the perceived illusoriness occurs without illusion. Therefore, within Supplementary Materials there are two illusions (figural and coloration effect) without illusoriness and illusoriness (the breath) without illusion.

The general emerging statements are now: the illusoriness is a phenomenal attribute independent from a mismatch between geometrical/physical and phenomenal domains; moreover, suggestions and cognitive processes can favor the emergence of both illusion and illusoriness.

The general statement related to the two juxtaposed contours and to the previous figures is as follows. Given two adjacent contours as in the watercolor illusion, the one with the higher luminance contrast in relation to the surrounding regions is seen as the boundary contours of the object, while the contour with the lower luminance contrast is seen as the color of the figure [12–14]. In more general terms, the assignment of the role is due to the relative contrast between the two adjacent contours. This explains the differences between Figures 1b and 1c previously described.

In Figure 4-top, subjects described the pattern of stimuli as an amoeboid shape. As in the previous conditions, the color was poorly mentioned (5 out of 12 subjects). The answer to the question "What about the color?" was "orange". However, the answer to a further question "What is the color of the inner region of the amoeboid shape?" was "white". "What kind of white?": "a white similar to the white of the background (6 out of 12), more dense and solid" (4 out of 12), "slightly darker than the white of the background" (5 out of 12), or ''white but slightly yellowish" (1 out of 12). These results suggest that, while the color of the amoeboid shape as a whole is orange, the color of its inner region is not orange. This contradiction suggests that part and whole colors can be different.

Figure 4. The illusion of discoloration.

None of the subjects perceived any sort of illusion or illusoriness. When Figure 4-bottom was shown to the same group of subjects, the outcomes were very different: they perceived a dark gray amoeboid shape. The results did not change significantly by asking the same questions to a new group of subjects. Again, not any kind of illusion or illusoriness was perceived.

First of all, this suggests that in Figure 4-top the inner orange surrounding the external boundaries of the shape was perceived as the color of the whole figure, although the inner region is white or slightly darker than the white of the background. So, a tiny chromatic portion of the figure assumes the role of color for the whole shape [15].

Second, the orange color of the boundaries of Figure 4-bottom was not mentioned at all. It was implicit and invisible. This is related to the basic assumption of the visual system, according to which two different regions of figure cannot assume the same role. Therefore, the color of the boundary contour and the one of the inner surface assume different roles, i.e., shape and color. The outer orange was in some way "discolored" and considered only as a contour and more specifically, as the boundary contour of the shape (on the different assignment of contour roles see also Figure 5).

Third, Figure 4-top shows the discoloration illusion [16] according to which the inner region appears white, although it is physically filled with the same gray of Figure 4-bottom. The gray is indeed phenomenally discolored. The same illusion occurs with chromatic inner coloration (not illustrated). The phenomenal discoloration can be considered as a phenomenon opposite to the coloration effect of the watercolor illusion.

The phenomenal attribute of the illusion of discoloration is different from the "discoloration" of the orange contours of Figure 4-bottom, where the "discoloration" is part of the visual/descriptive assignment of roles. The discoloration of Figure 4-top is, in fact, visually effective and "real" as it "really" occurs. This implies that the kind of discoloration of Figure 4-bottom is part of the perceptual syntactical organization of shape and color. In Figure 4-top, the discoloration is rather related to an earlier kind of visual process and perceptual organization that occurs before the syntactical one, in which the visual language operates and from which the spoken language picks up its atoms and components.

A different kind of illusory coloration can emerge under totally different geometrical and phenomenal conditions as shown in the following variations of Picasso's *Yellow Cock* (Figure 5). The three figures were described simply as roosters. When the color was asked, Figure 5a,b were described as yellow roosters, while Figure 5c was seen as a black rooster. Therefore, the black contours were perceived as the boundaries of the rooster's body, while the yellow contours were defined as its color. When the subjects were asked to compare Figure 5a,b, the consequence of the role assignment was that the rooster of Figure 5b was described as much slimmer than the one illustrated in Figure 5a. This is related to the fact that the black contours are now included within the yellow ones. Since the black contours were perceived as the boundaries of the yellow rooster, the rooster should also be perceived as slimmer than the one in Figure 5a, in which the black contours include the yellow ones. By replacing the yellow contours with black ones, the rooster was perceived as black and fat (Figure 5c). These results demonstrate that it is not the spatial position of the contours that determines their roles (boundary or color) but their difference in the luminance contrast as it occurred in the watercolor illusion.

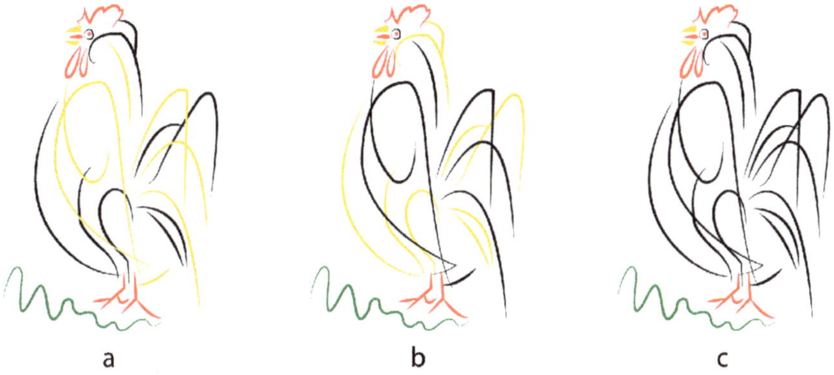

Figure 5. Variations of Picasso's *Yellow Cock*: *the fat-slim illusion*. (**a**) Yellow rooster; (**b**) slim yellow rooster; (**c**) fat black rooster.

It is worthwhile noticing that, although the coloration does not fill the whole inner edge of the shape, the roles are assigned anyway by the visual system. The general statement can be synthesized as follows: the perception of the color of an object does not require the full and perfect coloration of its inside edges [12,13].

In spite of the huge discrepancy between the geometrical/physical pattern of stimuli and the phenomenal results, the subjects did not report any illusion nor illusoriness. The outcomes appeared obvious and trivial. Only after a full and deep explanation of the differences were the meaning of the illusion and the illusoriness noticed, although they was mostly related to what apparently seems to be a fat vs. a slim rooster (the fat-slim illusion).

The illusoriness was perceived more saliently in Figure 6, in which a variation of the watercolor illusion without the figurality effect is illustrated. The red and light-blue colorations spreading and filling in the space among the circles (in reversed order between the two patterns) were spontaneously reported as "apparent", "illusory", "as an impression", or "as a result of brain processes". According to subjects' comments, this coloration was perceived as illusory, since it does not appear as belonging to a surface but because it is more similar to a light, a fog, or "a strange" color. Under these conditions the illusoriness emerged immediately revealing the sense of strangeness, deception, singularity, mendacity, and oddity.

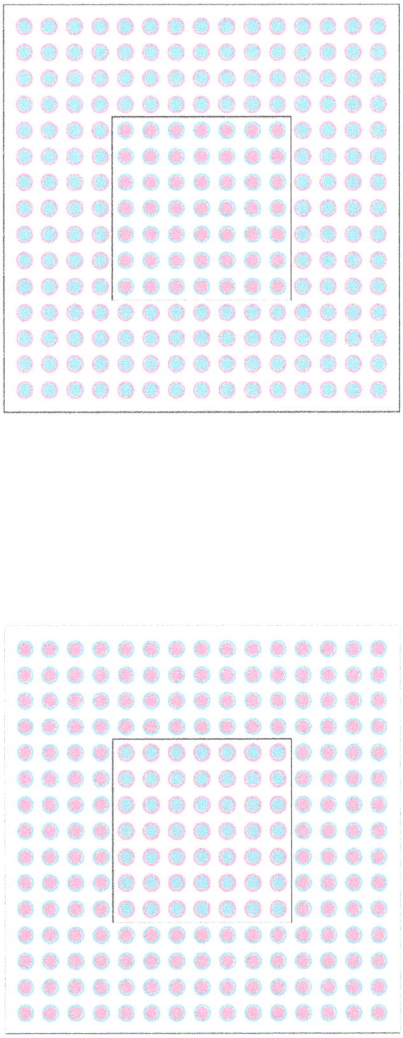

Figure 6. Reversed chromatic variations of the watercolor illusion.

This is also the case in Figure 7a,b, in which new variations of the neon color spreading [10] are illustrated. Under these conditions, the inner green/orange and green/black spreads with phenomenal qualities are analogous to those of Figure 6. These qualities induce again the emergence of the illusoriness as a sense of strangeness and the singularity of the color that is unexpected and different from the way it appears most of the time.

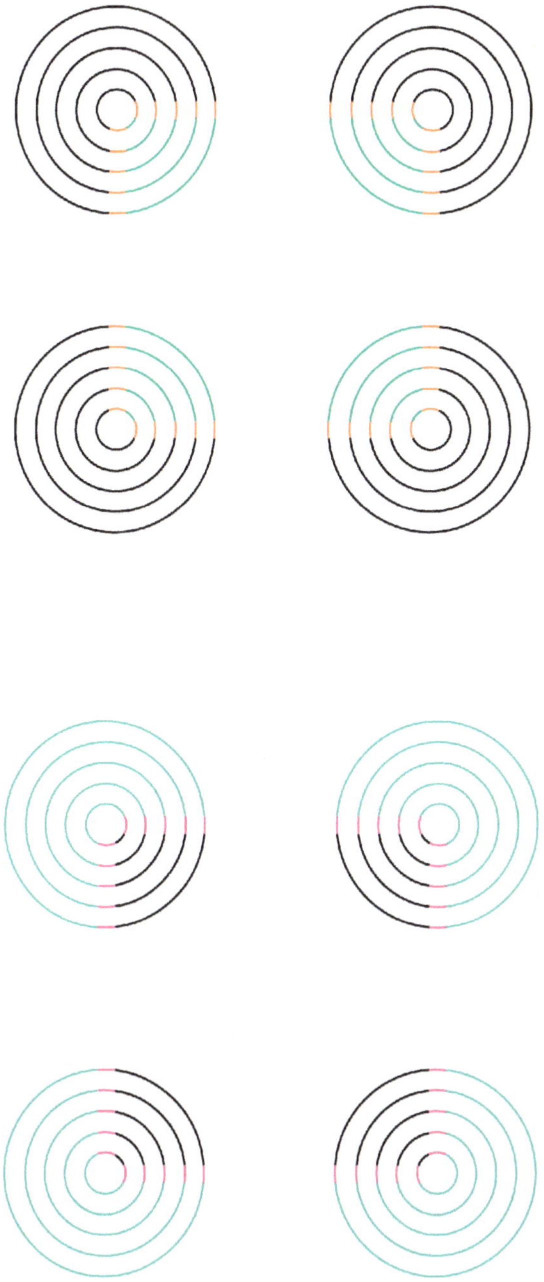

Figure 7. Variations of the Neon Color Spreading. (**a**) green-orange spreading; (**b**) black-pink spreading.

The phenomenology of the coloration effect in both illusions of Figures 6 and 7 manifests the following properties: (i) the color appears as a diffusion of a small amount of pigment of the embedded chromatic contours; (ii) the coloration is transparent like a light, a shadow, or a fog; (iii) the appearance of the color is diaphanous, and it appears similar to a veil that glows like a light upon the background, like a transparent layer, or like a dirty, shadowy, foggy, or muddy filmy sheet.

While the watercolor illusion occurs through the juxtaposition of parallel lines with different color, the neon color spreading is elicited by the continuation of one segment with another of a different color (continuation vs. juxtaposition). Pinna & Grossberg [10] showed that both illusions can be reduced to a common two-dots limiting case, based on nearby color transitions.

The chromatic illusion and illusoriness were fully and immediately perceived in Figure 8. Differently from the previous conditions, the answer to the question "what is it?" promptly revealed the chromatic illusoriness. The centers of the radial arrangement of lines were perceived of as a vivid and saturated green self-luminous, and as flashing with eye or stimulus movement. The green flashing effect disappears when the center of the radial lines is perceived directly (foveally). Under these conditions, the inner disk appeared gray. At the same time, all the other disks (extrafoveally) were perceived like flashing greens floating out of their confines, pulsing or oscillating. This is the flashing anomalous color contrast [17] fully perceived as an illusion with illusoriness.

Figure 8. The flashing anomalous color contrast.

3.2. Chromatic and Achromatic Whites

In the previous section, not only the chromatic components of the stimuli but also the whites of the objects and the background were considered. The illusoriness was shown to be related to these whites. More particularly, one of the illusions without illusoriness, not mentioned in the previous section, is the phenomenal difference between these whites as, for example, it happens in Figure 1a. This illusion was not mentioned as such and likely it was not even noticed by the reader accurately, as it does not manifest any illusoriness. This occurs in spite of the fact that it was a basic problem for the Gestalt psychologist. According to Rubin [18,19], the white of the undulated object is perceived of as full like an opaque surface and denser than the same physical white of the surrounding background that is perceived instead as empty and totally transparent. While the former manifests a surface color property with a chromatic paste seen as solid, impenetrable, and epiphanous like an opaque surface, the white of the background is, on the other hand, perceived as empty, penetrable, and diaphanous, like a void [20]. As soon as subjects were made aware of this difference they perceived the illusoriness, but they did so very weakly and took it for granted.

The distinction between the two Rubin's whites can be appreciated in Figure 9, although it is much stronger than the one illustrated in Figure 1a,b. Now, the word ARTE (Art in English) can be promptly read [21]. The R and the E letters manifest a strong surface and epiphanous white in contraposition to the diaphanous void of the background. The illusoriness of this condition is quite clear, and it is related to the illusion/mismatch that does not need to be revealed to be perceived as such. In spite of this clear effect, the illusoriness was considered by our subjects as derived from the readiness and past experience of the black letters. This naïve explanation weakens the illusoriness. This is not the case of the next condition.

Figure 9. The illusory word ARTE.

A white with a more salient illusoriness is illustrated in Figure 10. Now, the inside edges of the circular arrangement of letters are perceived as white surfaces and the effect here is much stronger that those described in Rubin's figures. The emerging illusory shape is promptly perceived as illusory and with a clear illusoriness. The white is seen much more prominently, like an independent surface, and is whiter than the white of the background that is perceived as empty and diaphanous. These illusory figures emerge in spite of the absence of any amodal completion and the incompleteness of the inducing elements [22,23].

The perceived illusoriness is about as strong as or slightly weaker than the condition in which these occlusion cues are introduced as shown in Figure 11. Here, the radial lines are perceived as amodally completed behind the illusory disks [24], whose white is similar to the one illustrated in Figure 10.

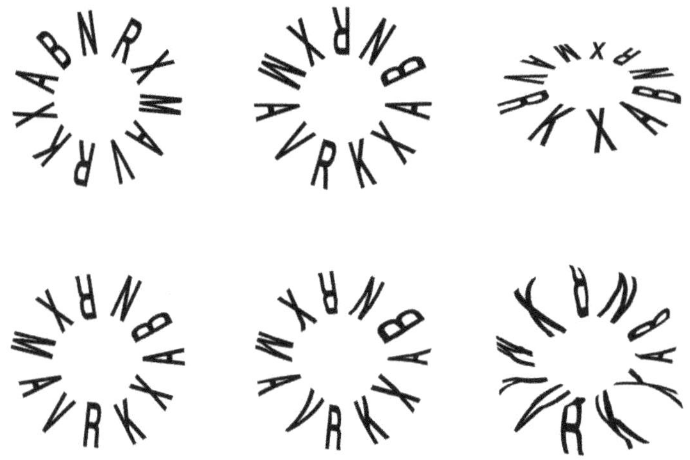

Figure 10. Circular arrangements of letters inducing inner illusory figures.

Figure 11. Radial lines inducing inner illusory disks.

The salience of the illusions and of the illusoriness of the whites illustrated in Figures 10 and 11 is strongly reduced by replacing the illusory contours of each disk of Figure 11 with black rings (see Figure 12). Now, although Rubin's whites are weaker than the previous ones, they are very similar for at least two important properties: they are all achromatic and opaque whites.

Figure 12. Radial lines and black rings showing achromatic whites.

By replacing each black ring with a purple one (Figure 13), the inner white becomes clearly chromatic. The resulting effect [23] is much brighter than in the Ehrenstein's illusion, and it appears like a surface colored with a dense paste of bright and quasi-luminous white color. The expression "chromatic white" is related to the fact that the white appears as if belonging not to the gray continuum but to the chromatic saturated colors. The phenomenon is clearly perceived as an illusion, and it manifests strong illusoriness attributes. All colors can produce these chromatic whites.

A control without radial lines is illustrated in Figure 14. Now the chromatic whites are absent and only Rubin's achromatic whites are perceived without any illusion and illusoriness.

The chromatic illusory whites can be enhanced as illustrated in Figure 15, in which the whites, although opaque and sharp in their contours, appear to be emanating their own light and self-luminous in a unique way that differs from a glare effect as shown in Figure 16a, in which each inner ring appears to shine with a dazzling apparent luminosity not necessarily brighter than the one of the background.

Figure 13. Radial lines and purple rings inducing chromatic whites.

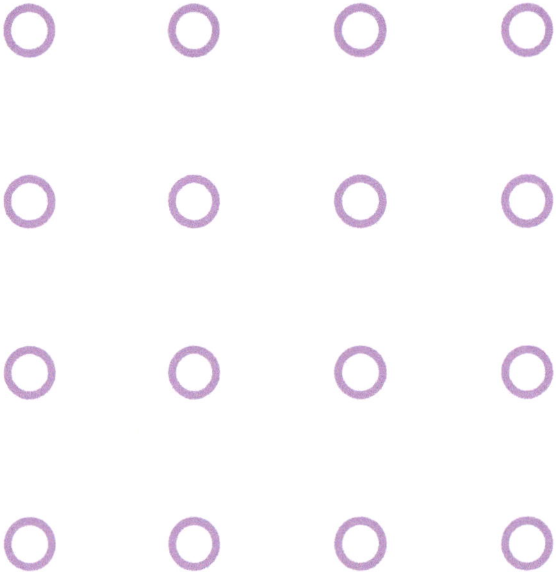

Figure 14. By removing the radial lines the chromatic whites are absent.

Figure 15. Two conditions showing chromatic whites enhanced and self-luminous.

It is worth noting, for our purposes, that the glare phenomenon, common in everyday life and not showing any illusoriness, as it appears, for example in Figure 16b, becomes illusory when the critical region is gradually zoomed as observable in the sequence of pictures of Figure 16b. In greater detail, the self-luminous white of Figure 16b, which is much brighter than the white of the background, gradually disappears when the picture zooms more and more inside that white. The sequence of photograms can annul or enhance the illusoriness depending on the starting point of the sequence, namely, from the beginning to the end or vice versa. This suggests that the perception of the illusoriness can change, not only by means of the awareness of the illusion but also through different ways of watching the stimulus.

(a)

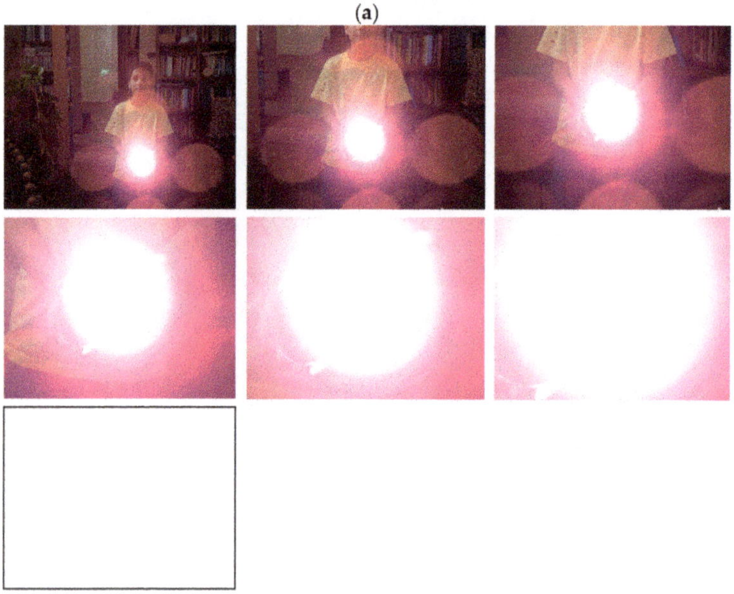

(b)

Figure 16. A glare effect of chromatic whites. (**a**) Dazzling apparent luminosity; (**b**) illusory glare due to the gradual zooming of the sequence of pictures.

This last point is important, since the visual world is full of illusory objects and elements that, at a first sight, appear strange, deceiving, *sui generis*, mendacious, and odd. Only after observation from a different point of view, in a different way, from a different angle, or with a simple prolonged sight, these objects decrease or increase their illusoriness. This is the case of the fish illustrated in Figure 17. By watching this figure, subjects were deceived for a short time by the two "eyes", although some of them (4 out of 12) needed a prolonged time (about 10 s) to discover the trick of the fish that the eyes are false. Therefore, in the beginning the illusoriness was absent and only after a relatively longer observation the illusoriness emerged.

Figure 17. A fish playing with illusions and illusoriness.

Other fishes, using colors to play with illusions and illusoriness, are illustrated in Figure 18. Again, spots, stripes, and false eyes create illusions that do not show, at first sight, any illusoriness. However, it emerges after a short time by revealing the trick.

Figure 18. Different fishes playing with illusions and illusoriness.

The illusoriness perceived by our subjects is not necessarily perceived by the peculiar preys and predators of these fishes. More generally, if a prey perceives the illusoriness then it survives; conversely, if a predator perceives the illusoriness, its hunting can succeed. In nature, animals have to deal with the play of illusion and illusoriness, according to which the perception of illusoriness is related to the adaptive fitness of an organism. In the Discussion section, these biological implications of the illusoriness will be explored in more detail.

4. Discussion

In contemporary vision science, illusions, which are becoming more and more popular, subsume two important issues that challenge scientists and deserve to be further investigated. The first issue is related to the mismatch between the geometrical/physical domain and the phenomenal one. The necessary condition to see an illusion is usually considered to be related to the perception of this mismatch. The second issue follows from the first, and it is related to at least two different visual levels that emerge from each illusion: the perception of the illusion and the perception of the illusoriness.

If in recent neuroscience all visual percepts are considered equally illusory, in this work we demonstrated that the phenomenology of the illusions and illusoriness is placed along a visual gradient of visibility with complex and useful interactions. More particularly, we demonstrated that (i) the mismatch between physical and phenomenal domains does not necessarily involve the perception of an illusion; (ii) besides the illusion is useful to study the illusoriness that is an independent visual attribute related to a sense of strangeness or oddity; (iii) the illusoriness emerges spontaneously or after a deeper observation and acknowledgment of mismatches; and, finally, (iv) the perception of an illusion does not necessarily require the perception of the illusoriness.

We also suggested that these issues are very useful to demonstrate new biological meanings and implications useful to understanding the complexity of the organization of the visual world and the interactions between visual and cognitive domains. As a matter of fact, the complex interactions and changes, studied in the previous sections, between explicitness and implicitness reflect the necessity to organize the visual world into a gradient of visibility, in which the most significant information in terms of adaptive fitness emerges in the foreground, to the detriment of the less important visual information placed in the background. This entails that, although all visual percepts can be considered equally illusory, it is not biologically useful to perceive everything as being illusory. In other terms, the perception of illusions and of illusoriness is not advantageous to the animal behavior that is required to act promptly without doubts and hesitations induced by illusions. Only under special conditions of illusoriness it is useful to stop, rearrange, and reorganize unexpected conditions.

Since illusoriness is related to a sense of strangeness, deception, singularity, mendacity, and oddity, its visual meaning plays a basic biological role. In fact, it is not a mere companion of illusions but a signal of a hidden structure invisible at a first sight, a structure that is related to the gradient of visibility and that can pop out thanks to a deeper investigation or cognitive discovery. More importantly, illusoriness plays a basic biological role by alerting and warning an organism about something wrong, unclear, ambiguous, tricky, or potentially dangerous within the visual world. As a matter of fact, all living organisms, from plants to animals, use visual strategies to deceive prey or predators, to attract, fascinate, or seduce conspecifics, or to reject, dissuade, or repulse them. In detail, each organism plays with illusions to change and manipulate their appearance and to deceive and influence other organisms to its will and purposes. This enhances the adaptive fitness. At the same time, organisms need to reduce the illusoriness that the represented illusion might manifest; otherwise, the illusory trick is uncovered and, immediately, discovered. Thus, to be effective, all the biological illusory strategies should elude the sense of illusoriness that could signal the deceiving behavior, structure, or coloration.

It is not coincident that colors and colorations are so commonly used to play with the illusory biological signals and identities. Color attributes are the best candidate to play with deceiving and illusory effects. Animal and plant colors play a basic role for their survival [25]. Color is used as a cue for species identification; it emphasizes and determines sexual dimorphism within the same species, and it also advertises the presence of animals and signals its dangerousness or poisonousness. Colors are also useful for identifying contenders, strength, and age. In addition, colors are used for cryptic camouflage and are seen as a defense against predation. In fact, cryptic camouflage [26–29] is a deceiving tool that is used to become imperceptible based on coloration. The opposite is the disruptive camouflage [26,30,31], whose purpose is to confuse by means of highly contrasted colorations and markings [32].

False eyes, dots (Figures 17 and 18), and a variety of patterns present in several animals' bodies are aimed to confuse by showing a disruptive masking and deceiving shape, and by hiding the most vital and important part of their body. The same defensive markings (diematic patterns) can also have the effect of startling or frightening potential predators. In butterflies, the presence of a decoy target, such as a false eye or other patterns, diverts the attack of possible predators away from the butterfly's body, towards the wings' borders.

To conclude, further studies of illusoriness could help to understand more deeply how adaptive fitness depends on the play of illusions and illusoriness, and how the biological meaning of the intense feedbacks and interconnections between organisms and between real and illusory visual conditions are essential to their survival.

Supplementary Materials: The following are available online at http://www.mdpi.com/2313-433X/4/2/30/s1, Video S1: By slowly reversing the contrast of adjacent contours, complementary peninsulas with inner illusory coloration emerge.

Aknowledgments: This work is supported by Fondazione Banco di Sardegna (finanziato a valere sulle risorse del bando Fondazione di Sardegna—Annualità 2015).

Author Contributions: Baingio Pinna and Daniele Porcheddu conceived and designed the experiments; Katia Deiana performed the experiments; Daniele Porcheddu and Katia Deiana analyzed the data; Baingio Pinna wrote the paper.

Conflicts of Interest: The authors declare no conflict of interest.

References

1. Coren, S.; Girgus, J.S. *Seeing is Deceiving: The Psychology of Visual Illusions*; Lawrence Erlbaum Associates: Hillsdale, NJ, USA, 1978.
2. Gregory, R.L. *Seeing Through Illusions*; Oxford University Press: Oxford, UK, 2009.
3. Robinson, J.O. *The Psychology of Visual Illusions*; Hutchinson: London, UK, 1972.
4. Gregory, R.L. Visual illusions classified. *Trends Cogn. Sci.* **1997**, *1*, 190–194. [CrossRef]
5. Metzger, W. *Psychologie: Die Entwicklung Ihrer Grundannhamen Seit der Einführung des Experiments*; Steinkopff: Dresden, Germany, 1941. (In German)
6. Pinna, B. New Gestalt principles of perceptual organization: An extension from grouping to shape and meaning. *Gestalt Theory* **2010**, *32*, 1–67.
7. Koffka, K. *Principles of Gestalt Psychology*; Harcourt, Brace & World: New York, NY, USA, 1935.
8. Pinna, B. Un effetto di colorazione. In Proceedings of the XXI Congresso Degli Psicologi Italiani, Arezzo, Italy, 28 September–3 October 1986; p. 158.
9. Pinna, B.; Brelstaff, G.; Spillmann, L. Surface color from boundaries: A new 'watercolor' illusion. *Vis. Res.* **2001**, *41*, 2669–2676. [CrossRef]
10. Pinna, B.; Grossberg, S. The watercolor illusion and neon color spreading: A unified analysis of new cases and neural mechanisms. *J. Opt. Soc. Am. A* **2005**, *22*, 2207–2221. [CrossRef]
11. Werner, J.S.; Pinna, B.; Spillmann, L. The Brain and the World of Illusory Colors. *Sci. Am.* **2007**, *3*, 90–95. [CrossRef]
12. Pinna, B. Perceptual organization of shape, color, shade and lighting in visual and pictorial objects. *Iperception* **2012**, *3*, 257–281. [CrossRef] [PubMed]
13. Pinna, B. The organization of shape and color in vision and art. *Front. Hum. Neurosci.* **2012**, *5*, 104. [CrossRef] [PubMed]
14. Pinna, B.; Reeves, A. Lighting, backlighting and watercolor illusions and the laws of figurality. *Spat. Vis.* **2006**, *19*, 341–373. [CrossRef] [PubMed]

15. Pinna, B.; Reeves, A. What is the purpose of color for living beings? Toward a theory of color organization. *Psychol. Res.* **2013**, *79*, 64–82. [CrossRef] [PubMed]
16. Pinna, B. The Discoloration Illusion. *Vis. Neurosci.* **2006**, *23*, 583–590. [CrossRef] [PubMed]
17. Pinna, B.; Spillmann, L.; Werner, J.S. Flashing anomalous colour contrast. *Vis. Neurosci.* **2004**, *21*, 365–372. [CrossRef] [PubMed]
18. Rubin, E. *Synsoplevede Figurer;* Glydendalske Boghandel: Kobenhavn, Denmark, 1915. (In German)
19. Rubin, E. *Visuell Wahrgenommene Figuren;* Gyldendalske Boghandel: Kobenhavn, Denmark, 1921. (In German)
20. Katz, D. *Die Erscheinungsweisen der Farben*, 2nd ed.; MacLeod, R.B., Fox, C.W., Eds.; The World of Color: London, UK, 1930. (In German)
21. Pinna, B. *Il Dubbio Sull'Apparire;* Upsel Editore: Padova, Italy, 1990. (In Italian)
22. Pinna, B.; Grossberg, S. Logic and phenomenology of incompleteness in illusory figures: New cases and hypotheses. *Psychofenia* **2006**, *9*, 93–135.
23. Pinna, B.; Spillmann, L.; Werner, J.S. Anomalous induction of brightness and surface qualities: A new illusion due to radial lines and chromatic rings. *Perception* **2003**, *32*, 1289–1305. [CrossRef] [PubMed]
24. Ehrenstein, W. Ueber Abwandlungen der L. *Hermannschen Helligkeitserscheinung. Z. Psychol.* **1941**, *150*, 83–91. (In German)
25. De Valois, K.K. The Role of Color in Spatial Vision. In *The Visual Neurosciences;* Chalupa, L.M., Werner, J.S., Eds.; MIT Press: Cambridge, MA, USA, 2004; pp. 1003–1016.
26. Cott, H.B. *Adaptive Coloration in Animals;* Methuen and Co. Ltd.: London, UK, 1940.
27. Edmunds, M. *Defence in Animals;* Longman Group Ltd.: Harlow, UK, 1974.
28. Endler, J.A. Interactions between predators and prey. In *Behavioural Ecology: An Evolutionary Approach*, 3rd ed.; Krebs, J.R., Davis, N.B., Eds.; Blackwell Scientific Publisher: Oxford, UK, 1991; pp. 169–196.
29. Ruxton, G.D.; Sherratt, T.N.; Speed, M.P. *Avoiding Attack: The Evolutionary Ecology of Crypsis, Warning Signals and Mimicry;* Oxford University Press: Oxford, UK, 2004.
30. Cuthill, I.C.; Stevens, M.; Sheppard, J.; Maddocks, T.; Párraga, C.A.; Troscianko, T.S. Disruptive coloration and background pattern matching. *Nature* **2005**, *434*, 72–74. [CrossRef] [PubMed]
31. Merilaita, S.; Lind, J. Great tits (Parus major) searching for artificial prey: Implications for cryptic coloration and symmetry. *Behav. Ecol.* **2006**, *17*, 84–87. [CrossRef]
32. Stevens, M.; Cuthill, I. Disruptive coloration, crypsis and edge detection in early visual processing. *Proc. Biol. Sci.* **2006**, *273*, 2141–2147. [CrossRef] [PubMed]

Journal of
Imaging

MDPI

Review

The Academy Color Encoding System (ACES): A Professional Color-Management Framework for Production, Post-Production and Archival of Still and Motion Pictures

Walter Arrighetti [1,2,3]

[1] CISSP, Media and Entertainment Technology Consultant, 00100 Rome, Italy; walter.technicolor@gmail.com
[2] Agency for Digital Italy (AgID), 00144 Rome, Italy
[3] Department of Mathematics and Computer Science, John Cabot University, 00165 Rome, Italy

Received: 24 July 2017; Accepted: 13 September 2017; Published: 21 September 2017

Abstract: The Academy of Motion Picture Arts and Sciences has been pivotal in the inception, design and later adoption of a vendor-agnostic and open framework for color management, the Academy Color Encoding System (ACES), targeting theatrical, TV and animation features, but also still-photography and image preservation at large. For this reason, the Academy gathered an interdisciplinary group of scientists, technologists, and creatives, to contribute to it so that it is scientifically sound and technically advantageous in solving practical and interoperability problems in the current film production, postproduction and visual-effects (VFX) ecosystem—all while preserving and future-proofing the cinematographers' and artists' creative intent as its main objective. In this paper, a review of ACES' technical specifications is provided, as well as the current status of the project and a recent use case is given, namely that of the first Italian production embracing an end-to-end ACES pipeline. In addition, new ACES components will be introduced and a discussion started about possible uses for long-time preservation of color imaging in video-content heritage.

Keywords: ACES; scene-referred; output-referred; color gamut; transfer characteristic; reference rendering transform; look modification transform; color look-up table; ColorLUT; Color Decision List; technicolor; primaries; illuminant; white point; film look; VFX; Input Transform; Output Transform; RRT; LMT; interoperable master format; CLUT; LUT; CDL; IMF

1. Introduction

As the acronym indicates, the Academy Color Encoding System (ACES) is a system championed by the Academy of Motion Pictures Arts and Sciences' Science and Technology Council (hereinafter referred to as AMPAS, or as simply "the Academy"), whose head organization is best known to the world for setting up the annual Academy Award, aka "the Oscars®". ACES is about Color management in a broad sense, as it is laid out as a series of technical and procedural documents describing how to generate, encode, process, archive imaging content (especially moving images) in an interoperable and standardized way [1–4].

The moving picture industry has only recently transitioned from a (mostly) film-based ecosystem, with certain "traditional" business logics and industrial scales, to a fully digital one ([5]) with all its disruptive methodologies regarding dematerialized content processing and security. This includes geographical, temporal, technological and marketing segmentation of industry services that are, nowadays, more akin to those of commercial/consumer digital markets. This also led both studios and indie productions (each with different timescales and budgets but converging toward the same solutions, as typical for this kind of economy) to turn to "pay-as-you-go" services and to the global market of service providers [6–8]. I am referring here to services deployed using Cloud business models, where almost no capital expenses are required (servers, storage, infrastructure, connectivity), in favor of metrics-driven operational costs, based on actual usage and production volume. Despite this shift raising understandable information-security concerns [9], this has already proved to be a key advantage, because it provides cost savings in production and postproduction as well, due to the increased flexibility, re-usability and global scope of digital technologies. Just to name two examples related to the present paper: the intent of the DoP (i.e., the cinematographer/director of photography or, as Storaro suggests calling it [10], the author of photography) can now be immediately checked "on-set", and the "look" of the film tailored since the beginning. This is achieved by means of a pre-grading workstation (with a calibrated reference monitor), without waiting weeks after the set is closed and look decisions are definitely moved into the post-production laboratory, Section 2.3 [11]. Second, due to the increase in global, faster Internet connectivity and secure file-transfer technologies, studios can easily distribute workloads for thousands of visual-effect shots (VFX), for each of their movies, among several companies around the globe, with the benefits of:

- sharing financial and content-security risks,
- reducing production times, and
- improving realistic outcome of the overall Computer-Generated Imaging (CGI), due to the differentiation of assets among several artists and VFX companies.

The downside of these new industrial-scale processes, worsened by the rapid inflation of digital tools emerged even before economically-feasible workflows (e.g., cloud media processing), is the lack of technology standards and procedures. These bred and evolved into an ecosystem with lots of vendors, producing proprietary technologies on one side (mostly incompatible with other vendors' products), and facilities using methodologies with radically different basic approaches on the other—sometimes even different nomenclature for the same things. As for that past in which fewer and bigger film laboratories owned patented frame formats and secret photochemical processes [12,13], this habit has continued in the digital age, leading to a plethora of new trade secrets and color processes, mostly incompatible with each other, and almost without any written formal procedures. This is especially true about color management (or "color science", as it is sometimes called, often abusing the scientific approach), which is a delicate process that, when not appropriately tackled with, easily leads to quick "patches" to overcome subtleties in things like camera matching, monitor/projector color-calibration, color-space conversions, inter-facility color pipelines, etc. All these patches lead to very specific color processes, rarely transportable across further shows, or translated for different engineers and/or software. As said at the beginning of the paragraph, while this may have had some business/marketing sense in the film era (in the role of Technicolor color consultants, for example, [12,13]), it is now just a threat to the average lean digital workflow.

It is exactly in this context [14], that ACES was born back in 2004 by an effort of the Academy which, like it did in the past many times with theatrical technology breakthroughs (panoramic film projection formats, stereo and multi-channel sound systems, etc., cfr. [12]), coveted a group of scientists, engineers, cinematographers, VFX technical directors and other industry professionals, to find an accessible and vendor-neutral solution to the problem of moving picture color management. This time for the first time, the Academy-proposed solution lives in the realm of "(color) metadata", is completely digital, open-source and supported by current internet/collaborative methodologies. The ACES project, initially called IIF (Interoperable Interchange Format), was then renamed "ACES" in 2012, when the author also officially joined the Academy internal group of experts. The first official version labeled 1.0 was released in December 2014 [1,15]. In fact, ACES ideas, software, formulas and techniques had been circulating among more experienced users and some vendors ever since earlier versions of it, so these are now usually addressed as "pre-release ACES", as they may involve legacy terminology and standards. The whole ACES framework will be introduced in Section 3.

The author has been mostly providing color-science consultancy to ACES' core internal components (cfr. color-spaces in Section 3.3 and the Reference Rendering Transform in Section 3.5) and—thanks to consolidated experience as technology executive for postproduction/VFX companies—technology expertise on the design of metadata integration of additional ACES components (LMT, CTL, CommonLUT, ACESclip in Sections 3.6–3.9) and file storage (Section 3.10). As of 2017, he is also cooperating with UK-based company The Foundry (London, UK) for the implementation of additional features introduced with ACES 1.0 (Sections 3.8 and 3.9) within their *Nuke* family of compositing, CGI texturing and image-processing software.

At the time of writing, more than 120 films (mostly theatrical, but including TV series/movies, documentaries and short films) were shot using ACES, with a similar figure estimated out of commercials and video art as well. Video games and virtual/augmented reality (VR/AR) are soon also expected in ACES, due to recent efforts to bring professional color management to those industries.

2. The Color Pipeline in the Post-Production and VFX Industry

In this section, several concepts in motion picture digital color management will be introduced or reviewed. They have a general validity even in contexts where ACES is not employed, but this chapter serves as both a background introduction for readers not acquainted with these industry-specific problems, and a pretext for introducing some colorimetry definitions and the terminology used throughout. Extensive, yet not exhaustive bibliographical references are given. The color issue introduced in Section 1 resides in three key areas of the below media production value-chain, cfr. Figure 1.

2.1. Input Colorimetry

Professional digital cameras usually record images using not only proprietary "raw" file formats (including compression and de-mosaicing algorithms [3,16]), but also proprietary imaging and colorimetries designed to account for camera sensors' characteristics and, possibly, counteract their technical weaknesses. This leads to a plethora of new color-spaces whose optical/digital definitions are, at best, only partially disclosed by camera manufacturers, e.g., ARRI LogC Wide Gamut, RED DRAGONColor*m* REDLogFilm*n*/Gamma*n*, Sony S-Log*m* S-Gamma*n* (m,n = 1,2,3 according to the camera manufacturers' color-science versions or generation of the sensors), Panasonic V-Log, GoPro CineForm/ProTune™, etc., cfr. Figure 2 and [17]. The native plates for most of the cinematic content are thus born digitally in these partially undefined spaces, but have one thing in common: they use a scene-referred colorimetry [3,18]. This means that each point-coordinate in the color-space—and the integer or floating-point code-value (CV) representing it in the dots of a digital raster—is related to a specific luminance value; its white-point CV is that of a perfect white diffuser [18,19] (as lit by the reference illuminant). The color-space's transfer characteristic [20] is, in this case, the transfer curve that maps all the luminance values from full-dark to full-white, into the (supposedly greyscale) CVs representing them.

2.2. Creative-Process Colorimetry

Postproduction software vendors are usually required to sign Non-Disclosure Agreements (NDAs) with the camera manufacturers in order to access codebase and information to process such formats; sometimes Software Development Kits (SDKs) are only provided to them, so all the "raw" processing is really black-box modeled further down the media-processing pipeline. Even without dealing with intellectual property (IP) and marketing mechanisms, the additional problem of preserving the original footage and its colorimetry (due to the obsolescence of proprietary colorimetries and raw file formats) is evident. Figure 3 shows the non-trivial, internal color-processing steps that average color-correction systems perform in the background, for when original camera colorimetry is processed to be correctly viewed. Real-world videos usually include cuts sourced from different cameras (not only film, TV, and advertising; photographic or animation; for instance documentaries and news); usually including partial- or full-CGI imaging composited over photographic plates [3]. Video editing, color-correction, compositing, as well as well as 3D CGI software must not only feature processing capabilities for as much of the above file formats and color-spaces [21], but also be flexible enough to accommodate for a wide range of color pipelines that were designed with many different kinds of constrains in mind (from budget, to filming locations/schedules, to hardware/software equipment used on-set and in postproduction, to artists/engineers' preferences). Color management is often problematic as not all the above solutions interoperate with each other—sometimes not even in nomenclature or user experience (UX). More on this will be detailed in Section 2.4.

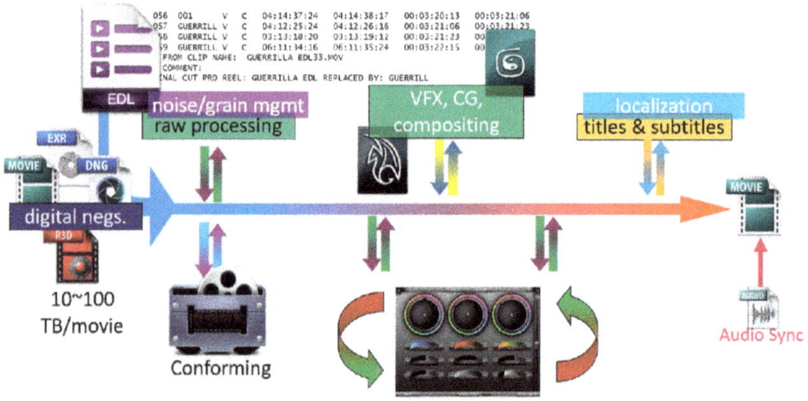

Figure 1. A very simple imaging postproduction pipeline (possibly involving several facilities).

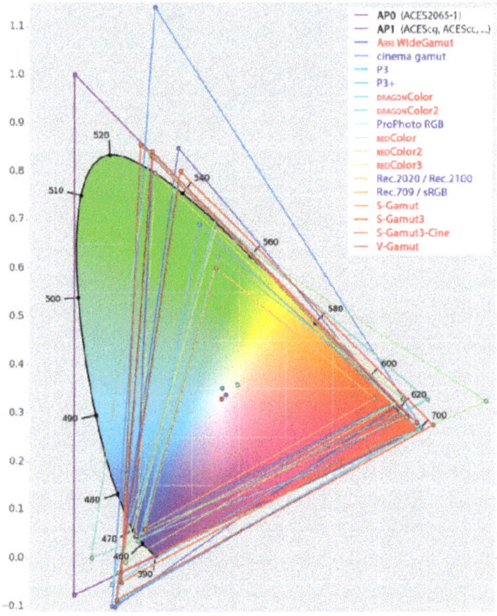

Figure 2. CIE 1931 chromaticity diagram (y axis shown) with comparison between ACES (black), professional digital motion picture cameras (red) and reference cinema/display (blue) color gamuts.

2.3. Output Colorimetry

Throughout all the imaging pipeline, from generation (camera or CGI) up to the delivery of masters, the same picture may not only be technically represented in different color-spaces and encodings according to the specific processing stage, but is actually viewed and creatively judged via different displaying devices, set in variegate viewing environments and employing a number of different optical mechanisms to reproduce the light. For this reason, the output colorimetry is called display-referred (or more generically, output-referred) [3,18]. Dually to scene-referred colorimetry, this means that color-space intensity and CVs relate to the luminance level as either directly produced by an additive-light device (monitors/displays), or by measuring the subsequent reflection of the projected light off a screen (theater projector); the transfer characteristic is physically the inverse of the one in Section 2.1; the above light source, at full power, is considered the "virtual scene illuminant" and the measured white is considered the color-space white-point [20]. For this, and accounting for other nonlinearities in the different color-reproduction physics, a change in the luminous intensity of a monitor or projector lamp, may change the output colorimetry, justifying peak luminance as another possible parameter for it. It is measured in SI unit cd/m^2; sometimes in US custom unit nits or foot-lamberts (fl), where 1 cd/m^2 $\equiv$ 1 nit = 0.2918 fl. In the case of projected light (e.g., in a digital cinema or grading theater), the transfer and the illuminant's spectral characteristics depend on those in the whole light path: lamphouse and lens first, the glass of the booth (if any), and especially the wide-screen which light is reflected off (effectively considered the 100%-white). The screen may, in fact, have micro-perforations or even metallic threads to preserve reflected-light polarization for some stereographic (S3D) theater systems. Environmental light comes also into play due to color adaption.

Figure 3. Toy-model of a simple, single-camera, non-ACES "basic" grading pipeline: the arrow on the right orders the "layer-based" representation of real-time image processing operations, as performed by the color-correction system in the background, while colorist works. Source footage in its camera-native colorimetry (**bottom**) is creatively modified by color transforms, either coming from the set (Color Decision Lists, cfr. Section 2.4) and applied in-theater (3-way CC'S); then a "viewing LUT" is applied on top to match the colorimetry with the output device's for viewing (**top**). A 3D LUT may represent the overall effect of a single primary grade (possibly lacking accuracy [17,20]), and may include the viewing LUT and all color-space conversions (courtesy: Hieronymus Bosch, *The Garden of Earthly Delights*, c.1495–1505).

From on-set reference monitors (dependent on the location's lighting, when not inside tents), to the reference monitor of the dailies' colorist (usually near-set), to the producers' tablets for review process, to the editor(s)' monitors, to the grading theater/room—original footage may undergo additional processing like "raw development", noise/grain management or pre-grading. Then, parallel to color-grading and happening in one or more phases (more on that in Section 2.4), pictures may receive

VFX or generic CGI, involving other artists creating rigged, animated, lit, and rendered 3D models and compositing them onto the plates, so that the different moving images realistically blend together [3]. As already said, this is usually demanded across several artists at the same time (sometimes at different studios), each having to evaluate the original look so that artificial pictures (and photographic plates) are all part of the same story. Last but not least, the imaging pipeline ends with the production of several masters, each for every end viewing typology (theatre, HDR/SDR TV, web, print, etc.) other than due to different localizations (audio, subtitles/closed-captions, cards/inserts, main/end titles); see Figure 1 for a diagram showing the steps of the post-production processes. All the above steps take place in different viewing conditions; color-critical operations are made at different times, by different subjects (artists but also imaging technicians); it is paramount that the initial creative intent (the DoP's) is preserved independently from viewing device and environment. The control of colorimetric process is a key requirement for this industry.

The color-correction, or (color) grading phase is the most delicate, cfr. [20,22], Section 2.4. It is set in either a dark-surround Digital Intermediate (DI) theatre equipped with a digital cinema projector in case of theatrical movies, or a TV grading room, dimly lit and equipped with one or more monitors (cfr. Figure 4a). Projectors and monitors (not only those in the grading theatre/room) are all color-calibrated according to their own reference color standards, dynamic ranges and viewing environments, cfr. Table 1 (and Figure 2). For multiple-delivery shows (e.g., theatrical, web, then SVoD and TV, both HDR and SDR), several color-grading sessions are needed, because there is not yet a universally-accepted HDR-to-SDR gamut-mapping algorithm [19], that is satisfactory from the creative standpoint—and automatically scales to different shows. The colorist first does a "master grade" (usually in the widest-gamut output/display color-space possible, so it is in fact an HDR-grade); this leads to a über-master, which derivative grades and subsequence masters descend upon [23]. The above is not yet a universally accepted standard practice (sometimes the HDR grade is a secondary grade), especially if several masters are needed for different HDR technologies.

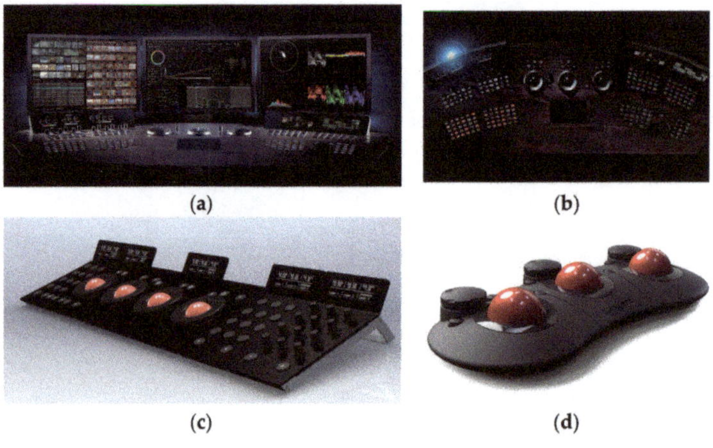

Figure 4. Example of color grading control surfaces: (**a**,**b**) The "Blackboard 2" panel connected to the high-end *Baselight* system via USB and DVI-D, featuring four 6" full-raster displays (800 × 240), 3 trackballs, 14 rotary encoders, 1 jog/shuttle, 220 backlit self-labeled buttons, plus integrated tablet and QWERTY keyboard (courtesy: FilmLight Ltd.); (**c**) The mid-end "Elements" modular panel, made up of four independent USB controllers featuring overall five 160-character displays, 4 trackballs/jogs, 12 rotaries, 45 buttons; (**d**) The low-end "Ripple" USB controller: 3 trackballs, 3 rotaries, 8 buttons; (**c**,**d**) are compatible with most commercial editing/color grading systems (courtesy: Tangent Wave Ltd.).

Table 1. Output colorimetric standards employed officially, or as *de-facto*, in (2017) media production, arranged according to viewing devices, environments and dynamic ranges; cfr. Table 2.

Dynamic Range	PC Monitor	HD TV	Cinema	UltraHD TV
SDR	sRGB [24]	BT.709 [3] [25]	DCI P3 [4] [26]	BT.1886 [27]
HDR [1]	— [2]	BT.2100 [3] [28]	ST2084 [5] [29]	BT.2020 [3] [30] and BT.2100 [3]

Note: [1] There exist many proprietary technologies not completely bound to standards, widely used in broadcast and sold in consumer TV: Dolby®Vision, Hybrid Log-Gamma (HLG), HDR10(+), etc.; [2] Market is pushing for DCI P3; [3] Informally known as Rec.709, Rec.2020 and Rec.2100 respectively; [4] Post labs often use D60 or D65 "creative white-points" instead of DCI's; [5] Defines "PQ curve" (Dolby®Vision and HDR10) and HLG transfer functions.

Table 2. CIE (x,y) chromaticity values for the ACES color-spaces' primaries/white-point, compared with a few known camera and reference-display color-spaces; cfr. Table 1.

Gamut Name	Red		Green		Blue		White-Point		
AP0	0.7347	0.2653	0.0	1.0	0.0001	−0.0770	D60	0.32168	0.33767
AP1	0.7130	0.2930	0.1650	0.8300	0.1280	0.0440	D60	0.32168	0.33767
sRGB/BT.709	0.6400	0.3200	0.3000	0.6000	0.1500	0.0600	D65	0.31270	0.32900
ROMM RGB	0.7347	0.2653	0.3000	0.6000	0.0366	0.0001	D50	0.34567	0.35850
CIE RGB	0.7347	0.2653	0.2738	0.7174	0.1666	0.0089	E	0.33333	0.3333
ARRI W.G.	0.6840	0.3130	0.2210	0.8480	0.0861	−0.1020	D65	0.31270	0.32900
DCI P3	0.6800	0.3200	0.2650	0.6900	0.1500	0.0600	DCI	0.31400	0.35105
BT.2100/2020	0.7080	0.2920	0.1700	0.7970	0.1310	0.0460	D65	0.31270	0.32900

2.4. Digital Color Grading Process

As introduced in Sections 2.2 and 2.3, when "conformed" (i.e., online-edited, high-resolution) footage is color-corrected, the clips in the timeline are each read from proprietary raw file formats, interpreting the "camera-native" color-spaces of their (possibly multiple) sources. Ideally these are the working color-spaces for grading as well: they are scene-referred and feature a pseudo-logarithmic transfer curve (Section 2.1): therefore, they are usually called "log" spaces. This nonlinearity is usually accounting for many optical and electronic phenomena, including the sensor's tone response curve, an optimization of its exposure range (measured in EV) into the CV space (i.e., the digital representation of the color-space [20,31]), plus the nonlinear effects of vision with respect to illumination which is, trivially by its own definition [32], photometrically "linear". To view and evaluate the look of the pictures, a color-space transformation to the output colorimetry (as described in Section 2.3) is automatically applied by the color-correction system, cfr. Figure 3.

The application of creative color transformations is operated with traditional computer aids (especially tablet) with the aid specific "control surfaces" (Figure 4); their trackballs and jogs allow for independent, either shadows/midtones/highlights, or RGB-components operations, thus called 3-way CC. Those applied on the whole frame of a video sequence are historically called primary grades, vs. the ones applied on specific subjects or areas of the frame and eventually moving with them along the timeline—which are called "secondaries" [22]. Sophistication of modern color-correction technology does neither depend only on pure color-handling tools—e.g., making secondaries' partial selections according to either geometries (windows), specific shades or gradations of other colors within the image (keys) or by hand-painted masks (mattes)—nor on the algorithms to automatically "follow" the corrections along with the motion picture flow (power windows and point-tracking in general); current color grading tools (and in this sense both I prefer the use of the noun "grading" instead of "correction") also take care of the image "look" in a broader sense, including what is referred to as creative finishing [23], including other creatively "dosed" imaging processes on the footage like motion blur, re-graining (i.e., adding artificial film grain—or adding it back after sensor noise or real film grain was removed/polished prior to VFX work [3]), texturing and glow effects, etc.

Color-metadata technologies like the American Society of Cinematographers' Color Decision Lists (ASC CDL [33,34]), allow the colorist to start from the creative "look" information as it was pre-graded during principal photography [35], thus channeling, if correctly transitioned in the color pipeline, the DoP's creative intent without the need of "burning" the pre-grades in additional, time- and storage-expensive renders of the whole footage: 10 floating-point values per grade (a single CDL), collected and transferred as either a sidecar XML file or embedded in other video-editing project files, are enough to preserve this color-critical, creative information as tiny-footprint metadata. This is a winning strategy that ACES uses as well [15,16,35]. The core of an XML-embedded CDL is like:

```
<ColorCorrection id="ClipName_or_CorrectionName">
    <SOPNode>
        <Slope> 1.06670 1.05820 1.06270</Slope>
        <Offset>-0.07250 -0.05260 -0.05060</Offset>
        <Power> 1.04020 1.06050 1.06350</Power>
    </SOPNode>
    <SatNode>
        <Saturation> 0.85000</Saturation>
    </SatNode>
</ColorCorrection>
```

The 10 numbers are $(\mathbf{s},\mathbf{o},\mathbf{p},\sigma) \equiv ((s_R,s_G,s_B),(o_R,o_G,o_B),(p_R,p_G,p_B),\sigma) \in [0,+\infty[^3 \times \mathbb{R}^3 \times [0,+\infty[^3 \times]-1,1]$, grouped in 3 groups of 3-tuples, plus 1 number; the 3 groups each store one parameter type for the color-correction Equation (1), which acts the same on each channel of input color $\mathbf{c} \in \mathbb{R}^3$. The first tuple stores "slope" values $\mathbf{s}$, the second "offset" values $\mathbf{o}$, the third tuple "power" values $\mathbf{p}$; the first coordinate of tuples applies to red, the second to green and the third to blue channel. The tenth number σ is a "saturation" modifier. Overall, the CDL—i.e., the Slope-Offset-Power Equation (1), plus saturation—is applied as Equation (2):

$$\mathbf{c} \equiv \begin{pmatrix} r \\ g \\ b \end{pmatrix}, \quad \mathrm{SOP}(\mathbf{c}) \equiv \mathrm{SOP}(r,g,b) := \begin{pmatrix} (s_R r + o_R)^{p_R} \\ (s_G g + o_G)^{p_G} \\ (s_B b + o_B)^{p_B} \end{pmatrix} = (\mathbf{sc} + \mathbf{o})^{\mathbf{p}}. \tag{1}$$

Overall, the CDL—i.e., the slope-offset-power Equation (1) plus saturation—is applied as Equation (2):

$$\begin{aligned} L(\mathbf{c}) &= 0.2126\,\mathrm{SOP}(r) + 0.7152\,\mathrm{SOP}(g) + 0.0722\,\mathrm{SOP}(b); \\ \mathrm{CDL}(\mathbf{c}) &:= \sigma\,\mathrm{SOP}(\mathbf{c}) + (1-\sigma)L(\mathbf{c})\mathbf{1}_3 = \begin{pmatrix} \sigma(s_R r + o_R)^{p_R} + (1-\sigma)L(\mathbf{c}) \\ \sigma(s_G g + o_G)^{p_G} + (1-\sigma)L(\mathbf{c}) \\ \sigma(s_B b + o_B)^{p_B} + (1-\sigma)L(\mathbf{c}) \end{pmatrix}. \end{aligned} \tag{2}$$

Inversion of a CDL is possible in closed form but, unless $\sigma = 0$ and $\mathbf{p} = \mathbf{1}_3 \equiv (1,1,1)$, it may not be a CDL itself: the four operations Equation (2) are to be applied in reverse order (inverse saturation, reciprocal power, opposite offset, reciprocal slope). Algebraically, the set of possible CDLs does not form a group.

Both technical color processing (e.g., color-space conversions) and primary grades applied for creative purposes may be approximated by ColorLUTs, cfr. [2,16], Figure 5: color look-up tables were sampled source CVs are mapped to target ones, while colors not in LUT are mapped by interpolation methods (usually trilinear and tetrahedral); among them, 3D LUTs allow better representation of both technically-complex and creative transforms [3] and Figure 6, e.g., some creative looks, or emulation of cross-talk in developed film dyes [13]. Coarser samplings improve accuracy, but not all the systems can handle larger LUTs and their interpolation in real-time, especially if they use floating-point instead of integer arithmetics (cfr. Section 3.3). Furthermore, source shaping the source color mesh along the aforementioned intrinsic nonlinearities of imaging processes (illumination, electronics, and vision [19]) optimizes performance by adding accuracy just where perceptually and numerically advantageous.

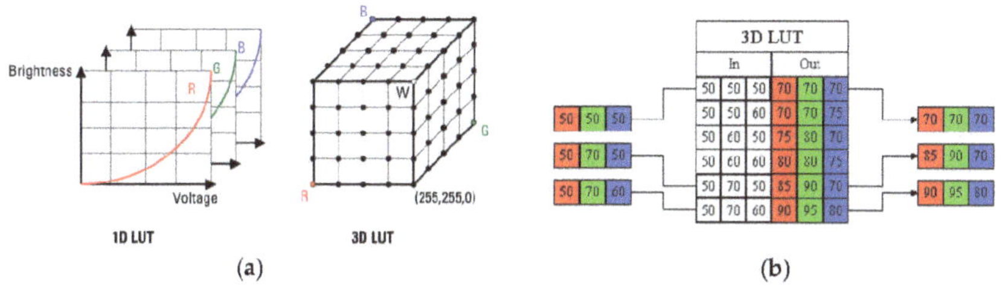

Figure 5. How ColorLUTs work (in 8 bits/channel color-spaces): (**a**) concept difference between a 1D LUT (discrete version of a color curve) and a 3D LUT (a discrete vector space [2], i.e., a 3rd-order tensor); (**b**) a 3D LUT maps a mesh of sampled RGB 3-tuples into target RGB 3-tuples; the other colors are mapped by interpolating points in both the source and target color-spaces (also cfr. Figure 6).

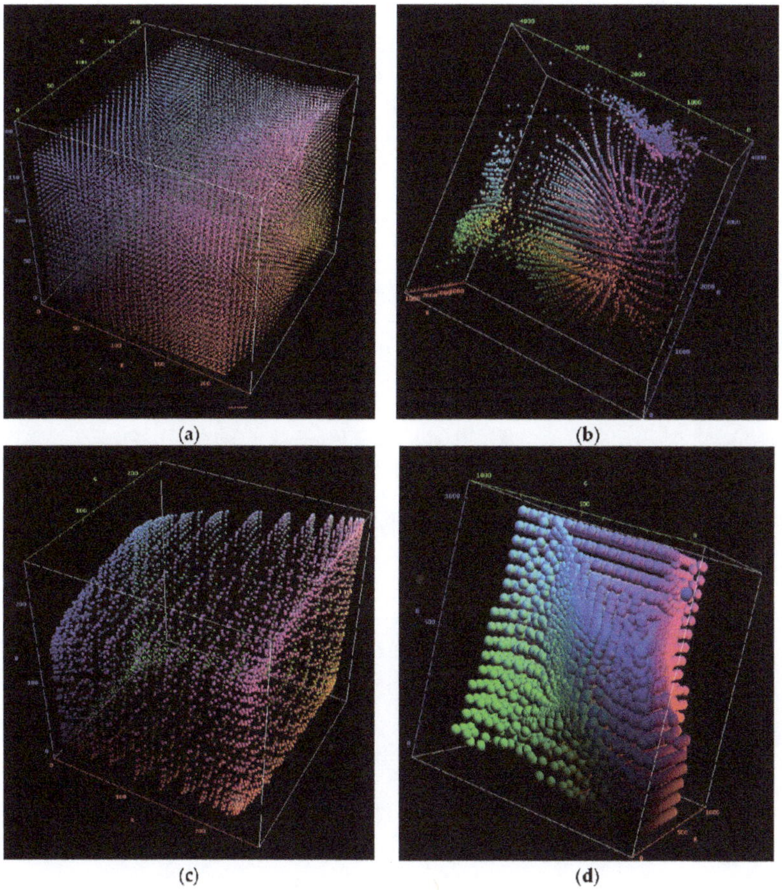

Figure 6. Vector-field representation of 3D LUTs [20]: source gamut is a regular cubic grid (not shown) whose points are mapped to an (irregular) target gamut. Each sphere in these graphs is a map between two RGB 3-tuples: the sphere's hue as source and its position as target. (**a**) Creative "musical" look obtained via mild primary grading; (**b**) extreme creative look obtained via clipped primary grading; (**c**) Kodak Vision print film emulation (PFE) to Rec.709; (**d**) Inverse PFE into a HDR color-space.

ColorLUTs and CDLs are just two examples of color metadata, as they can describe most color transforms in a non-destructive way (as original images are preserved) and without computational and storage footprint due to rendering new images with those transformed "burned" over the raster.

To tackle with all this, color science comes into play, and before ACES were adopted, no single solution existed to be re-usable show after show. Color scientists had thus to provide effective methods and workaround to overcome lots of different optical, psychovisual and digital issues at the same time. This was feasible only in larger imaging/postproduction companies (usually those that had naturally been dealing with color consistency for "traditional" Digital Intermediate (DI) process [31], starting with digital 16/35 mm film scans and ending with 35/70 mm film-out) that had expertise and resources, for example, to engineer ColorLUTs. Smaller production had to rely on non-tailored processes with little to no control. This boosted the industry though, forcing creation of a common vocabulary first [32,36], then standards to refer to display colorimetries [18] and correctly measure their deviation [37]. Overall, color science ensures uniformity of the color pipeline (possibly "from set to screen") by controlling processes like:

- profiling and cross-calibration of all the output devices accounting for different light emission/reflection spectral densities [19,38], colorimetry standards [24–30,39];
- accounting for the different viewing environments (including surrounding lighting), thus including chromatic adaption aspect and other effects to the viewers [19,40];
- using mathematics, physics and IT concepts throughout [21,41], to keep a broader understanding of both fundamental and engineering processes involved, in order to provide the best viewing experience [42], and craft quantitatively accurate color transforms (either technical and creative).

3. ACES Components

First of all, ACES "core" components, i.e., documentation, code and reference images, can be downloaded from both AMPAS' website [4] and its GitHub page [43]. The Academy also created a specific website called ACES Central [44], which is, at the same time, a repository for all the above, a showcase for all the news related to ACES (e.g., new events and lists of projects and products using it) and, last but not least, a forum-style network where people can ask questions and exchange feedbacks and tricks with both ACES experts and among themselves.

Figure 7 shows a gross bird's eye view to an end-to-end production/postproduction/VFX workflow, with relevant ACES component names [45] specified for each phase of it.

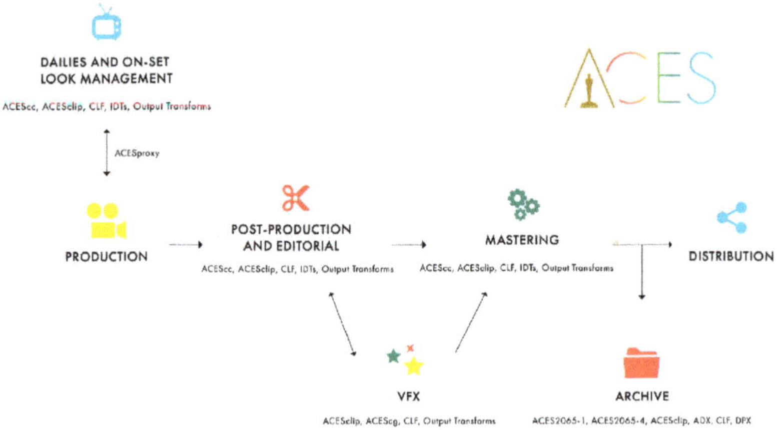

Figure 7. Overall view of an ACES 1.0 end-to-end pipeline; the ACES logotype (Section 3.2) is shown at the top-right corner (courtesy: AMPAS).

3.1. Reference Implementation

The "reference implementation" is the ACES corpus, containing everything needed to check that applications of this framework, as well as products built to use parts of it, have reproducible results that are interoperable with those of other products and applications; its components are listed below.

- The ACES documentation is mostly public domain [1,45–60], plus a group of standards by Society of Motion Picture and Television Engineers (SMPTE) standards which were assigned the ST2065 family name [61–65], plus an application of a different SMPTE standard [66] and two external drafts hosted on GitHub [67,68]. The SMPTE documents are technical standards mainly for vendors and manufacturers to build ACES-compliant hardware and software products; they are not needed by creatives and engineers that just want to use ACES in their workflows and pipelines across production, postproduction and VFX, as this framework was designed with low- and mid-level budget productions in mind. ACES requires neither major-studio budgets, nor top-notch engineers [15].
- The development codebase [43] is an open source implementation of ACES color encodings (in CTL language, Section 3.7), metrics and file formats. It is OS-neutral (although meant to be primarily compiled on Linux and macOS), written in C++ language, plus a few scripts in Python, and depends on a few additional open source libraries [69]. The executables are not intended for production uses—they are neither optimized for performance or batch/volume usage, nor have ergonomic interfaces (being mostly command-line utilities)—but rather for validating third-party products compliance with a reference, as specified at the beginning of the paragraph.
- Collection of reference images as a deck of still photographs about several categories of subjects in diverse lighting conditions, encoded in different ACES color-spaces (cfr. Section 3.3) and using elective file formats like OpenEXR, TIFF and DPX [69–71]. Together with the above codebase, ACES vendors and users are expected to test their products and workflows on them and compare them with their own rendered pictures for accuracy in different conditions.

Among the ACES documentation there are three important papers that provide a key to an effective reading of the whole corpus: two [45,47], provide nomenclature and versioning naming convention for the other components (particularly for color transformations in [43]); another one [46], is instead a guide for both users and implementers of ACES as regards usability/user experience (UX) design: ACES needs to be interoperable and vendor-neutral, thus suggestions are made neither about implementation, for example, of products' user interface (UI), nor on internal algorithms; however a consistent UX is still needed so that users of different products may confidently switch from one to another without having to invent missing parts of workflow that glues them or having to guess equivalent component names across products.

It is in this setting that Sony ImageWorks Inc. created in 2003 an open-source project called OpenColorIO (abbreviated to OCIO) as unified color management frameworks dedicated to VFX [72]. It is supported by most CGI software vendors and also integrates ACES color transforms (despite not all its usability conventions [45,46] are currently respected), so every product supporting OpenColorIO (e.g., The Foundry *Nuke*, Autodesk *Maya*®, SideFX *Houdini*™), basically handles ACES color science as well, at least as far as color-space conversions are concerned.

3.2. Product Partners and the Logo Program

In addition to the above, and to foster interoperability across different products and manufacturers, the Academy created a categorization of products potentially using ACES, a list of product partners (i.e., vendors officially providing products supporting it [33]), and a "Logo Program" where vendors can apply their products to and undergo a certification path for the categories that each products belongs to, eventually receiving an ACES logo for it (drawn at the top-right corner of Figure 7). During the certification, the Academy ascertains that, and the vendor provides evidence for, the product complies with the reference implementation of Section 3.1. This assures to the users that any products with the ACES logo, whatever vendors they are from, are always fully interoperable with each other, so end-user color pipelines can be freely designed without any fear of "lock-in" or color-accuracy problems when switching products among each other.

3.3. ACES Color Spaces

The basics of the ACES framework revolve around a set of "core" color-spaces for which complete and invertible mathematical formulas are specified so that conversions may be applied without any loss of accuracy or precision. In this subparagraph, the color-spaces will be defined and in detail and their generic use introduced; the last color-space, ADX, will be introduced in Section 3.11 as it devoted to photochemical-film color management. Full understanding of ACES color science is delayed to Section 4. Among these color-spaces the first one—ACES2065-1 by the SMPTE Standard [61] defining it—is the most important because it works as principal connection space (or PCS, cfr. [38]) for the whole pipeline; it was also introduced first, well before the others were organically defined. For this reason, ACES2065-1 may still be referred as "ACES Linear" or as "the ACES color-space" in many pre-release implementations of the framework; currently, it is so in the first and former SMPTE standard as well [61], erroneously. Common to these color-spaces (except or ADX, cfr. Section 3.11) is that they are based on the RGB model, scene-referred (cfr. Section 2.1) and the effective white-point is equivalent to that of a D60 standard illuminant (i.e., a 6000 K coordinated color temperature); their gamuts can be defined to be either of two sets of color primaries, called AP0 and AP1 respectively, whose chromaticities are reported in Table 2 and shown in Figure 8 for comparison with other color-spaces.

It is important to stress that, being all scene-referred and wide-gamut color-spaces, they are never meant to be viewed directly on any display device. As usual in traditional motion picture imaging science, they need a color-transform to be applied on top of the color pipeline, just before the output.

The AP0 gamut (cfr. Figure 8a), much like CIE XYZ or CIE RGB [19] is defined to be the smallest triangle enclosing the whole CIE 1931 Standard observer chromaticity diagram, therefore every chromatic stimulus perceivable by the human eye can be represented in this gamut. This choice future-proofs the imaging content of assets stored in ACES colorimetry, no matter how better capture and display devices will improve. The only three caveats in using AP0 are:

- RGB color primaries in AP0 (as well as in DCI $X'Y'Z'$ [26]) are substantially "tilted" with respect to other RGB spaces' and to natural-cone primaries (LMS [40]), thus imaging operations that affect chromaticity, hue or saturation (e.g., color-grading), produce unnatural color-shifts.
- A significant area of the AP0 gamut falls outside of average observer's (imaginary colors), therefore many CVs are "lost" when these primaries are used in a color-space; besides, most of the in-gamut chromaticities cannot be captured/displayed by current technologies (as of 2017). Thus, a higher bit-depth may be needed to retain the same CV density within a "usable" gamut.
- Since no colorimetric cinema camera exists (yet), and ACES colorimetry is based on this, the correspondence between real tristimuli captured from a scene and the recorded CVs (even before conversion into ACES colorimetry), depends on the manufacturer-chosen sensitometry of today's real cameras (or on the emulated physics of the CGI application's lighting engine).

The AP1 gamut (Figure 8) was introduced in ACES 1.0 to overcome the first two drawbacks above; it is a subset of AP0, it does not cover the whole standard observer's, yet it is "HDR wide gamut", like ARRI WideGamut, DCI P3 and Rec.2020 (Table 2). In fact, AP1 contains P3 and Rec.2020 gamut used in D-Cinema and HDR reference displays (Figure 2). It represents colors that are viewable on displays and projectors, including HDR, without noticeable color skew. This is the right gamut to apply the internal mathematics of color-grading and CGI applications. It is important to note, however, that colors from AP0 but outside AP1 are gamut-mapped to CVs outside the [0.0, 1.0] range: this is expected behavior and ACES-compliant applications shall handle negative CVs.

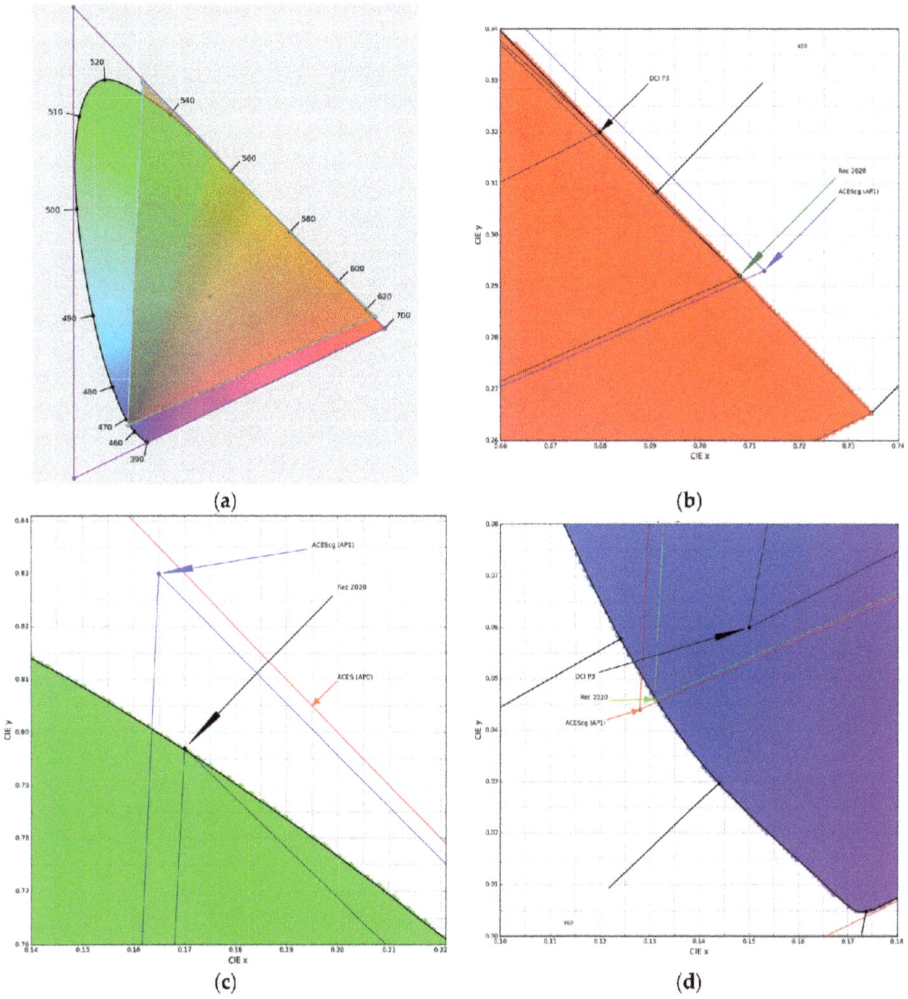

Figure 8. Comparisons for ACES gamuts: (**a**) AP0 (purple) vs. AP1 (cyan, yellow-filled); (**b**–**d**) Details of several gamuts (AP0, AP1, P3, BT.2100) cross-coverage near the edges of the three color primaries.

In ACES, the colorimetric stance is preservation and future-proofing of moving picture creative intents and, even if it works equally well for "full-CGI" features (e.g., animation), traditional films color-accuracy starts with principal photography, and follows on color-grading. ACES colorimetry is thus scene-referred by design (cfr. Section 2.1) and based on the CIE standard illuminant D60 (which is widely used as "creative white-point", cfr. [26], Table 1 note 4). Besides, to faithfully reproduce the widest possible dynamic range and extrapolate its interpretation from todays' imaging technologies, a photometrically linear transfer characteristic (i.e., "gamma 1.0" or color-space linearity) is chosen. Such use of "linear color-spaces" is also consistent with the physics of light, thus notably simplifying image lighting operations on CGI textures, including animation and VFX. Refer to Table 2 for references to documents defining the color-spaces below.

- **ACES2065-1** is the main color-space of the whole framework and it is the one using AP0 primaries, since it is meant for short-/long-term storing as well as file archival of footage. It has a linear transfer characteristic and should be digitally encoded with floating-point CVs of at least 16 bits/channel precision according to [73].

ACES terminology introduced in [61] defines a Reference Image Capture Device (RICD) as an ideal/virtual colorimetric camera whose sensor's spectral sensitivities record scene luminance directly in the above color-space (i.e., linearity between scene relative exposure and CVs) [35]. RICD is defined void of system noise, with 0.5% camera flare off a perfect reflecting diffuser, whose reflected light from a standard D60 illumination source (i.e., a 100% reference-white) is scaled to CV (1.0, 1.0, 1.0), whereas the same recorder light off an 18% grey card maps to CV (0.18, 0.18, 0.18). The viewing environment is always considered observer-adaptive to D60, with 1600 cd/m^2 minimum adapted luminance and 0% viewing flare. The next color-spaces are intended as working-only, not at all for storage: compliant appliances shall internally convert back and forth between ACES2065-1 to the sole extent and for the duration of specific purposes only (lighting, compositing, color-grading, video transport).

$$C_{\mathrm{AP0}}^{\mathrm{AP1}} := \begin{pmatrix} 1.4514393161 & -0.2365107469 & -0.2149285693 \\ -0.0765537734 & 1.1762296998 & -0.0996759264 \\ 0.0083161484 & -0.0060324498 & 0.9977163014 \end{pmatrix} \in \mathrm{GL}_3(\mathbb{R}). \tag{3}$$

- **ACEScg** was specifically designed as a working color-space for CGI applications [74], which it should be the standard working color-space for internal operations that still need linear-to-light transfer characteristic for physics-/optics-/lighting-based simulations. It is different from ACES2065-1 only due to the use of AP1 primaries; conversion to it is done via an isomorphism represented by Equation (3). Encoding uses 16 or 32 bits/channel floating-point CVs ("floats").

$$\log_{\mathrm{cc}}^{\mathrm{ACES}}(x) := \begin{cases} -0.358447, & x \leq 0, \\ 0.554795 + 0.0570776 \log_2 (2^{-16} + \frac{x}{2}), & x < 2^{-15} \\ 0.554795 + 0.0570776 \log_2 x, & x \geq 2^{-15} \end{cases} \tag{4}$$

$$\log_{\mathrm{cc}}^{\mathrm{ACES}-1}(y) := \begin{cases} 2^{17.52y-8.72} - 2^{-15}, & y \leq -0.30137 \\ 2^{17.52y-9.72}, & y < 1.468 \\ 65504, & y \geq 1.468 \end{cases} \tag{5}$$

- **ACEScc** was designed to help with color-correction applications, where a specifically crafted spline-logarithmic transfer function of Equation (4), $\log_{\mathrm{cc}}^{\mathrm{ACES}}$ whose inverse is Equation (5), supports color-grading operators; it applies indistinctly to all RGB channels after a color-space conversion to AP1 via Equation (3). Digital encoding for FPU or GPU processing [34], is in either 16 or 32 bits/channel floats.

$$\log_{\mathrm{cct}}^{\mathrm{ACES}}(x) := \begin{cases} 0.0729055341958355 + 10.5402377416545x, & x \leq 0.0078125 \\ 0.554795 + 0.0570776 \log_2 x, & x > 0.0078125 \end{cases} \tag{6}$$

- **ACEScct** is an alternate color-grading space to ACEScc, specifically designed with a different linear/logarithmic spline curve (6) instead of (4), resulting in a distinct "milking" look on shadows, due to additional toe added in that range; this additional characteristics was introduced following many colorists' requests to have a "log" working space more alike those used in traditional film color-grading and have a similar and vendor-neutral feeling/response when manipulating control surfaces, cfr. Section 2.4 and Figure 3. ACEScc and ACEScct are identical above CV$_{\mathrm{ACES2065}}$ 0.0078125, although their black pedestal is different (cfr. Table 3 and Figure 9a).

$$\log^{\mathrm{ACES}}_{\mathrm{proxy}10|12}(x) := \begin{cases} 64|256, & x \leq 2^{-9.72} \\ \mathrm{rect}^{940|3760}_{64|256}[\![550|2200 + \log_2 x]\!], & x > 2^{-9.72} \end{cases} \tag{7}$$

- **ACESproxy** is introduced to work with either devices transporting video signals (with integer CV encoding), or with intermediate hardware that supports integer-based arithmetic only (instead of floating-point) [34]. These include video-broadcast equipment based on Serial Digital Interface (SDI) among the former category; LUT boxes and references monitors among the latter. Such professional encodings are implemented in either 10 or 12 bits/channel, therefore two isomorphic flavors exist: ACESproxy10 and ACESproxy12. This is the elective encoding as long as it is used only for transport of video signals to endpoint devices (and processing finalized for such intents only), with no signal or data ever stored in, or re-converted back from ACESproxy. By design, it is an integer epimorphism of ACEScc (WARNING: *not* of ACEScct); it also scales CV to video-legal levels [34] for compatibility with broadcast equipment, as shown in Figure 9b, as they may include legalization or clipping across the internal signal paths. The conversion from ACES2065-1 is done applying (3) first, followed by either one of the two functions in Equation (7) $\log^{\mathrm{ACES}}_{\mathrm{proxy}} : \mathbb{R} \rightarrow \mathbb{N}_0$ (red for 10-bits/channel or blue for 12-bits/channel).

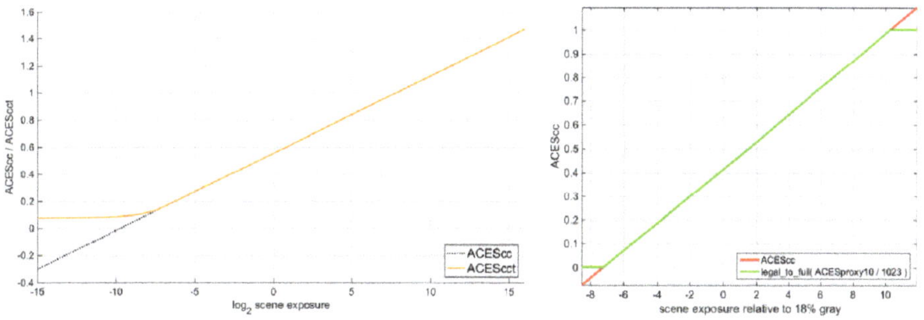

(a) **(b)**

Figure 9. ACEScc transfer curve in a "EV/CV plot", i.e., the $\log_2$ of scene exposure relative to 18% grey in abscissa (photometric stops) vs. CV range in ordinate. Compare with: (**a**) ACEScct; (**b**) ACESproxy.

The color-space conversion formulas between any of the above spaces, except ACESproxy, are exactly invertible because so are matrix (3), $C^{\mathrm{AP0}}_{\mathrm{AP1}} = C^{\mathrm{AP1}-1}_{\mathrm{AP0}}$ and Functions (4)–(6). Conversion from ACESproxy back to ACES2065-1 is numerically possible introducing quantization errors, but this is not allowed as ACESproxy, by design, reserved for transport of color-data to end devices, or within "last-mile" processing inside such devices (with inferior computational performance) just before final output. In no way should ACESproxy code-stream be recovered, recorded, or even stored in files.

- **ADX** is different from all the other color-space, is reserved for film-based workflows, and will be discussed in Section 3.11.

Color transformations between ACES color-spaces, and between ACES and non-ACES color-spaces, can be implemented by any product, application and workflow according to their specific architectures and pipelines: ACES does not require or suggest any solutions, for the sake of increasing its adoption, and encouraging each actor to freely use the best methodologies and technologies that are considered appropriate, without any limits. In order to verify adherence to the reference implementation (thus assuring interoperability, cfr. Section 3.1), the Academy communicates any color conversion formulas—regardless of its creative or technical intent—via scripts written in a C-style language and both (the language and, generically, the scripts themselves) are called CTL; more on that in Section 3.7. CTL can be used to represent pre-baked color transforms such as ColorLUT, although there is an elective file format for them as well (cfr. Section 3.8). The CTLs coming from product partners (cfr. Section 3.2) are integrated in the next ACES release, thus are automatically retrieved from [43,44] (and, eventually, from [72] as well): new or updated color-spaces get automatic integration into users' workflows and applications, without needing of full software upgrades.

Table 3. Technical characteristics of the six ACES color-spaces.

	ACES2065-1	ACEScg	ACEScc/ACEScct	ACESproxy	ADX
primaries	AP0	AP1	AP1	AP1	APD
white-point	D60	D60	D60	D60	~Status-M
gamma	1.0	1.0	$\log_{cc/cct}^{ACES}$	$\log_{cc}^{ACES}$	$\log_{10}$
arithmetic	floats 16b	floats 16/32b	floats 16/32b	int. 10/12b	int. 10/16b
CV ranges:	Full	Full	Full	SDI-legal	Full
legal (IRE)	[−65504.0, 65504.0]	same [1]	[−0.358447,65504.0] [2]	[64, 940] [3]	[0, 65535] [4]
±6.5 EV	[−0.0019887, 16.2917]	same [1]	[0.042584, 0.78459] [2]	[101, 751]	— [5]
18%, 100% grey	0.180053711, 1.	same [1]	0.4135884, 0.5579	426, 550 [3]	— [5]
Purpose	file interchange; mastering; archival	CGI; compositing	color grading	real-time video transport only	film scans
Specification	[1,48,61]	[49]	[50]/[52]	[53]	[54,62,63]

Note: [1] ACEScg has the same full/legal ranges and 18%-grey/100%-white CVs as ACES2065-1. [2] This is for ACEScc only; ACEScct uses legal range [0.072906, +∞] and ±6.5 EV range [0.093867, 0.78459]. [3] These are 10-bit CVs; The full range/18%-grey/100%-white 12-bit CVs are [256, 3760], 1705 and 2200. [4] This is for the 16-bits variant; the 10-bits has [0, 1023]. [5] ADX has no set CVs for certain EVs.

3.4. Entering ACES

The RICD introduced in Section 3.3 is an idealized colorimetric camera, and the closest "real" thing to it are virtual cameras used in CGI software for previsualization, animation and rendering of assets (although in this case the working space shall be ACEScg). It is, therefore, up to camera vendors to provide conversion formula from their sensors' native colorimetries into ACES, cfr. Figure 2 and Section 2.1. Any color transformation that takes a non-ACES color-space into ACES2065-1 is called an (ACES) Input Transform (also abbreviated in IDT, whose acronym comes from its deprecated, pre-release name: input device transform). Some manufacturers provide several colorimetries according to different sensor models (or camera firmware), but due to sampling/quantization algorithms getting different firmware-based optimizations as the sensitometry is influenced by user-selectable optical and electronic settings, there may be often several Input Transforms per model, depending on one or more camera settings, like:

- sensitivity, measured in EI (ISO exposure index),
- correlated color temperature (CCT), measured in Kelvin or, equivalently,
- generic shooting illumination conditions (e.g., daylight, tungsten light, ...),
- presence of special optics and/or filters along the optic path,
- emulation of the sensor's gamut of some cameras (e.g., REDColor, DRAGONColor, S-Log),
- creative preference on a contrast curve (Section 3.1, e.g., LogC, REDLogFilm, REDGamma, S-Gamma).

At application level UIs can be designed such that the selection of the right Input Transform becomes completely transparent to the user: for example, most color-grading software read the source colorimetry and camera settings from the metadata embedded into the raw file headers and use the correct Input Transform accordingly, thus greatly reducing a color-science effort by the users.

The recommended procedure for creating IDTs for digital cameras is described in [55]. Input Transforms can be created for other color-spaces as well, even from output color-spaces (in case one has to bring footage calibrated for an output device into ACES) like sRGB [24], which is the default colorimetry for web-based imaging, as well as those assumed, within motion picture industry, for most of non-color-managed images (e.g., stored in file formats like PNG, GIF, MPEG2, or in JPEG files without an embedded ICC profile).

Any source image can be brought into an ACES pipeline by just applying the Input Transfer corresponding to its colorimetry. This should be the first process and the only place where Input Transforms are used; after that, every image is brought into ACES2065-1 color-space, that is to a photometrically equivalent, scene-referred footage coming from one ideal camera, the RICD.

Product partners manufacturing cameras may provide their own Input Transforms to the Academy for inclusion in the next version of ACES release, so that every other logo-program product can implement those transforms with guaranteed accuracy and without any re-certification, cfr. Section 3.2.

3.5. Viewing and Delivering ACES

As hinted in Section 3.3, ACES colorimetry is always scene-referred, thus no images shall be viewed without some conversion to an output-referred colorimetry first. In short, stored RGB CVs in any ACES color-spaces are not proportional to any intensities for display/projector color primaries. Conversion from ACES2065-1 to any non-ACES colorimetry is done—dually with respect to input, cfr. Section 3.4—via an (ACES) Output Transform, be it just for viewing or for mastering/delivery purposes. Every Output Transform is the composition of a fixed conversion from scene- to output-referred colorimetry, by means of a Reference Rendering Transform (RRT), whose formula closely resembles that of a tonecurve, plus an output device transform (ODT) which maps the output-referred CVs out of the RRT into specific colorimetry standards, cfr. Section 2.3 and Figure 2.

Since the RRT's shape definition involves mostly technical but also a few creative decisions: it should be a neutral tonecurve, yet pleasant to look at by default, before any look development starts, plus is should be functional enough so that all color-grading operations can be viewed through it. Due to this, the RRT may change in future versions of ACES (current version as of 2017: 1.0.1); the Academy will always provide the official forward/inverse RRT as individual CTLs as part of [43].

Users just need to make sure that every viewing pipeline has an Output Transform applied before any ACES image is viewed, or any footage is rendered/transcoded into non-ACES colorimetry. No Output Transforms need to be used when footage is rendered to "stay" within ACES' color-managed pipeline, also because the involved ODTs are, usually, highly non-invertible.

Product partners manufacturing viewing devices such as monitors and projectors can provide their own ODTs to the Academy for inclusion in the next version of ACES release, so any logo-program product, cfr. Section 3.2, automatically includes the relevant Output Transforms, as soon as the updated versions of either ACES and OpenColorIO are re-linked from or replaced into the color-managed applications. That may be a bit more complex for some kinds of HDR monitors that adapt the output dynamic range in real time, based on the exposure range of currently displayed/projected content; this process may involve transport of metadata describing changes in the gamut [75]; in this case a separate Output Transform is required, at application level, for every dynamic-range interval. UIs, though, can make this additional metadata path completely transparent to the user/viewer.

It is important to point out the difference between an Output Transform (logically, inverse function of an Input Transform, Section 3.4) and an ODT: the former is the combination of the RRT plus an ODT; for this reason, the former is informally called "RRT + ODT", as in Figure 10. These are unessential technicalities, but subject to the improper nomenclature used in pre-release ACES.

3.6. Creative Intent in ACES: The Transport of "Color Look" Metadata

ACES colorimetry is intended for all color-critical processes and, being based on standardized color-spaces (Section 3.3) with fully invertible cross-conversion formulas, the interoperability is guaranteed as long as images are transported in this colorimetry; creativity is not hindered by technical constrains posed by the use of different products, toolsets available to colorists (via product partner products) are ACES-agnostic, so artists can decide exclusively based on creative evaluations, while engineers may remove many technical constrains off the color pipeline. As defined in [56], any "image-wide creative 'look' to the appearance of ACES images [...] that precedes the" Output Transform is called a Look Modification Transform (LMT); more abstractly, a LMT is any color transformation mapping ACES2065-1 CVs into themselves and applied in between the Input and the Output Transforms (i.e. before the RRT), like in Figure 11a,b. Again from [56]: "LMTs exist because some color manipulations can be complex, and having a pre-set for a complex look makes a colorist's work more efficient. [...] The LMT is intended to supplement—not replace—a colorist's traditional tools for grading."

Figure 10. Schematic of a linear and very simple ACES color pipeline; the black-edged triangles indicate ACES components. Left-to-right: Camera footage enters ACES scene-referred colorimetry via an IDT; creative color decisions are applied as Looks (LMT); final output is carried over through the Output Transform (RRT + ODT) combination, leading to a display-referred colorimetry ready for output to standard output device (a theatrical projector in this case).

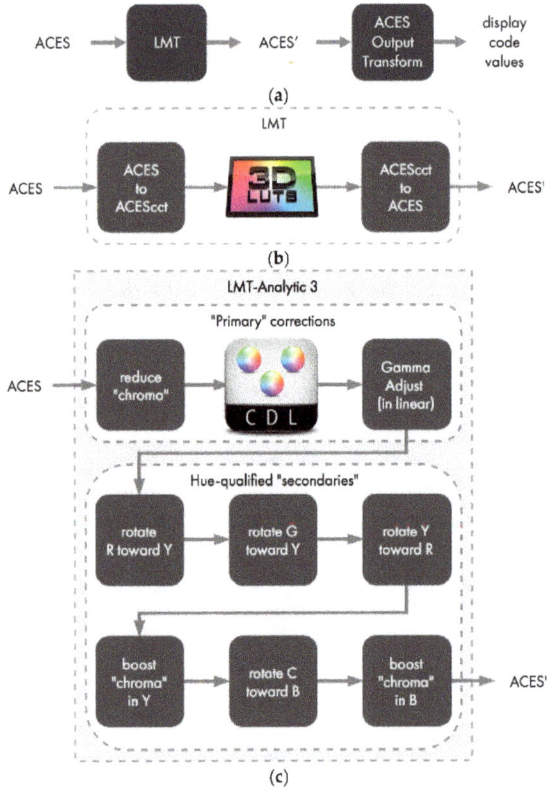

Figure 11. Diagrams showing LMT-based color pipeline samples: (**a**) "Black-box" LMT as mapping ACES2065-1 CVs ACES into ACES2065-1 CVs ACES′; (**b**) Empirical LMT including a color-grading operation (set, for example, as a 3D LUT acting within ACEScct color-space); (**c**) Analytic LMT acting as primary color-correction first (as a CDL in ACEScc), followed by concatenation of hue shifts.

Technically, it may include color-space transformations in between so that the overall action is done within the ACES2065-1 colorimetry, like shown in Figure 11b. As far as creative color modifications are concerned, primary color-grading operations (including those carried over by ColorLUTs or by CDLs, cfr. Section 2.4) can be part of an LMT; closed-form mathematical formulas can be as well. In this case one talks about an "analytical LMT"; when the LMT is mostly originated from a pre-existing technical or creative color mapping (e.g. in the form of a ColorLUT), one talks about an "empirical LMT" instead. Sometimes the grades represent the early or tentative setting for the "film look" [2], coming from on-set grading sessions (and called, in this context, pre-grading [11]); sometimes they represent the complete finished look of a scene; most of the times, the look is just an intermediate passage before more refined finishing is applied to fine-static and moving details. The possibility to save this look as an ACES key component and make it part of the interoperability framework is, however, essential. For this reason, LMTs can be represented, at the low level of the ACES reference implementation (Section 3.1), as CTLs, but they are usually stored using the color-correction software's own algorithms as well. The Academy provides two interchange formats, for looks/LMTs as well as for generic color transformations, that will be introduced in the next paragraphs, cfr. Sections 3.7 and 3.8.

3.7. Color Transformation Language

There are two AMPAS-sponsored projects that describe new syntaxes, detailed in the current and next subparagraphs, respectively. The former is the Color Transformation Language (CTL) [67,68], a C-style language (`.ctl` file extension), that was born with ACES, to describe the widest possible range of color transformations in algorithmic form (i.e., by means of algorithms and numeric formulas); it is organized in "library modules" where CTL files containing higher-level functions may include references to other CTLs defining lower-level functions. A CTL can contain either generic functions for manipulating data (even as abstract as matrix algebra) and production-ready color transformation algorithms; it may or may not reference additional CTLs.

CTL is not intended for production use, because neither the language nor its interpreter, `ctlrender` (which accepts an image plus a color transform specified as CTL, rendering out that image with the CTL applied, or "baked" in), were designed with parallel processing in mind. For similar reasons, most production-ready hardware and software systems, should not implement direct processing of CTL files; they serve as references which actual ColorLUTs can be built upon and later on processed faster, especially by real-time systems.

The CTL system is not only modular, but also extensible: Academy-provided CTLs, being part of the reference implementation (cfr. Section 3.1), are structured using the filesystem hierarchy under `transforms/ctl` folder; those of the core transforms in Sections 3.3–3.6 can be found in the corresponding subfolders (and grouped therein using subfolders named along with camera vendors or output color-spaces), whereas a generic subfolder `./utility` contains the "library" of lower-level mathematical and colorimetric functions commonly used by the other ODTs.

For example, the official `LMT.Academy.ACES_0_7_1.a1.0.1.ctl` file in [43], containing a legacy sample LMT, does not depend on any library CTLs, as it is essentially a pre-baked 3D LUT, cfr. Sections 2.4 and 3.6. Let instead the `ODT.Academy.P3D60_ST2084_4000nits.a1.0.1.ctl` file in [43] be considered (Section 3.5); its filename can be interpreted using [45] as a map from ACES colorimetry version 1.0.1 to a very specific, output-referred color-space for technical mastering/viewing called "P3@D60 (i.e., P3 primaries [26] with D60 white-point (Table 1 note 4)) with SMPTE ST2084 transfer curve [29] and 4000 cd/m^2 peak luminance"—cfr. Section 2.3 for technical insights about this output colorimetry. This CTL depends on four CTLs in the `../../utility` subfolder, progressively defining lower level formulas:

- used in most ODTs (e.g., white roll, black/white points, dim/dark surround conversion);
- the RRT itself, including generic tone-scale and spline functions contained together with it;
- basic color-space linear conversion formulas (plus colorimetry constants);
- functions from linear algebra and calculus (plus a few mathematical constants).

A single ODT file may have a (commented) XML tag **`<ACESTransformID>`** whose element is the ODT's original filename itself—but it may be some kind of universally-unique identifier (UUID) or, in the future, some hash-digest or other message integrity code (MIC). Color-management applications may honor this tag in order to properly index and identify CTLs within larger databases, possibly retrieve legacy ones from a registrar/repository—whereas only newer versions may be shipped along with the reference implementation—and, in case of hash/MIC, validate the transforms to double-check they were not altered or corrupted (which is essential if CTLs are all remotely stored, for example, for IP protection reasons—cfr. Section 2.2).

3.8. *CommonLUT Format*

The latter file format/language, sponsored by both the Academy and the ASC, is the CommonLUT Format (CLF) [51,68], which is a XML dialect for storing both exact and approximated color transforms (like ColorLUTs [2,16] and CDLs [33], cfr. Section 2.4), as well as their combinations. This is suggested as elective interchange file format for ColorLUTs in ACES workflows, and an open SDK in Python is also available [43] as reference implementation for bi-directional conversions between CLF and other open or proprietary ColorLUT file formats used by current commercial software. CLF also has the following features:

- combined, single-process computation of ASC CDL, RGB matrix, 1D + 3D LUT and range scaling;
- algorithms for linear/cubic (1D LUT) as well as trilinear/tetrahedral (3D LUT) interpolations;
- support for LUT shapers (cfr. Section 2.4) as well as integer and floating-point arithmetics.

More features will be implemented in the next versions of CLF, like other technical and authoring metadata (to better characterize source and target color-spaces), new color transform paradigms (including direct CTL referencing), new interpolation methods, and integration with Academy-provided color transforms (via `ACESTransformID` tag, Section 3.6). Expansion of the CommonLUT format is also one of the author's main contributions to ACES, particularly on color-metadata extensions and on interpolation algorithms using advanced Algebraic Topology techniques based on simplicial geometry, cfr. [20,76,77]. As of 2017, CLF has received mild support by a few commercial color-correctors (e.g. Light Illusion *LightSpace*), mainly because most of the algorithms behind its main process nodes can hardly be applied in real time on footage being played back. Yet, some vendors adopted CLF, and then further extended it, into real-time proprietary file formats for their color-grading, finishing or CGI modeling software: CTF file format for Autodesk® *Smoke®/Maya®* and DCTL file format for Blackmagic Design *DaVinci Resolve*.

3.9. *ACESclip: A Sidecar for Video Footage*

Version 1.0 of ACES introduced a very important addition to existing color management frameworks for Media and Entertainment, in the form of the (ACES) Clip-level metadata file, briefly called "ACESclip" [58]. Despite a current scarce implementation by product partners, this "sidecar" file is meant to be generated for each video clip entering (even just formally) an ACES pipeline, and stay along with it any futur re-renders, at least for as long as the clip logically resides in an ACES-color-managed environment. In this context, a video clip may be either a "frame sequence" (i.e., a sequence of still-picture files with a consistent, consecutive and uninterrupted enumeration in their filenames, where each file progressively represents one frame of a locally-referenced video sequence—e.g., {`frame-0001.ari`, `frame-0002.ari`, `...` , `frame-0859.ari`} for a sequence or ARRIRAW files [78], containing one clip recorded by an ARRI Alexa motion picture camera), or a single video file (e.g., the file `A050_C029_0803HB_001.R3D` containing one clip recorder by a RED DRAGON camera, cfr. Section 2.1). An ACESclip is an XML-language file that persists for every such clip, acting as both a manifest file, i.e., storing metadata about the clip that may or may not be stored within the clip's file(s) header, and as a sidecar file, i.e., ACESclip's locality-of-reference with its clip shall be maintained by storing it in the same folder as the clip and with the clip's own filename, plus `.ACESclip.xml` extension appended, cfr. Figure 12.

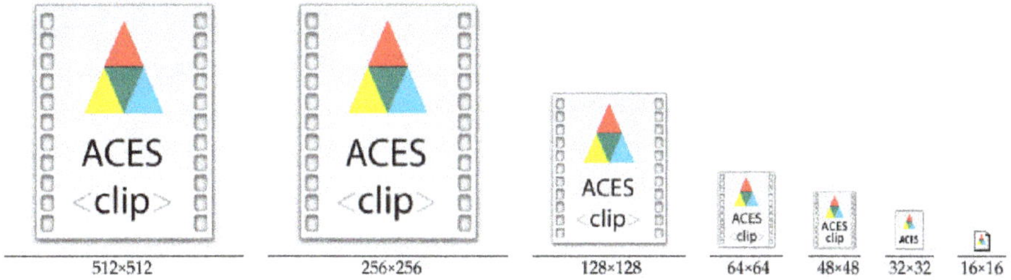

Figure 12. File-type icon study for ACESclip (`.ACESclip.xml` extension), sized 512 down to 16 pixels.

In the former, frame-sequence case above, that is `frame-.ari.ACESclip.xml`; in the latter, single-file case, it is `A050_C029_0803HB_001.R3D.ACESclip.xml`.

ACESclip also has an XML namespace reserved to it: **aces**. This component may be needed in contexts where either the clip's file-format does not support fields where relevant color-related metadata can be unambiguously stored, or this association cannot be enforced (e.g., such fields exist in the file headers but not all applications honor them or expect such information as written according to a standard encoding, etc.), or the conversion of the clip to file format allowing such metadata in the header is impractical (due to time, storage and/or computing constrains). Keeping the (original camera) footage in its native file format(s) and colorimetries, while adding missing information in an ACESclip, is always advantageous for the same parenthesized reasons. Despite the ACES framework specifies file formats for storage of ACES footage, read Section 3.10 and [64–66], such formats are not mandatory in order to be ACES color-managed; original camera footage, for example, may be usually stored in its original raw format (Section 2.1), untouched; ACESclip comes in handy especially in this context, where a record of the colorimetry and color transformations that the clip has either undergone (in the past), represented in its present state, and likely to be interpreted (in the future) for viewing/mastering purposes may be advantageous. Information should also be present in ACESclip to unambiguously re-associate it to its original clip in case the above locality is lost (e.g., the ACESclip file is separated from the clip itself), although ACESclip should stay along with its content throughout its entire lifecycle (until archival and/or destruction). Information stored in the sidecar file includes:

- reference to the clip itself by means of its filename(s) and/or other UIDs/UUIDs;
- reference to the Input Transform either used to process the clip in the past or intended for entering the clip into an ACES pipeline in the future;
- reference to LMTs that were applied during the lifecycle of the clip, with explicit indication whether each is burnt on the asset in its CVs, or this a metadata association only;
- in case of "exotic" workflows, the Output Transform(s) used to process and/or view the clip.

References to the above color transformations like Input/Output Transforms and LMTs are made by either linking the CTL's `ACESTransformID` (if any), or directly writing the ColorLUT or ASC CDL data as XML extensions, or by using ACESclip's own XML tags. Future versions of ACESclip will have more features (which is one of the author's current activities within this project), also thanks to the use of XML, like:

- the clip's color pedigree, i.e., full history of the clip's past color-transformations (e.g., images rendered in several passages and undergoing different technical and creative color transforms);
- a more specific correspondence, for a selected number of professional imaging file formats or wrappers (e.g., ARRIRAW [78], REDCODE, QuickTime, MXF, CinemaDNG [79], TIFF [70],

DPX [71], Kodak Cineon™), between own UID-like file-header fields and ACESclip's `ClipID` element;

- extending the ACESclip XML dialect with other production metadata potentially useful in different parts of a complete postproduction/VFX/versioning workflow (frame range/number, framerate, clip-/tape-name, TimeCode/KeyKode, frame/pixel format, authoring and © info, ...).

An example of ACESclip XML file, reference for namespace and most of its tags, is found in [58].

3.10. Storage and Archival

As written in Section 3.3, storage of ACES footage is allowed in ACES2065-1 color-space only. Nevertheless, files are not always stored or transported in ACES colorimetry and, even if not endorsed in ACES guidelines, many VFX facilities prefer to encode footage, and temporarily store it, in ACEScg color-space for several practical reasons [80]—ACESclip can come in handy to store color parameters that will be used to do both things properly, cfr. Section 3.9.

However, a small group of file formats are defined as elective choices for storing footage in ACES colorimetry (ACES2065-1): within the SMPTE family of standards about ACES, metadata constrains are defined for filesystem implementation as either frame sequences and single video files, cfr. Section 3.9. Stored as file-per-frame sequences (ST2065-4 [54,64]) the open-source OpenEXR still-picture format ("EXR") is chosen [69]—uncompressed, scanline-based, floating-point 16 bits/channel encoding—plus other specific constrains. Stored as video files (ST2065-5 [65]) the SMPTE MXF multimedia container [81], wraps a series of frames, each encoded as per ST2065-4, plus additional constrains on the MXF internal structure. Table 4 lists all the principal constrains for the above formats.

Table 4. Technical constrains and mandatory metadata for ST2065-4/5 compliant ACES content.

OpenEXR (ST2065-4)	**MXF (ST2065-5)**
version/endianness/header: 2.0/little/≤1 MiB	essence container: MXF Generic Container
attribute structure: name, type name, size, value	mapping kind: ACES Picture Element
tiling: scanlines (↑↓ order), pixel-packed	content kind: Frame- or Clip-wrapped as ST2065-4
bit-depth/compression: 16 bpc floats [73]/**none**	MXF Operational Pattern(s): any
channels: (`B`,`G`,`R`) or (α,`B`,`G`,`R`) or S3D: ([α],`B`,`G`,`R`,[left.α],left.`B`,left.`G`,left.`R`)	content package (frame-wrapped): in-sync and unfragmented items of each system/picture/audio/data/compound type
color-space: ACES2065-1 [61]	Image Track File's top-level file package: RGBA Picture Essence
raster/ch./file size: ≤4096 × 3112/ ≤8 / >200 MB	channels: (`B`,`G`,`R`) or (α,`B`,`G`,`R`)
Mandatory metadata: `acesImageContainerFlag=1`, `adoptedNeutral`, `channels`, `chromaticities`, `compression=0`, `dataWindow`, `displayWindow`, `lineOrder`, `pixelAspectRatio`, `screenWindowCenter`, `screenWindowWidth` (stereoscopic images: `multiView`)	Mandatory essence descriptors: Frame Layout = `0` = `full_frame`, Video Line Map = [0,0], Aspect Ratio = abs(`displayWindow[0]`· `pixelAspectRatio)/displayWindpw` [1], Transfer Characteristic = `RP224`, Color Primaries = AP0 [61], Scanning Direction = `0`

There has also been a growing interest by studios and OTT companies in encoding mezzanine/master formats preserving the original ACES colorimetry, so that interchange and archival copies are stored in one shared colorimetry. The *de-facto* standard for this purpose has become the SMPTE Interoperable Master Format (IMF), which has a dedicated family of standards, ST2067 [82]. In short, IMF prescribes that a content—either video, audio, and timed-text (TT) data—to be either exchanged among facilities, long-term archived, or simply prepared for mastering, as per Section 2.3, is organized in a IMP (interoperable master package), i.e., a sub-filesystem structure that may have one or more of:

- video tracks, (monoscopic or S3D stereoscopic);
- sound groups, as separate sets of audio tracks, each with possibly multiple channels (e.g., a sound group may have 3 audio tracks: one has "5.1" = 6 discrete channels, one 2 discrete "stereo" channels and the other a Dolby-E® dual-channel—all with different mixes of the same content);
- TT tracks (e.g., subtitles, closed captions, deaf-&-hard-of-hearing, forced narratives, ...);
- one Packing List (PKL) as the inventory of files (assets) belonging to the same IMP, listing their filenames, UUIDs, hash digests and sizes;
- one Composition Play-List (CPL) describing how PKL assets are laid out onto a virtual timeline;
- Output Profile List(s) (OPLs) each describing one output format rendering the IMP into a master.

PKL, CPL and a few other files in an IMP are XML sidecar files, whereas video and audio assets are wrapped in MXF files [81]. The separation between content essences (video/audio/TT clips) and their temporal/logical organization within a timeline (the IMP's virtual timeline) allows easy interchange of single components for versioning purposes (e.g., localization, assets' distribution, censorship, etc.). An IMP can, in fact, depend on other IMPs (which it is a "supplemental" package of), so it is easy to build a new cut/edit/mix of a piece of content by having a new CPL referencing essences from several PKLs. As well as MXF standards include "Operational Patterns" that add additional technical and operating extensions to the file format (as regards essence codecs, metadata, internal file structure, interoperability and, above all, usage workflows), IMF standards include numbered "Applications" for specific requirements where, for example, IMPs must obey one or more rules regarding either naming conventions, compulsory metadata, package and virtual timeline structure, etc.

Use of ACES colorimetry with the very versatile and extensible IMF is being standardized in Application #5 [66], including author's contributions, which prescribes video content encoding in ST2065-5 standard (MXF-wrapped EXR frames in ACES2065-1 colorimetry), plus additional metadata describing the version of ACES and of the core transforms from Sections 3.4–3.6. Optional reference still-frames, encoded via Output Transform(s) and included in the IMP as either PNG or TIFF files, are linked at specific frames of the virtual track, so that content mastering (at least in the same color-space(s) as the reference's) can be visually evaluated for consistency in case the content is "restored" in more-or-less unknown conditions. This is a very important (yet optional) feature, as IMF Application #5 deals with packages in ACES2065-1 and purposes them for interchange and even long-term archival, delegating to otherwise unreferenced Output Transforms the task to produce any viewing colorimetry for the content. In a sense, IMF Application #5 is a complement and an extension to ACESclip specifically targeting interchange, versioning, mastering and even long-term archival, as further detailed in Section 6.

3.11. ACES Integration with Photochemical Film Process

A last-but-not-least part of the ACES framework is dedicated to workflows involving film scanning and film-out (i.e., film printing) processes; this part was one of the earliest to be designed—with involvement of digital film-processing labs [23,33]—and is increasingly relevant to film preservation, as remarked in Section 6. A multi-device, yet simple ACES workflow for a DI pipeline, from scan to film-out, is shown in Figure 13. For a complete description of photochemical film-based processes cfr. [12,13]; for a "traditional DI" workflow instead, cfr. [31]. First, a new densitometry for photochemical negative and internegative films is defined, the Academy Printing Density (APD), replacing other standard optical density metrics, like Kodak Cineon™ Printing Density (CPD). For a film printer, APD depends on the overall spectral power distribution (SPD) along the light path (i.e., light source, spectral transmission/reflectance/absorbance of optics, film medium). It has 3 spectral components $(\overline{r}_{\mathrm{APD}}, \overline{g}_{\mathrm{APD}}, \overline{b}_{\mathrm{APD}}) \in \mathbb{R}^3$, one per spectral responsiveness of the printer's RGB light sources, is numerically defined in [62], and modeled after the Bell & Howell Model C® with

Kodak Wratten® filter No. 2B applied—resulting in a variant of Status-M densitometry [83]. It is thus a vector field $\boldsymbol{\Pi}_{\text{APD}} \in [0, +\infty[^3$ depending on the film sample's spectral transmittance $T(x,y,\lambda)$:

$$\boldsymbol{\Pi}_{\text{APD}}(x,y) := -\log_{10} \int_{360 \cdot 10^{-9}}^{730 \cdot 10^{-9}} T(x,y,\lambda) \begin{pmatrix} \overline{r}_{\text{APD}}(\lambda) \\ \overline{g}_{\text{APD}}(\lambda) \\ \overline{b}_{\text{APD}}(\lambda) \end{pmatrix} \mathrm{d}\lambda. \tag{8}$$

An APD minimum value of 0 means 100% transmittance, thus $\boldsymbol{\Pi}_{\text{APD}} = \mathbf{0}_3 \equiv (0,0,0)$ corresponds, once re-printed to positive stock (as APD natively refers to negative films), to the deepest printable black. As regards the representation of negative/internegative film scans as digital data, a densitometric color-space called Academy Density Exchange (ADX) is introduced [63], with logarithmic transfer characteristic and integer CVs (10-/16-bits per channel), encoding image data as the APD values of a color negative film that will be printed on recent, average motion picture print stock (i.e., a mix of Kodak Vision®, Fujifilm Eterna® and F-CP®). ADX thus represents quantized film density scanned from a negative film, as it is expected to be printed by a reference printer. ADX is output-referred to a reference print film: all in all, it encodes a "film-referred" color-space.

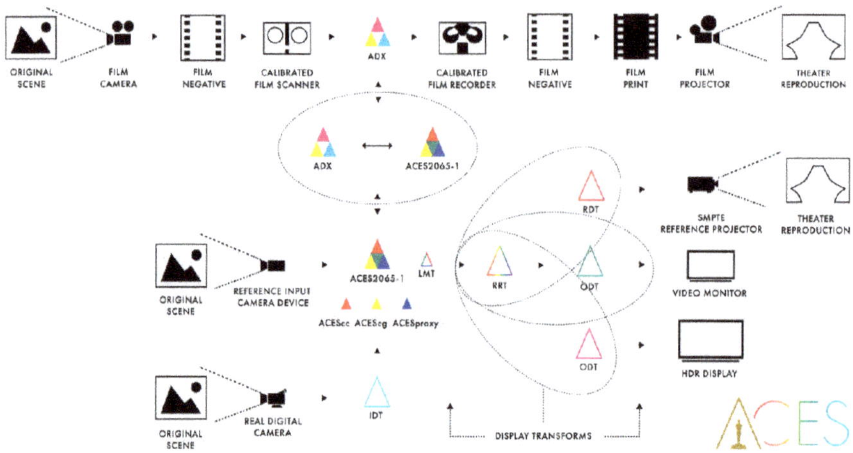

Figure 13. Comprehensive (yet simple) ACES pipeline involving multiple sources: film scans (Section 3.11), a real digital camera, an ideal RICD (Section 3.4), and multiple deliveries (film + digital projection, SDR + HDR TV, Section 3.6).

To compute an ADX CV $\mathbf{c}_{\text{ADX}} \in \mathbb{N}^3$, the optical density of a film receiving 0 EV exposure, $\mathbf{D}_{\min} \in \mathbb{R}^3$ (i.e., that very negative film's clear-base density), is subtracted from the APD triple $\boldsymbol{\Pi}_{\text{APD}}$ of a particular point on the film, then it is quantized according to Equation (9):

$$\mathbf{c}_{\text{ADX}10|16} = \text{rect}_{0}^{1023}\Big|_{0}^{65535} \left[5|80 \begin{pmatrix} 100 \\ 92 \\ 95 \end{pmatrix} (\boldsymbol{\Pi}_{\text{APD}} - \mathbf{D}_{\min}) + 95|1520 \begin{pmatrix} 1 \\ 1 \\ 1 \end{pmatrix} \right]. \tag{9}$$

A sequence of DPX frames [71], can be used to store film scans in ADX color-space; it is recommended to ensure that the scanner software fills as many DPX Film-Area metadata as possible, including printing-density type (APD), clear-base $\mathbf{D}_{\min}$ and, above all, perforation-wise KeyKode™ (if available). The Academy provides Output Transforms from ADX to ACES2065-1 and documentation on how to calibrate the laboratory's printing process to ACES colorimetry, cfr. [59], which means matching APD with all the films stocks used in the laboratory's scanners and printers. PFEs may also be generated out of this—as usual in traditional DI workflows—and become "empirical LMTs", as described in Section 3.6 and Figure 14.

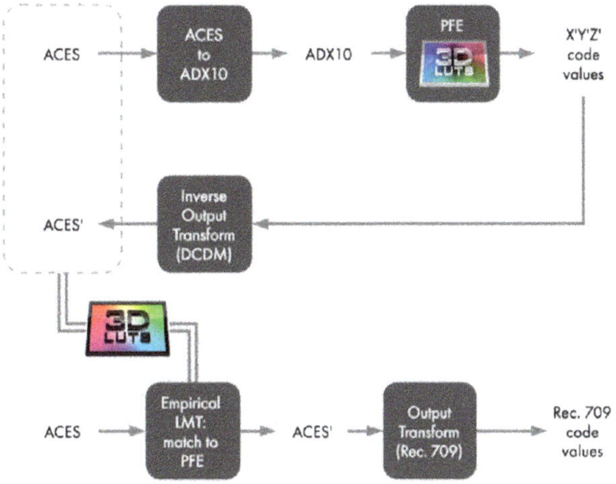

Figure 14. Example of creation and use of an empirical LMT from a film workflow (cfr. Section 3.6): after applying a conversion to ADX color-space and concatenating a pre-existing PFE (cfr. Section 3.11 and Figure 6c) from a traditional DI color pipeline (used to mimic a specific film stock into a DCI $X'Y'Z'$ output colorimetry, cfr. Section 6), a conversion back to ACES2065-1 is done to map to ACES′ CVs. This overall transformation may be burned into another 3D LUT acting as empirical LMT to achieve the same look (ACES′ CVs) for any other footage—including those not from film scans.

4. The ACES Color Pipeline, from Theory to Practice

Now that all the ACES components have been defined and general concepts for their use were explained, a more detailed, end-to-end workflow based on ACES 1.0 will be explained (Figure 15), including general processes already introduced Section 2.

In a typical on-set or near-set scenario, footage from the camera is either saved as is, in its camera-native file formats and colorimetries, or can be sometimes converted to ST2065-4-compliant EXRs (cfr. Section 3.10); latter choice is rarely employed because it requires additional storage and computing power which is so critical for on-set operations. Sometimes though, the camera-native format may be unsupported by some of the tools, so *a priori* conversion is preferable, as long as it does not reduce the footage quality. Viewing of the footage is possible through monitors equipped with view LUTs from the camera-native colorimetry to the on-set monitor colorimetry, but in case of ACES workflows, the monitor reference output from the camera is expected to provide a video signal in ACESproxy to an ACES-compliant monitor that reads that video feed and, once detected, displays it in its right colorimetry. Most professional cameras by RED, ARRI and Sony Cinealta™ have this monitor-out SDI port with ACESproxy colorimetry (so the conversion is internally done by the camera), while the footage is sent to the recorder or to SSD/flash storage in its native colorimetry. If CDLs are applied on-set, the footage should be converted to either ACEScc, ACEScct or, in case a true floating-point capable color-corrector is not available on-/near-set, to ACESproxy. This conversion is necessary because the CDL Equations (1) and (2) encode meaningful color-correction metadata only if operating on CVs of a "log" color-space.

Figure 15. ACES-based color grading (cfr. Figure 3), with the arrow on top ordering this "layer-based" representation of real-time image processing. Source footage in its camera-native colorimetry (**left**) technically enters the color pipeline by means of Input Transforms (camera A is represented, but different IDTs may be used for cameras B, C, …). After everything is converted in ACES2065-1 CVs, creative grading is done via same operators as in the non-ACES case, i.e., CDLs, 3-way CC's, creative LUTs (both LMT and secondary grades), with implicit conversion to either ACEScc or ACEScct as working color-space and, after all CC layers, back. Viewing on the reference monitor/projector (**right**) is done via RRT + ODT (courtesy: Hieronymus Bosch, *The Garden of Earthly Delights*, c.1495–1505).

By taking Figure 13 as a bird's-eye view reference for the video post-production phase, Figure 15 concentrates on color grading. In this case, again, the footage is internally converted via Input Transform(s) adequate to the camera footage, into ACES2065-1; sometimes this passage was done before—either near-set or in the postproduction lab during a preparation phase like data management, digital negative "development", noise/grain management. Internally, the color corrector applies most the color-grading operations by pre-converting the footage to ACEScc or ACEScct, then applying the transforms (CDLs, 3-way CC, printer lights, lift/gamma/gain, "X-versus-Y", etc. [22]) and finally post-converting the footage back to ACES2065-1. As said in Section 3.3 and specified in [46], this conversion should be almost (if not completely) invisibly to the user. A very important caveat, though, is that the choice of working color-space to use for color-grading (among ACEScc or ACEScct) should be the same across departments, otherwise the CDL coordinates cannot represent the same color-correction when moving from the set to the post-production lab.

For example, if ACESproxy are employed on-set for CDLs, then ACEScc may be safely used in postproduction because, as per Section 3.3, formula (7) is the "integer-arithmetics analogue" to (4), ACES-compliant CDLs use $\log_{cc}^{ACES}$ as transfer characteristic (provided the difference between the video-legal range of ACESproxy is taken care as well). If ACEScct is used in post-production instead (usually on specific colorists' creative request on the default look of the footage), CDLs generated on-set on top of ACEScc or ACESproxy cannot be used at all: there is no integer-arithmetics, transport color-space analogue to ACEScct.

Primary color-correction is usually applied globally as a LMT, cfr. Section 3.5, so ACESclip sidecar files may be enough to describe and represent such creative operations on the footage. However, color-grading is usually much more complex (as seen in Section 2.4) therefore highly dependent on the color-correction application's tools and file formats to store project metadata. Although secondary grades and more complex operations are, in fact, not meant for LMTs, they are powerful instruments to help preserve or build more complex looks starting from pre-existing LMTs or other kind of simpler "primary" color operations; for example, empirical LMTs can be built out of pre-existing LMTs—e.g., out of a Print Film Emulation LUT (PFE) created in a film laboratory—and hard-coded ("baked") as an individual 3D LUT, to replicate advanced color effects and transport them like black boxed in a node-based color pipeline (cfr. Figure 15).

5. Use Case of an End-to-End ACES Workflow in a Full-Feature Film

A real-world use case where ACES was effectively and successfully used, will now be explained, concentrating on the pre-production tests, with the intent to show not only how ACES actually "works", but that it is a solid and viable option for films with limited budget as well. It may take a while to make everyone comfortable working with ACES color management (from camera to editorial, to VFX, to finishing departments) but—especially if ones does not own an expensive color science department—it really pays off in the end and increases everyone's trust in what is seen on monitors while jobs are carried on.

During the author's service as Chief Technology Officer (CTO) of Frame by Frame Italia (a leading theatrical and commercials post-and-VFX company based in Rome, Italy), this company provided video postproduction and VFX services for Indigo Film's superhero movie sequel *Il Ragazzo Invisibile: Seconda Generazione* (G. Salvatores, 2017 [84]). After a pre-production meeting with all the above departments involved (including the camera department from Kiwii Digital, the colorist, and the VFX supervisor), it was decided to use a workflow integrally based on ACES version 1.0.3 (as regarded color-science), thus spanning photography, postproduction and VFX. The decision was due to the expected use of ARRI Alexa XT cameras for principal photography (shooting in ARRIRAW frame sequences, uncompressed, 2.8K [78]), plus additional RED DRAGON (R3DCODE raw clips, JPEG2000-compressed, 6K) and GoPro (clips in ProTune™ color-space, H.264-compressed, 4K) cameras, thus leveraging on the system's multi-input compatibility. The author, as both an active color-scientist contributing to the ACES project in first person [51], and for having already conducted camera-comparison sessions for ACES 1.0 compliance before [85], designed two different pre-production tests dedicated, respectively, to the camera department (Kiwii Digital, Section 5.2) and the VFX department (Frame by Frame, Section 5.3), to individually interoperate with the finishing department. The company's DI theater, equipped with FilmLight *Baselight* color-correction system (cfr. Figure 4a,b) and a Christie 2K reference projector (DCI P3 color-space, with 14 fl full-white and dim surround color-adaption), as a "hub" for color-critical evaluation. As said in Section 2.4, in fact, the DI theater is where the cinematographer and the colorist spend most of the time developing and finishing the "look" of the movie, in a lighting and viewing environment that is a reference for all movie theaters where film will be screened to the audiences.

The choice of tools across the various phases, and their total compliance with ACES 1.0.3, made this test very straightforward: *a fortiori*, it simply confirmed that all possible production roundtrips for the film's color pipeline produce mathematically color-identical results. In the end, no specific "show LUT" was crafted to use as a empirical LMT, because the cinematographer and the colorist liked the default look given by ACES' default RRT (version 1.0.1, Section 3.5); the only technical 3D LUT that was exchanged with camera department was the Input Transform for the ProTune™ color-space (as GoPro was not an ACES product partner, Section 2.2, at the time of the tests). All cross-department data and metadata paths undergoing ACES processing were therefore tested, as described in Sections 5.2 and 5.3, with positive outcome.

5.1. *Production Workflow and Color Metadata Path*

For the above movie, a workflow very like that described in Section 4 was used. The on-set/camera department provides on-set grading, footage backup copies and generate dailies for editorial (using Avid® *Media Composer®*). For grading the camera department uses ACEScc, as their on-set equipment has full floating-point processing capabilities (ACESproxy is not needed for this workflow), exporting CDLs on top of this space. Dailies are MXF-wrapped DNxHD files, 1080p24@120 Mb/s, encoded in Rec.709 color-space with pre-grades burnt in; Avid project files ("bins") with all the footage metadata, including the CDLs, are sent from camera to editorial directly. Kiwii Digital also prepares the backup of all original footage (ARRIRAW, R3DCODE, H.264), audio, CDLs other metadata into 6th-generation digital LTO tapes (2.8 TiB per tape) using interoperable LTFS format. The fact that the postproduction facility provides both finishing and VFX services, and hosts the external editorial department (2 editing suites), all within the same premises, helps a lot for intercommunication.

Editorial department produces Avid bins containing references of footage needed for compositing (VFX plates), i.e., the shot scenes that will receive composited VFX over. From the above bins only the relevant footage is restored from LTO6s (plus 15 head/tail handle-frames for each clip), converted into ST2065-4-compliant EXRs, and transferred to the VFX storage. The reason for restoring just the needed clips is economy on the storage footprint (9 MiB per frame); reason for converting from camera-native formats is explained at the end of Section 5.2. At the same time, using in-house Python code written by the author, each VFX plate from the above bin is converted into a new *Nuke* projects that already include specific high-level automations (further customizable on a show basis), like:

- filesystem naming-convention enforced for input (VFX plate footage) and output files (renders);
- OpenColorIO set as color-management system, with ACES 1.0.3 configuration, cfr. Section 3.1;
- all color-spaces for input footage, optional CGI models, CDLs and render files, automatically set;
- customizations for the specific artist assigned to the job (name, OS, workplace UI, ...);
- customizations/metadata for the specific job type (rotoscoping, "prep", compositing, ...);
- set up of a render node that writes both the full-resolution composited plates (as a ST2065-4-compliant EXR sequence), and a reference QuickTime-wrapped Apple® ProRes 4444, with informative slate at the beginning, to use for previz of VFX progress in calibrated UHD TV room;
- set up of viewing path that includes the CDL for each plate node (with ACEScg to ACEScc implicit conversion), just before the `Viewer` node (implicit Output Transform to the monitor);
- set up of alternative viewing path using a *Baselight for Nuke* plugin, to previsualize the specific secondary color-corrections from the finishing department (read below).

For the last bullet-point it was decided that, for a few selected shots with complex grading and/or compositing requirements, it is necessary to simulate the full secondaries for exact matching (e.g., a different CC applied to the plate and to the composite, that needs to change as the VFX is prepared, both as CGI and compositing). In this case, FilmLight provides a metadata format called BLG (BaseLight Grade) that exports the full CC of a single shot, as done in *Baselight*, later to be imported by *Nuke* or *Media Composer®* via a node-based plugin. For those selected shots, the in-house tools automatically replace the CDL node with the latter plugin node (referencing the relevant BLG file).

Regarding the color-correction phase, the editorial department sends an Avid bin with the final cut of the movie, and the editing information embedded as metadata (including the CDLs) to the postproduction lab; the conforming phase (cfr. Figure 1) rebuilds, in *Baselight*, the complete timeline using either the original footage selectively retrieved from LTO6's, or with the rendered versions of VFX plates (if available, even as temporary versions). As long as the project "reels" are prepared, the colorist and the DoP work on the film look and finishing. Color grading starts from the CDLs and internally works in ACEScc color-space, but produces ACES2065-1 files in the end.

At the end of the project (still ongoing at the time of writing) Frame by Frame provides the finished footage in several delivery formats, including those for long-term archival in the form of both ST2065-4-compliant OpenEXR sequences (cfr. Section 3.10) and DCDM-compliant TIFF sequences, to be archived in 7th-generation LTO tapes (5.6 TiB per tape, LTFS).

5.2. Camera Department Test

First of all, the reference monitors used by the camera department during principal photography are all color-calibrated to Rec.709 (gamma 2.4), and cross-calibration with the reference monitors used in all the other departments was double-checked. The camera department does not alter the original camera footage (that is archived and always worked on in its native color-space, i.e., mostly ARRI LogC WideGamut, into ACES2065-1 via official Input Transforms), while pre-grading and dailies (with burnt-in CDLs) are color-handled on-/near-set. For this reason, during the usual pre-production camera and lens tests, Kiwii shoots a few clips for each used camera type and a few different lighting settings; then processes the footage and renders each of the clips in four variants, {1–4}, differing by color-space and file formats (cfr. Figure 16a for a sample from ARRI Alexa XT camera):

- EXR sequence (ST2065-4-compliant, original resolution), without {1} and with {2} baked CDL;
- DPX sequence, resampled to 2K and Rec.709 color-space, without {3} and with {4} baked CDL.

The finishing department then generates all the variants {1–4} from scratch, via *Baselight*, from the original footage in their camera-native color-spaces; for each clip, the samples generated in the two facilities are compared both visually (in the DI theater), then technically by pixelwise subtraction of one sample from above the other, as shown in Figure 16.

Comparison of the original footage with {1} and {3} respectively verifies the consistency of the Input Transforms for the ARRI Alexa camera, and of the Output Transform for Rec.709; the above comparisons were indeed not necessary, as they are standard components used across logo-partner products, Section 3.2. Comparison of the original footage with the EXR variant with CDL {2} verifies the implicit ACEScc color-space conversions used by the color-grading processes (both on-set and DI); that with variant {4} verifies the whole viewing pipeline consistency (Input Transform + LMT + Output Transform), i.e., that the on-set pre-grades look the same throughout to finishing.

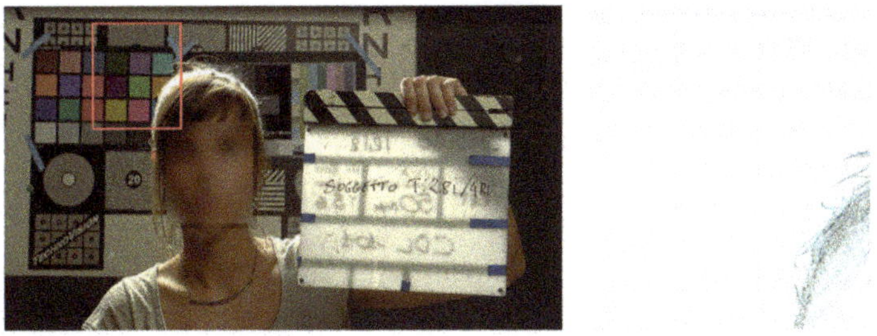

(a) (b)

Figure 16. Camera-department tests: (**a**) camera-test footage, as processed by on-set equipment (no CDL), with detail zone in red; (**b**) differences with the same footage and same processing by finishing department (white areas mean the two sources sum to 0). High-frequency discrepancies are evident.

The tests showed no visible differences, thus confirming effective interoperability. Difference tests were done in floating-point arithmetics, naturally supported by both Foundry *Nuke* (32 bits/channel) and Adobe® *Photoshop*® CC (16 bits/channel), and using ACEScg as working color-space—naturally implemented in *Nuke* via OpenColorIO, and configured in *Photoshop*® via an ICC profile that was previously compared for accuracy by means of additional software by Light Illusion.

The only detected—yet expected—inconsistency was present, in all variants, in high-frequency and/or high-saturation details of ARRIRAW footage, as seen in Figure 16b. This is not due to color-management disparities though, but rather to the different de-mosaicing algorithms in software products used by the two teams to "remove" the Bayern-pattern, that is the color-grid pattern that is written in each ARRIRAW file that replicates the layout and geometry of photosites in the Alexa camera's Alev-V sensor, as described in [75]. Demosaicing can be implemented in many ways (e.g., by either CPU or GPU processing [34]), and different accuracy/quality goals; since it is a computationally-intensive process, some color-correction systems employ higher-speed but lower-quality algorithms to prioritize real-time computation of CC operators along with footage playback, while switch to higher-accuracy/higher-quality algorithms for non-real-time operations like rendering the final files. Despite camera vendors either provide SDKs or disclosure of their certified exact algorithms, nor all the products use them, in favor of customized algorithms. The difference is sometimes hardly visible in moving scenes, but can be appreciated in stills, and technically detected in high-detail/high-saturation areas, like in Figure 16b. This is also why, during postproduction, plates sent to the VFX department are pre-converted to EXRs (thus also de-mosaiced) by the same *Baselight* software used for finishing, cfr. Section 5.1. Apart from this consideration though, all the above tests were successful, confirming full interoperability between the on-set and the postproduction workflows. By direct comparison of {2–4} with {1} instead, with all systems using the same EXR footage de-mosaiced by FilmLight software (either *Baselight* or its on-set relative, *Prelight*), no practical differences are detectable: less than 0.0004 CVACES$_{cg}$ delta for all pixels within ±6 EV normal-exposure range.

5.3. VFX Department Test

As first thing for correct 3D work, it was verified that input colorimetries maintain photometric linearity throughout CGI and compositing (ACES2065-1 and ACEScg are both "linear" in this sense, cfr. Section 3.3) so that all lighting, shading and effects based on Physics (optics) simulation are consistent; as usual in VFX methodology [3], principal photography of VFX shots reflection balls on set.

For consistency, a full-CGI 3D asset was also modeled, rigged, shaded and rendered in Autodesk® *Maya*® (as a sequence of EXRs with alpha channel encoding transparency)—also set to use OpenColorIO for ACES [72]—with lighting parameters read from those present in the original footage where it would be composited over using The Foundry *Nuke*-X, which was also used to run this test. Three different versions of each clip were prepared:

- EXR sequence (ST2065-4-compliant), exported by *Baselight* {5}, with and without baked CDL {6};
- *Nuke* render of sequence {5} with composited 3D asset {7}, and with baked CDL {8}.

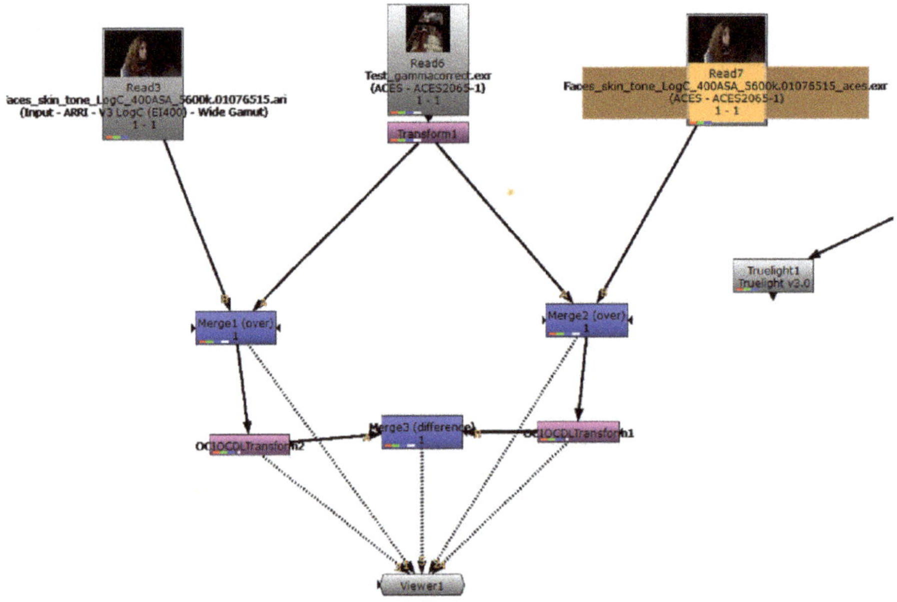

Figure 17. Node-based graph from The Foundry *Nuke*'s CGI/VFX comparison project. `Read` nodes apply the corresponding Input Transforms to footage (if needed, as in this case, for ARRIRAW `.ari` files). The `OCIOCDLTransform` nodes interpret CDLs from the XML metadata to compare with versions with baked-in CDLs. `Merge` nodes composite different images together or compute pixelwise subtraction to verify the consistency. The `Truelight` node is the *Baselight for Nuke* plugin when a BLG file is used to transport a scene's color correction instead of a CDL. The `Viewer` node is where different viewing paths are sent for application of the Output Transform (RRT+ODT) to compositor's PC monitor.

Both the accuracy of the CDL color-path and algorithms, by comparing {6} with {8}, and the accuracy of the compositing operations and of neutrality of OpenEXR render engine, by comparing {5} with {7}, are successfully verified. The node-graph of the *Nuke* project for comparisons is shown in Figure 17, with all internal processing done in ACEScg, as per Section 3.3. There are 0.0 $\mathrm{CV}_{\mathrm{ACES_{cg}}}$ differences between {5} and {7}; not even for the luminance, within ±6 EVs of artificial 3D objects that were generated in *Maya*® (using scene-referred data inferred from the original footage), then rendered and composited over the plates in *Nuke*.

6. Conclusions

As introduced in Section 1, at the time of writing (2017), more than 120 films were produced using ACES components in the workflow, with a similar figure estimated out of commercials and video art (including music videos) [14,44,86]. A significant number comes from independent-studio productions, confirming that ACES does not require major-film-studio budgets [15]. The use-case in Section 5 of the first Italian movie with a complete, end-to-end ACES pipeline includes not only a description of the color workflow, but also a report of the successful pre-production tests, necessary to put the involved teams on the same track with this new "tool". In the author's opinion, much has been done with ACES, to have such a deep worldwide impact in the industry (thanks to the initial effort by the Academy and the early proponents of ACES, but also to SMPTE as a standardization body); much still needs to be done, in fact, to foster interoperability beyond the current preparatory steps. As many technological innovations in Computer Science, ACES is now thought more of as a "process" that is continuously improving, and while some components are mature enough, some other (especially CommonLUT, ACESclip and the storage-related standards from Sections 3.8–3.10) need to be exploited at their full potential, and leverage on automation features that come with wise metadata management (e.g., the color-pedigree and UUID-based identification features described in Section 3.9).

Storage is another key area that has lots of improvement margins—and needs them the most. Industry, governments, and society do not just need to store and exchange content for short-term commercial development. Preservation of video-based cultural heritage (both film-based from the past, and current digital-born assets) is a necessity where technology can and shall help by providing automated and efficient, metadata-aware workflows. The purpose of such a digital transformation is two-fold: effective preservation on one side, and business optimization on the other. Costs for archival of Media and Entertainment content is constantly growing, being sometimes an outstanding show-stopper to many film archives. Technologies like IMF, for example, help aggregate, rationalize, and globally manage the exponentially increasing quantity and size of assets [87]. Disney Studios championed its inception and others, like Netflix on top of the list, were early adopters of IMF within their production pipeline, now gaining huge benefits form it [88]. IMF is a technical means to save Content Delivery Network (CDN) storage footprint, as the same content needs to be either available and archived in tens or hundreds of different versions (due to audio/video/TT localization, special editions, censorship, A\V technology formats like HDR, HFR, immersive audio, etc.). Such versions proliferate according to outstanding power-laws, then get stale at a faster rate than they are indexed, used, or than the technologies storing them obsolesce. In a word, this industry-wide problem has been nicknamed "versionitis" [87].

IMF's compared anatomy reveals how it stemmed from Digital Cinema packaging system (DCP—[89], also defined by SMPTE standards), borrowing supplemental packaging features from it, but without the cryptographic and operational considerations required, in that case, for the secure distribution of finalized theatrical content to theaters, using DRM-style public-key cryptography [90]. IMF is a business-to-business (B2B) tool assuming agreed-upon content-security measures are already in place among the exchanging parties. IMF Application #4 [91], was specifically designed for film distribution and possibly archival but, at the time of writing, still lacks features that can be effective and provide cost-savings to existing preservations or interchange between film archives. For example, it has no standardized ways to archive preservation metadata, logs and complete frame-by-frame versioning information within the IMP structure (despite IMF supports textual metadata within its XML sidecar files). Secondly, IMF Application #4 stores images in output-referred color-spaces, whereas from a preservationist's standpoint, digital archival of film scans should preserve film density information of the film as illuminated by the scanner's lighting and captured by its sensor, thus using scene-based colorimetry (the scene being the density on each film frame). This future-proofs the scans from the choice and progress of output/display technologies. Third and foremost, IMF's multiple and incremental versioning system would benefit processes where, for example, different facilities scan and restore aged photochemical films, or scans are subsequently integrated when missing parts of a show are found at later ages, or different versions are used for a unique restoration process. For example, the original version of a film missing one whole reel (likely lost or destroyed) can be later integrated when a different version of the reel is found in a private collection, yet is scanned at a different lab. As another example, a better-preserved, colored 35 mm print version with analog stereo sound to be used in place of the original black and white camera negative reels for selected frames ranges where aging prevented scanning or restoring to a decent quality.

On the contrary, DCP is being chosen by many film archives and studios as a *de-facto* long-term archival format, despite this poses technological and operational problems. For example, DCPs are encoded in DCI $X'Y'Z'$ color-space, whose wide gamut is fine, but not so is its display-referred state. Besides, they are visually-lossless, yet lossy-compressed with an algorithm optimized for real-time playback, without absolute image quality rationale. Also, encrypted DCPs (frequently used as archived masters) can be played back on pre-determined devices only, and pending certificates' expiration, with the risk of preventing generation of any new decryption keys, unless a resilient preservation chain for the whole Public Key Infrastructure (PKI) is established and maintained. Digital Cinema was designed for theatrical distribution and playback; IMF is designed for interchange and archival.

The other large cultural problem related to color preservation in digital imaging (which ACES may very well be part of a technological method for, but that is still lacking operational solutions) is even more "occult" because it has not been identified as a problem yet. In the contemporary world, there are lots of dispersed imaging devices capable of capturing either still and moving images. Most of the devices apply image-enhancing and often creative color-correction operations to the imagery, often discarding the original image while preserving the result, and without any chance to disable or even manually discern whether there can be advantage/benefit at all to them. From the heritage point of view, we, as a community, are losing the grip with the original look of untouched images—not only because non-professional imaging devices have poor color reproduction capabilities, but also because of such image-enhancing and color-manipulating features do not preserve the colors of original pictures. Social networks' cloud storage also preserves just the final images and not the originals. In a few years' time, most of our visual historical documents' may be "altered" because of this. Apart from storing both the original and the color-corrected version—which, without compression, may affect storage footprint up to 200%—a possible solution would be that all image- and color-enhancing operations are globally, non-destructively applied, i.e., as metadata only (just alike methodologies introduced in Sections 2.4 and 4), so that the original CVs are archived and, thus, preserved.

Acknowledgments: Figures 1, 3, 12 and 15 designed by W. Arrighetti. Figures 4 and 5 sourced from public internet. Figures 2 and 8 generated using *Colour* Python API [92]; Figure 8 courtesy of H.P. Duiker [74]. Figure 6 rendered with W. Arrighetti's toolset written in Python, WebGL and Wolfram *Mathematica*®. Figures 7, 9–11, 13 and 14 courtesy of AMPAS. Figure 17 GUI screenshot of a *Nuke-X* 10.5v3 (courtesy: The Foundry) project.

Author Contributions: W. Arrighetti conceived and designed two ACES experiments: the 2015 camera test [93] and the 2016 end-to-end pre-production test for *Il Ragazzo Invisibile: Seconda Generazione* [85], Section 5. They were independently performed in studios, together with E. Zarlenga and F.L. Giardiello respectively. In both cases W. Arrighetti analyzed the results at Frame by Frame Italia post-production facility, with the support of colorist C. Gazzi. W. Arrighetti wrote the papers [2,16] and, together with F.L. Giardiello, was guest speaker of SMPTE professional development webcast [15], which also quickly introduces the experiment [85].

Conflicts of Interest: The author declares no conflict of interest.

Abbreviations

AMPAS stands for Academy of Motion Pictures Arts and Sciences, SMPTE stands for Society of Motion Picture and Television Engineers, IEEE stands for Institution of Electrical and Electronics Engineers, ITU stands for International Telecommunication Union, ISO stands for International Standards Organization, CIE stands for Commission Internationale de l'Éclairage, MPAA stands for Motion Picture Association of America.

References

1. AMPAS. *Academy Color Encoding System (ACES) Documentation Guide*; Technical Bulletin TB-2014-001; AMPAS: Beverly Hills, CA, USA, 2014. Available online: http://j.mp/TB-2014-001 (accessed on 20 September 2017).
2. Arrighetti, W. Motion Picture Colour Science and film 'Look': The maths behind ACES 1.0 and colour grading. *Color Cult. Sci.* **2015**, *4*, 14–21. Available online: jcolore.gruppodelcolore.it/numeri/online/R0415 (accessed on 20 September 2017). [CrossRef]
3. Okun, J.A.; Zwerman, S.; Rafferty, K.; Squires, S. *The VES Handbook of Visual Effects*, 2nd ed.; Okun, J.A., Zwerman, S., Eds.; Focal Press: Waltham, MA, USA, 2015, ISBN 978-1-138-01289-9.
4. Academy Web Page on ACES. Available online: www.oscars.org/science-technology/sci-tech-projects/aces (accessed on 20 September 2017).
5. Pierotti, F. The color turn. L'impatto digitale sul colore cinematografico. *Bianco Nero* **2014**, *580*, 26–34.
6. MPAA. *Theatrical Marketing Statistics*; Dodd, C.J., Ed.; MPAA Report; Motion Picture Association of America: Washington, DC, USA, 2016. Available online: www.mpaa.org/wp-content/uploads/2017/03/MPAA-Theatrical-Market-Statistics-2016_Final-1.png (accessed on 20 September 2017).
7. Anonymous; U.S. Availability of Film and TV Titles in the Digital Age 2016 March; SNL Kagan, S & P Global Market Intelligence. Available online: http://www.mpaa.org/research-and-reports (accessed on 15 September 2017).
8. Mukherjee, D.; Bhasin, R.; Gupta, G.; Kumar, P. And Action! Making Money in the Post-Production Services Industry, A.T. Kearney, 2013. Available online: http://www.mpaa.org/research-and-reports (accessed on 15 September 2017).
9. Arrighetti, W. La Cybersecurity e le nuove tecnologie del Media & Entertainment nel Cloud. *Quad. CSCI* **2017**, *13*, in press.
10. Storaro, V.; Codelli, L.; Fisher, B. *The Art of Cinematography*; Albert Skira: Geneva, Switzerland, 1993, ISBN 978-88-572-1753-6.
11. Arrighetti, W. New trends in Digital Cinema: From on-set colour grading to ACES. In Proceedings of the CHROMA: Workshop on Colour Image between Motion Pictures and Media, Florence, Italy, 18 September 2013; Arrighetti, W., Pierotti, F., Rizzi, A., Eds.; Available online: chroma.di.unimi.it/inglese/ (accessed on 15 September 2017).
12. Salt, B. *Film Style and Technology: History and Analysis*, 3rd ed.; Starword: London, UK, 2009, ISBN 978-09-509-0665-2.
13. Haines, R.W. *Technicolor Movies: The History of Dye Transfer Printing*; McFarland & Company: Jefferson, NC, USA, 1993, ISBN 978-08-995-0856-6.
14. Arrighetti, W. (Frame by Frame, Rome, Italy); Maltz, A. (AMPAS, Beverly Hills, CA, USA); Tobenkin, S. (LeTo Entertainment, Las Vegas, NV, USA). Personal communication, 2016.
15. Arrighetti, W.; Giardiello, F.L. ACES 1.0: Theory and Practice, SMPTE Monthly Educational Webcast. Available online: www.smpte.org/education/webcasts/aces-1-theory-and-practice (accessed on 15 September 2017).
16. Arrighetti, W. The Academy Color Encoding System (ACES) in a video production and post-production colour pipeline. In *Colour and Colorimetry, Proceedings of 11th Italian Color Conference, Milan, Italy, 10–21 September 2015*; Rossi, M., Casciani, D., Eds.; Maggioli: Rimini, Italy, 2015; Volume XI/B, pp. 65–75, ISBN 978-88-995-1301-6.
17. Stump, D. *Digital Cinematography*; Focal Press: Waltham, MA, USA, 2014, ISBN 978-02-408-1791-0.
18. Poynton, C. *Digital Video and HD: Algorithms and Interfaces*, 2nd ed.; Elsevier: Amsterdam, The Netherlands, 2012, ISBN 978-01-239-1926-7.
19. Reinhard, E.; Khan, E.A.; Arkyüz, A.O.; Johnson, G.M. *Color Imaging: Fundamentals and Applications*; A.K. Peters Ltd.: Natick, MA, USA, 2008, ISBN 978-1-56881-344-8.

20. Arrighetti, W. Colour Management in motion picture and television industries. In *Color and Colorimetry, Proceedings of 7th Italian Color Conference, Rome, Italy, 15–16 September 2011*; Rossi, M., Ed.; Maggioli: Rimini, Italy, 2011; Volume VII/B, pp. 63–70, ISBN 978-88-387-6043-3.
21. Arrighetti, W. Kernel- and CPU-level architectures for computing and A\V post-production environments. In *Communications to SIMAI Congress*; DeBernardis, E., Fotia, G., Puccio, L., Eds.; World Scientific: Singapore, 2009; Volume 3, pp. 267–279. Available online: cab.unime.it/journals/index.php/congress/article/view/267 (accessed on 15 September 2017). [CrossRef]
22. Hullfish, S. *The Art and Technique of Digital Color Correction*, 2nd ed.; Focal Press: Waltham, MA, USA, 2012, ISBN 978-0-240-81715-6.
23. Doyle, P.; Technicolor, London, UK. "Meet the Colourist" Video Interview, FilmLight, 2017. Available online: www.filmlight.ltd.uk/customers/meet-the-colourist/peter_doyle_video.php (accessed on 20 September 2017).
24. IEC. *Colour Measurement and Management in Multimedia Systems and Equipment—Part 2-1: Default RGB Colour Space—sRGB*; Standard 61966-2-1:1999; IEC: Geneva, Switzerland, 1998.
25. ITU. *Parameter Values for the HDTV Standards for Production and International Programme Exchange*; Recommendation BT.709-6; ITU: Geneva, Switzerland, 2015. Available online: www.itu.int/rec/R-REC-BT.709-6-201506-I (accessed on 20 September 2017).
26. SMPTE. *D-Cinema Quality—Reference Projector and Environment for the Display of DCDM in Review Rooms and Theaters*; Recommended Practice RP431-2; SMPTE: White Plains, NY, USA, 2011.
27. ITU. *Reference Electro-Optical Transfer Function for Flat Panel Displays Used in HDTV Studio Production*; Recommendation BT.1886; ITU: Geneva, Switzerland, 2011. Available online: www.itu.int/rec/R-REC-BT.1886 (accessed on 20 September 2017).
28. ITU. *Image Parameter Values for High Dynamic Range Television for Use in Production and International Programme Exchange*; Recommendation BT.2100; ITU: Geneva, Switzerland, 2016. Available online: www.itu.int/rec/R-REC-BT.2100 (accessed on 20 September 2017).
29. SMPTE. *Dynamic Range Electro-Optical Transfer Function of Mastering Reference Displays*; Standard ST 2084:2014; SMPTE: White Plains, NY, USA, 2014.
30. ITU. *Parameter Values for Ultra-High Definition Television Systems for Production and International Programme Exchange*; Recommendation BT.2020-2; ITU: Geneva, Switzerland, 2014. Available online: www.itu.int/rec/R-REC-BT.2020 (accessed on 20 September 2017).
31. Throup, D. *Film in the Digital Age*; Quantel: Newbury, UK, 1996.
32. CIE. *Colorimetry*; Standard 11664-6:2014(E); ISO/IEC: Vienna, Austria, 2014.
33. Pines, J.; Reisner, D. *ASC Color Decision List (ASC CDL) Transfer Functions and Interchange Syntax*; Version 1.2; American Society of Cinematographers' Technology Committee, DI Subcommittee: Los Angeles, CA, USA, 2009.
34. Clark, C.; Reisner, D.; Stump, D.; Levinson, L.; Pines, J.; Demos, G.; Benitez, A.; Ollstein, M.; Kennel, G.; Hart, A.; et al. Progress Report: ASC Technology Committee. *SMPTE Motion Imaging J.* **2007**, *116*, 345–354. [CrossRef]
35. Arrighetti, W. Moving picture colour science: The maths behind Colour LUTs, ACES and film "looks". In *Color and Colorimetry, Proceedings of 8th Italian Color Conference, Bologna, Italy, 13–14 September 2012*; Rossi, M., Ed.; Maggioli: Rimini, Italy, 2012; Volume VIII/B, pp. 27–34, ISBN 978-88-387-6137-9.
36. CIE; International Electrotechnical Commission (IEC). *International Lighting Vocabulary*, 4th ed.; Publication 17.4; CIE: Geneva, Switzerland, 1987.
37. CIE. *Chromaticity Difference Specification for Light Sources*, 3rd ed.; Technical Note 1:2014; CIE: Geneva, Switzerland, 2004.
38. Tooms, M.S. *Colour Reproduction in Electronic Imaging Systems: Photography, Television, Cinematography*; Wiley: Hoboken, NJ, USA, 2016, ISBN 978-11-190-2176-6.
39. SMPTE. *Derivation of Basic Television Color Equations*; Recommended Practice RP 177:1993; SMPTE: White Plains, NY, USA, 1993.
40. Westland, S.; Ripamonti, C.; Cheung, V. *Computational Colour Science*; Wiley: Hoboken, NJ, USA, 2012, ISBN 978-0-470-66569-5.

41. Arrighetti, W. Colour correction calculus (CCC): Engineering the maths behind colour grading. In *Colour and Colorimetry, Proceedings of 9th Italian Color Conference, Florence, Italy, 19–20 September 2013*; Rossi, M., Ed.; Maggioli: Rimini, Italy, 2013; Volume IX/B, pp. 13–19, ISBN 978-88-387-6242-0.
42. Sætervadet, T. *FIAF Digital Projection Guide*; Indiana University Press: Bloomington, IN, USA, 2012, ISBN 978-29-600-2962-8.
43. AMPAS Software Repository on GitHub (Version 1.0.3). Available online: http://github.com/ampas/aces-dev (accessed on 20 September 2017).
44. ACES Central Website. Available online: http://ACEScentral.com (accessed on 20 September 2017).
45. AMPAS. *ACES Version 1.0 Component Names*; Technical Bulletin TB-2014-012; AMPAS: Beverly Hills, CA, USA, 2014. Available online: http://j.mp/TB-2014-012 (accessed on 20 September 2017).
46. AMPAS. *ACES Version 1.0 User Experience Guidelines*; Technical Bulletin TB-2014-002; AMPAS: Beverly Hills, CA, USA, 2014. Available online: http://j.mp/TB-2014-002 (accessed on 20 September 2017).
47. AMPAS. *ACES—Versioning System*; Standard S-2014-002; AMPAS: Beverly Hills, CA, USA, 2014. Available online: http://j.mp/S-2014-002 (accessed on 20 September 2017).
48. AMPAS. *Informative Notes on SMPTE ST2065-1—ACES*; Technical Bulletin TB-2014-004; AMPAS: Beverly Hills, CA, USA, 2014. Available online: http://j.mp/S-2014-004 (accessed on 20 September 2017).
49. AMPAS. *ACEScg—A Working Space for CGI Render and Compositing*; Standard S-2014-004; AMPAS: Beverly Hills, CA, USA, 2014. Available online: http://j.mp/S-2014-004 (accessed on 20 September 2017).
50. AMPAS. *ACEScc—A Working Logarithmic Encoding of ACES Data for Use within Color Grading Systems*; Standard S-2014-003; AMPAS: Beverly Hills, CA, USA, 2014. Available online: http://j.mp/S-2014-003 (accessed on 20 September 2017).
51. Houston, J.; Benitez, A.; Walker, D.; Feeney, R.; Antley, R.; Arrighetti, W.; Barbour, S.; Barton, A.; Borg, L.; Clark, C.; et al. *A Common File Format for Look-up Tables*; Standard S-2014-006; AMPAS: Beverly Hills, CA, USA, 2014. Available online: http://j.mp/S-2014-006 (accessed on 20 September 2017).
52. AMPAS. *ACEScct—A Quasi-Logarithmic Encoding of ACES Data for Use within Color Grading Systems*; Standard S-2016-001; AMPAS: Beverly Hills, CA, USA, 2016. Available online: http://j.mp/S-2016-001 (accessed on 20 September 2017).
53. AMPAS. *ACESproxy—An Integer Log Encoding of ACES Image Data*; Standard S-2013-001; AMPAS: Beverly Hills, CA, USA, 2013. Available online: http://j.mp/S-2013-001 (accessed on 20 September 2017).
54. AMPAS. *Informative Notes on SMPTE ST2065-4 [64]*; Technical Bulletin TB-2014-005; AMPAS: Beverly Hills, CA, USA, 2014. Available online: http://j.mp/TB-2014-005 (accessed on 20 September 2017).
55. AMPAS. *Recommended Procedures for the Creation and Use of Digital Camera System Input Device Transforms (IDTs)*; Procedure P-2013-001; AMPAS: Beverly Hills, CA, USA, 2013. Available online: http://j.mp/P-2013-001 (accessed on 20 September 2017).
56. AMPAS. *Design, Integration and Use of ACES Look Modification Transforms*; Technical Bulletin TB-2014-010; AMPAS: Beverly Hills, CA, USA, 2014. Available online: http://j.mp/TB-2014-010 (accessed on 20 September 2017).
57. AMPAS. *Alternative ACES Viewing Pipeline User Experience*; Technical Bulletin TB-2014-013; AMPAS: Beverly Hills, CA, USA, 2014. Available online: http://j.mp/TB-2014-013 (accessed on 20 September 2017).
58. AMPAS. *ACES Clip-Level Metadata File Format Definition and Usage*; Technical Bulletin TB-2014-009; Academy of Motion Picture Arts and Sciences: Beverly Hills, CA, USA, 2014. Available online: http://j.mp/TB-2014-009 (accessed on 20 September 2017).
59. AMPAS. *Informative Notes on SMPTE ST2065-2 [62] and SMPTE ST2065-3 [63]*; Technical Bulletin TB-2014-005; AMPAS: Beverly Hills, CA, USA, 2014. Available online: http://j.mp/TB-2014-005 (accessed on 20 September 2017).
60. AMPAS. *Informative Notes on SMPTE ST268:2014—File Format for Digital Moving Picture Exchange (DPX)*; Technical Bulletin TB-2014-007; AMPAS: Beverly Hills, CA, USA, 2014. Available online: http://j.mp/TB-2014-007 (accessed on 20 September 2017).
61. SMPTE. *Academy Color Encoding Specification (ACES)*; Standard ST 2065-1:2012; SMPTE: White Plains, NY, USA, 2012.
62. SMPTE. *Academy Printing Density (APD)—Spectral Responsivities, Reference Measurement Device and Spectral Calculation*; Standard ST 2065-2:2012; SMPTE: White Plains, NY, USA, 2012.

63. SMPTE. *Academy Density Exchange Encoding (ADX)—Encoding Academy Printing Density (APD) Values*; Standard ST 2065-3:2012; SMPTE: White Plains, NY, USA, 2012.
64. SMPTE. *ACES Image Container File Layout*; Standard ST 2065-4:2013; SMPTE: White Plains, NY, USA, 2013.
65. SMPTE. *Material Exchange Format—Mapping ACES Image Sequences into the MXF Generic Container*; Standard ST 2065-5:2016; SMPTE: White Plains, NY, USA, 2016.
66. SMPTE. *IMF Application #5—ACES (Working Draft)*; Standard ST 2067-50; SMPTE: White Plains, NY, USA, 2017.
67. Kainz, F.; Kunz, A. Color Transformation Language User Guide and Reference Manual, 2007. Available online: http://github.com/ampas/CTL (accessed on 20 September 2017).
68. Houston, J. A Common File Format for Look-up Tables, Version 1.01; Starwatcher Digital, 2008. Available online: http://github.com/starwatcherdigital/commonlutformat; http://github.com/ampas/CLF (accessed on 15 September 2017).
69. Kainz, F.; Bogart, R.; Stanczyk, P.; Hillman, P. OpenEXR Documentation; Industrial Light & Magic, 2013. Available online: www.openexr.com (accessed on 20 September 2017).
70. ITU-T Specification, TIFF™, Revision 6.0; Adobe Systems Incorporated, 1992. Available online: www.itu.int/itudoc/itu-t/com16/tiff-fx/ (accessed on 15 September 2017).
71. SMPTE. *File Format for Digital Moving Picture Exchange (DPX)*; Standard ST 268:2014; SMPTE: White Plains, NY, USA, 2014.
72. OpenColorIO Project Website and GitHub Page; Sony ImageWorks, 2003. Available online: http://www.opencolorio.org; http://github.com/imageworks/OpenColorIO (accessed on 15 September 2017).
73. IEEE Computer Society Standards Committee; American National Standard Institute. *Standard for Floating-Point Arithmetic*; IEEE Standards Board. Standard P754-2008; IEEE: New York, NY, USA, 2008.
74. Duiker, H.P.; Forsythe, A.; Dyer, S.; Feeney, R.; McCown, W.; Houston, J.; Maltz, A.; Walker, D. ACEScg: A common color encoding for visual effects applications. In Proceedings of the Digital Symposium on Digital Production (DigiPro'15), Los Angeles, CA, USA, 8 August 2015; Spencer, S., Ed.; ACM: New York, NY, USA, 2015; p. 53. [CrossRef]
75. SMPTE. *Mastering Display Color Volume Metadata Supporting High Luminance and Wide Color Gamut Images*; Standard ST 2086:2014; SMPTE: White Plains, NY, USA, 2014.
76. Arrighetti, W. Topological Calculus: From algebraic topology to electromagnetic fields. In *Applied and Industrial Mathematics in Italy*, 2nd ed.; Cutello, V., Fotia, G., Puccio, L., Eds.; World Scientific: Singapore, 2007; Volume 2, pp. 78–88. [CrossRef]
77. Arrighetti, W. Mathematical Models and Methods for Electromagnetics on Fractal Geometries. Ph.D. Thesis, Università degli Studi di Roma "La Sapienza", Rome, Italy, September 2007.
78. SMPTE. *ARRIRAW Image File Structure and Interpretation Supporting Deferred Demosaicing to a Logarithmic Encoding*; Registered Disclosure Document RDD 30:2014; SMPTE: White Plains, NY, USA, 2014.
79. Adobe Systems Incorporated. *CinemaDNG Image Data Format Specification*, Version 1.1.0.0. Available online: www.adobe.com/devnet/cinemadng.html (accessed on 20 September 2017).
80. Arrighetti, W. (Technicolor, Hollywood, CA, USA); Frith, J. (MPC, London, UK). Personal communication, 2014.
81. SMPTE. *Material Exchange Format (MXF)—File Format Specifications*; Standard ST 377-1:2011; SMPTE: White Plains, NY, USA, 2011.
82. SMPTE. *Interoperable Master Format—Core Constraints*; Standard ST 2067-2:2016; SMPTE: White Plains, NY, USA, 2016.
83. National Association of Photographic Manufacturers, Inc.; American National Standards Institute. *Density Measurements—Part 3: Spectral Conditions*, 2nd ed.; Standard 5-3:1995(E); ISO: London, UK, 1995.
84. Internet Movie Database (IMDb) Page on "Il Ragazzo Invisibile: Seconda Generazione Movie". Available online: www.imdb.com/title/tt5981944 (accessed on 15 September 2017).
85. Arrighetti, W. (Frame by Frame, Rome, Italy); Giardiello, F.L. (Kiwii Digital Solutions, London, UK); Tucci, F. (Kiwii Digital Solutions, Rome, Italy). Personal communication, 2016.
86. ShotOnWhat Page on "Il Ragazzo Invisibile: Seconda Generazione Movie". Available online: http://shotonwhat.com/il-ragazzo-invisibile-fratelli-2017 (accessed on 15 September 2017).
87. The IMF (Interoperable Master Format) User Group Website. Available online: https://imfug.com (accessed on 15 September 2017).

88. Netflix Technology Blog on Medium. IMF: A Prescription for "Versionitis", 2016. Available online: http://medium.com/netflix-techblog/imf-a-prescription-for-versionitis-e0b4c1865c20 (accessed on 15 September 2017).
89. DCI Specification, Version 1.2; Digital Cinema Initiatives (DCI), 2017. Available online: www.dcimovies.com (accessed on 15 September 2017).
90. Diehl, E. *Securing Digital Video: Techniques for DRM and Content Protection*; Springer: Berlin, Germany, 2012, ISBN 978-3-642-17345-5.
91. SMPTE. *IMF Application #4—Cinema Mezzanine Format*; Standard ST 2067-40:2016; SMPTE: White Plains, NY, USA, 2016.
92. Mansecal, T.; Mauderer, M.; Parsons, M. Colour-Science.org's Colour API. Available online: http://colour-science.org; http://github.com/colour-science/colour (accessed on 15 September 2017). [CrossRef]
93. Arrighetti, W.; Gazzi, C. (Frame by Frame, Rome, Italy); Zarlenga, E. (Rome, Italy). Personal communication, 2015.

MDPI
St. Alban-Anlage 66
4052 Basel
Switzerland
Tel. +41 61 683 77 34
Fax +41 61 302 89 18
www.mdpi.com

Journal of Imaging Editorial Office
E-mail: jimaging@mdpi.com
www.mdpi.com/journal/jimaging

www.ingramcontent.com/pod-product-compliance
Lightning Source LLC
Chambersburg PA
CBHW051730210326
41597CB00032B/5675